Engineering
Problem Solving
with MATLAB®

Engineering Problem Solving with MATLAB®

D. M. Etter

Department of Electrical and Computer Engineering
University of Colorado, Boulder

MATLAB® Curriculum Series

PRENTICE HALL, Englewood Cliffs, New Jersey 07632

Library of Congress Cataloging-in-Publication Data

Etter, D. M.
    Engineering problem solving with MATLAB / D.M. Etter.
        p.   cm.
    Includes bibliographical references and index.
    ISBN 0-13-280470-0
    1. Engineering--Data processing.   2. MATLAB.   I. Title.
TA331.E88 1993
    620'.00285'5369--dc20   92-34148
                                                            CIP

Acquisitions Editor: Don Fowley
Editor-in-Chief: Marcia Horton
Production Editor: Bayani Mendoza de Leon
Copy Editor: Brenda Melissaratos
Marketing Manager: Tom McElwee
Design Director: Florence Dara Silverman
Designer: Linda J. Den Heyer Rosa
Cover Designer: Lisa A. Domínguez
Prepress Buyer: Linda Behrens
Manufacturing Buyer: Dave Dickey
Editorial Assistant: Phyllis Morgan

 © 1993 by Prentice-Hall, Inc.
A Paramount Communications Company
Englewood Cliffs, New Jersey 07632

Printed in the United States of America

10 9 8 7 6 5 4 3 2

ISBN 0-13-280470-0

Prentice-Hall International (UK) Limited, *London*
Prentice-Hall of Australia Pty. Limited, *Sydney*
Prentice-Hall Canada Inc., *Toronto*
Prentice-Hall Hispanoamericana, S.A., *Mexico*
Prentice-Hall of India Private Limited, *New Delhi*
Prentice-Hall of Japan, Inc., *Tokyo*
Simon & Schuster Asia Pte. Ltd., *Singapore*
Editora Prentice-Hall do Brasil, Ltda., *Rio de Janeiro*

TRADEMARK INFORMATION

MATLAB is a registered trademark
of MathWorks, Inc.

In memory of my dearest Mother, Muerladene Janice Van Camp

MATLAB® CURRICULUM SERIES

Cleve Moler, The MathWorks, Inc.
Editor

The Student Edition of MATLAB® Student User Guide

The Student Edition of MATLAB® for Macintosh Computers

The Student Edition of MATLAB® for MS-DOS Personal Computers—
with 5-1/4" Disks

The Student Edition of MATLAB® for MS-DOS Personal Computers—
with 3-1/2" Disks

Engineering Problem Solving with MATLAB®, D. M. Etter

Contents

PART III   NUMERICAL TECHNIQUES   213

8 Solutions to Systems of Linear Equations   215

9 Interpolation and Curve Fitting   231

Foreword

This is an exciting and unusual textbook. It is exciting because it represents a new approach to an important aspect of engineering education. It is unusual because it combines topics from what are traditionally three or four different courses into a single, introductory course. This new course, intended for freshman or sophomore students in engineering and science, covers:

- Elementary applied mathematics
- Basic numerical methods
- Computer programming
- Problem solving methodology.

Where do you first see complex numbers? Where do you first learn about 3-by-3 matrices? When do you begin to do useful mathematical computations? How do you combine these ideas into the solution of practical engineering and scientific problems? This book provides answers to such questions early in the collegiate career.

One exercise in the book illustrates this multi-faceted approach:

Write a MATLAB expression for the resistance of three resistors in parallel.

Here, in one problem, we have some basic electrical engineering, some elementary mathematics, and a little bit of computer programming.

Ten years ago, Delores Etter wrote two popular textbooks on computer programming and numerical methods. She chose to use Fortran which then was clearly the most widely used language for technical computing. Coincidentally, it was also about ten years ago that MATLAB began to be used outside of the matrix computation community where it originated.

Today, there is a wide variety of languages and environments available for technical computing. Fortran is certainly still important; but so are Basic, C, and Pascal. There are sophisticated programmable calculators, spread sheets and mouse- and menu-based systems. And there are several commercial mathematical languages. In our opinion, MATLAB is the right choice for courses such as this because it is:

- Easy to learn and use
- Powerful, flexible and extensible

- Accurate, robust and fast
- Widely used in engineering and science
- Backed by a professional software company.

At the MathWorks and at Prentice-Hall, we are committed to the support of MATLAB's use in education. In the future, you will see new features added to the Student Edition and many new books available in the MATLAB Curriculum Series.

A friend of ours, who is a professor of electrical engineering and an expert on signal processing, says "MATLAB is so good for signal processing because it wasn't designed to do signal processing—it was designed to do mathematics."

Our friend's observation is also the basis for this book. Mathematics, and its embodiment in software, is a foundation for much of modern technology. We believe you will enjoy, and benefit from, this introduction.

—*Cleve Moler*
The MathWorks, Inc.
Natick, Massachusetts

Preface

Engineers use computers to solve a variety of problems ranging from the evaluation of a simple function to solving a system of nonlinear equations. This text was written to introduce engineering problem solving with the following objectives:

- to develop a consistent methodology for solving engineering problems,
- to present the capabilities of MATLAB, the premier software package for interactive numeric computation, data analysis, and graphics, and
- to illustrate the problem solving process with MATLAB through a variety of engineering examples and applications.

To accomplish these objectives, Chapter 1 presents the five-step process that is used consistently for solving engineering problems. The rest of the chapters present the capabilities of MATLAB for solving engineering problems. Throughout all the chapters, we present a large number of examples from many different engineering disciplines. The solutions to these examples are developed using the five-step process and MATLAB.

Organization

The text is divided into three parts: Introduction, Fundamental Engineering Computations, and Numerical Techniques. Part I contains an introductory chapter on engineering problem solving and an introductory chapter on MATLAB. Part II contains five chapters of MATLAB material that is fundamental to basic engineering computations and graphics. Part III contains numerical techniques that are used in solving special types of engineering problems. A recommended plan for using this text in an introductory course would be to cover all of Parts I and II, and to present selected material from Part III. In particular, we would recommend the following topics from Part III: solutions of systems of linear equations, interpolation and curve fitting, and roots of equations.

Prerequisites

No prior experience with the computer is assumed. The mathematical background needed for Parts I and II is algebra and trigonometry; more advanced mathematics is needed for some sections of Part III.

Grand Challenges Theme

Chapter 1 introduces you to the exciting area of engineering through two sets of applications—the ten top engineering achievements of the last 25 years and a set of grand challenges currently facing engineers. We particularly focus on the following grand challenges:

- prediction of weather, climate, and global change
- speech recognition and understanding
- machine vision
- vehicle performance
- superconductivity
- enhanced oil and gas recovery
- nuclear fusion

These grand challenges are used as a general theme throughout the text. Each chapter will begin with a photograph and a discussion of some aspect of one of these grand challenges. We also refer to these grand challenges in many of the examples and applications within the chapters.

Five-Step Problem Solving Methodology

An important emphasis in this text is on developing problem solving skills. A five-step problem solving process is introduced in Chapter 1, and then used to solve engineering problems throughout the remaining chapters. These five steps are:

1. Define the problem clearly.
2. Describe the input and output.
3. Work a simple example by hand or with a calculator.
4. Develop a MATLAB solution.
5. Test the solution with a variety of data.

Each step is clearly outlined each time that a MATLAB solution is developed for an engineering application. One way to reinforce the development of problem solving skills is to provide a large number of examples that use a consistent set of steps in approaching and solving problems. A key part of this process is the development of solutions that are simple and straight-forward to understand. Therefore, in all examples in the text, emphasis is placed on creating readable and simple solutions.

Graphics

Engineers (and engineering students) use plots to analyze, interpret, and evaluate data. Therefore, it is important to learn to use a powerful graphics tool so that plots can be easily generated in a variety of formats. MATLAB allows you to generate x-y plots, polar plots, bar graphs, contour plots and 3-D plots. Chapter 7 covers these types of plots and the options that can be used with them.

Text Diskette

Since the text includes a large number of examples that illustrate the various engineering applications, a diskette is included that contains all the examples and data files so that students can run these examples without having to enter the MATLAB commands or the data files. In addition, a number of engineering data files is included in the diskette that are referenced in the problems at the end of the chapters. These data files include climatology data, speech signals, robot manipulator arm paths, engine temperature data, and oil well production data.

Instructor's Manual

An Instructor's Manual is available that contains complete solutions to all Practice exercises and all end-of-chapter problems. In addition, data files are suggested for all problems that do not include sample data. The end-of-chapter problems and the new data files are included in ASCII format in a diskette that accompanies the Instructor's Manual.

Acknowledgments

I am very appreciative of the support of Cleve Moler (Chairman of The MathWorks, Inc.) and Pete Janzow (former Senior Engineering Editor) during the initial development of this text, and of Don Fowley (Senior Engineering Editor) during the final stages of the text. I also want to acknowledge the enthusiastic encouragement of many people at Prentice Hall, including Marcia Horton, Bernard Goodwin, Gary June, Tom McElwee, Bayani de Leon, Alice Dworkin, and Mike Sutton. This has been an exciting project to develop, and I appreciate the suggestions and encouragement from a number of reviewers, including Randall Janka, The MITRE Corporation; Professor John A. Fleming, Texas A&M; Professor Jim Maneval, Bucknell University; Professor Helmuth E. Worbs, University of Central Florida; Professor Hüseyin Abut, San Diego State University; Professor Richard Shiavi, Vanderbilt University; Captain Randy Haupt, U.S. Air Force Academy; Professor Zoran Gajic, Rutgers University; Professor Robert F. Stengel, Princeton University; Professor William Beckwith, Clemson University; Professor Thomas H. Sloane, Bucknell University; Professor Virginia Stonick, Carnegie-Mellon University; and Professor Juris Vagners, University of Washington. I would also like to acknowledge the assistance of Mark Summers, a project engineer at Lawrence Livermore Laboratory, for his assistance in obtaining information and photographs of the Nova project. A final note of gratitude goes to my husband Jerry Richard, a mechanical engineer, for his help in developing the material on control systems.

Delores M. Etter
Department of Electrical/Computer Engineering
University of Colorado, Boulder

Engineering
Problem Solving
with MATLAB®

PART I

Introduction

In Chapter 1 we introduce you to the challenges of engineering. After presenting the Ten Top Engineering Achievements of the last 25 years, we present a group of Grand Challenges—fundamental problems in science and engineering that have broad applications and that will require technological breakthroughs to solve them. These grand challenges are used as a theme for the engineering applications used in the chapter openers and in the examples. Since problem solving is such an important part of engineering, we present a problem solving methodology for solving technical problems. This process has five steps, and we will illustrate problem solving with these five steps throughout the text. Since many engineering problem solutions require the use of a computer, we then turn to a brief introduction to computing systems in Chapter 2. After a quick summary of the key components in computer hardware and computer software, we then address MATLAB. MATLAB (MATrix LABoratory) has become the premier software package for interactive numeric computation, data analysis, and graphics, and is thus an ideal choice for introducing you to engineering computing. We illustrate the use of MATLAB with a simple program that allows you to easily generate a plot of laboratory data. With this capability, you will be ready to prepare graphs for your next homework in Physics, Calculus, or Chemistry.

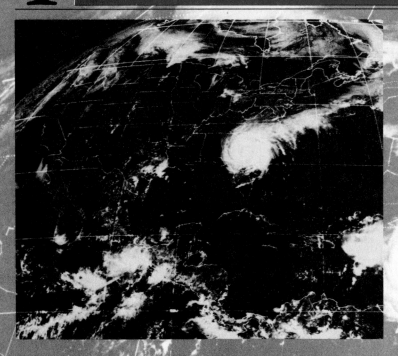

GRAND CHALLENGE: Weather Prediction

Weather satellites provide a great deal of information to meteorologists who attempt to predict the weather. Large volumes of historical weather data can also be analyzed and used to test models for predicting weather. In general, we can do a reasonably good job of predicting the overall weather patterns; however, local weather phenomena such as tornadoes, water spouts, and microbursts are still very difficult to predict. Even predicting heavy rainfall or large hail from thunderstorms is often difficult. While Doppler radar is useful in locating regions within storms that could contain tornadoes or microbursts, the radar detects the events as they occur and thus gives little time for issuing appropriate warnings to populated areas or aircraft. Accurate and timely prediction of weather and associated weather phenomena is still an elusive goal.

An Introduction to Problem Solving

■■ Introduction

■■ Summary

Introduction

Although most of this text is focused on teaching you to use MATLAB, which has become the premier software package for interactive numeric computation, data analysis, and graphics, we begin by discussing problem solving. Much of engineering centers on problem solving, and it is important to develop a consistent approach to solving problems. We also introduce you to the types of problems that engineers solve. We begin by discussing 10 outstanding engineering achievements over the past 25 years. We then present a group of grand challenges—problems that require technological breakthroughs for their solution. One of the grand challenges includes the prediction of weather, which we used in the chapter opening discussion. With this framework of interesting engineering problems that have been solved, and important engineering problems remaining to be solved, we then present a problem-solving methodology that we will use throughout the text. This methodology has five steps for describing a problem and then developing a solution. Finally, we return to the problem of weather prediction and discuss some of the different types of weather data that are currently being collected. These data are critical for developing the understanding and intuition needed to develop a mathematical model to predict weather. These data are also important because they can be used to test hypothetical models.

1.1   ENGINEERING CHALLENGES

Engineers solve real-world problems using scientific principles from disciplines that include mathematics, physics, computer science, chemistry, and biology. It is this variety of subjects, and the challenge of real problems, that makes engineering so interesting and so rewarding. In this section we discuss 10 outstanding engineering and scientific achievements of the past 25 years. We then present a group of grand challenges—important problems that need to be solved by engineers and scientists.

Challenges Met

National Academy
of Engineering

Top 10
achievements

In celebration of its 25-year anniversary, the National Academy of Engineering recently selected what it considers to be the 10 most important engineering achievements during the past 25 years. Each of these achievements represents a major advance or breakthrough in engineering and a significant contribution to human welfare. The following brief discussions [1, 2] present each of these 10 achievements and some of the interesting engineering challenges that were met. They are examples of the many areas that are included in engineering, and they demonstrate how engineering has improved our lives and expanded the possibilities for the future.

Moon Landing   The moon landing is probably the most complex and ambitious engineering project ever attempted. Major technological breakthroughs occurred in a number of areas supporting the Saturn/Apollo system that consisted of the Apollo spacecraft, the lunar lander, and the three-stage Saturn V rocket. A new inertial navigation system was developed using gyroscopes and accelerometers that sensed change in direction and speed. The lunar lander was the first manned vehicle designed to fly solely in space. Its small ascent engine, which had no backup, was used to return the astronauts to the mother ship. The spacesuits consisted of a three-piece suit and backpack, which together weighed 190 pounds. The first piece of the suit was a cooling undergarment of knitted nylon spandex with a network of plastic tubes filled with circulating water. The next piece was a rubber-coated bladder between a cloth lining and a nylon cover. The final piece was a protective outer garment of 17 layers; 6 of the 17 layers were made of a fireproof fabric of fiberglass threads coated with Teflon. An individual moon flight required the coordination of more than 450 persons in the launch control center and nearly 7,000 others on nine ships, in 54 aircraft, and at stations located around the earth.

Application Satellites   Satellite systems are used to send weather photographs, relay communication signals, map uncharted terrain, and provide navigational information. Satellites in a geostationary orbit rotate the earth 22,300 miles above the equator; they complete one revolution every 24 hours and thus remain stationary relative to the surface of the earth. Other groups of satellites use oblong orbits looping over northern latitudes. Landsat satellites were first launched in 1972 and provide information on the entire earth from near-polar orbits, observing each spot at the same local time every 18 days. Landsats use multispectral sensors that transmit data that are processed into images yielding information unavailable in standard images. For example, infrared satellite sensors use thermal information to provide images showing heat differentials that can be used to analyze geologic formations, to monitor pollution, and to observe volcanic activity. Navstar is a Global Positioning System (GPS) that will have 18 satellites plus three spares in orbit when it is complete. Each satellite contains an atomic clock that loses or gains one second every 33,000 years. A GPS receiver on the earth receives signals from four satellites, each telling the time it was emitted and the position of the satellite. The receiver can then determine its location to within 10 meters by calculating how far each signal traveled.

Microprocessors   A microprocessor is a tiny computer smaller than your fingernail. It can be programmed to perform specific tasks, such as operating video cassette recorders or controlling a home security system. If extra memory and input/output devices are added, a microprocessor can also be used as a general-purpose computer in hand-held calculators or personal computers. Technology continues to provide smaller, and yet more powerful, microprocessors at lower prices. As a result, microprocessors are being used in electronic equipment, household appliances, toys, and games, as well as in many components of cars and airplanes. Applications are also being developed that use many microprocessors at once, with each one given a specific task that can be performed simultaneously with other tasks. These massively parallel computer systems can be designed to solve problems that cannot feasibly be solved with a single computer.

Computer-Aided Design and Manufacturing   Computer-aided design (CAD) and computer-aided manufacturing (CAM) have brought another industrial revolution to the world by increasing the speed and efficiency of many types of manufacturing processes. CAD typically refers to the use of the computer to generate designs, particularly using computer graphics. CAD programs combine the steps to take an idea through the initial sketch (which can also be done on the computer) to the completed design with schematics, parts lists, and computer simulation results. CAM uses design parameters (sometimes produced by CAD systems) to control machinery or industrial robots in manufacturing parts automatically, assembling components, and moving them to the desired locations. CAM is especially useful in operations that would be hazardous to humans or that require high precision in tooling or milling.

CAT Scan   A computerized axial tomography (CAT) scanner is a machine that generates three-dimensional images or slices of an object using X-rays generated from many angles, encircling the object or patient. Each X-ray measures a density at its angle, and by combining these density measurements with sophisticated computer algorithms, an image can be reconstructed to give clear, detailed pictures of the inside of the object. CAT scans have been used to identify medical problems such as tumors, infections, bleeding, and blood clots that would not have been identified by regular X-rays or other noninvasive procedures. Brain abnormalities that relate to schizophrenia, alcoholism, manic-depressive illness, and Alzheimer's disease can also be identified from CAT scans. CAT scanners are available in most hospitals, and the U.S. Army is developing a rugged, lightweight one that can be transported to medical stations in combat zones.

Advanced Composite Materials   Composites consist of materials that can be bonded together in a manner such that one material is reinforced by the fibers or particles of another material. Examples include using straw in mud bricks to make the bricks stronger and using steel rods in concrete to reinforce bridges and buildings. Particle board and plywood are composite materials that are strong and resistant to warping. Advanced composite materials were developed to provide lighter, stronger, and more temperature-resistant materials for aircraft and spacecraft. The first commercial advanced composite materials appeared in sporting goods. Downhill snow skis use layers of woven Kevlar fibers to increase their strength and reduce weight. Golf club shafts of graphite/epoxy are stronger and lighter than the steel in conventional shafts. Composite materials are now incorporated in new designs of fishing rods, bicycle frames, and race-car chassis.

Jumbo Jets   The origins of the jumbo jet (747, DC-10, L-1011) came from the Air Force C-5A cargo plane that began operational flights in 1969. Much of the success of the jumbo jets can be attributed to the high-bypass fanjet engine that allowed the planes to fly farther, with less fuel, and with less noise than previous jet engines. The core of the engine operates like a pure turbojet, in which compressor blades pull air into the engine's combustion chamber. The hot expanding gas thrusts the engine forward, and at the same time spins a turbine that in turn drives the compressor and the large fan on the front of the engine. The spinning fan provides the bulk of the engine's thrust. The turbines in the new engines tolerate burner-exit gas temperatures up to 2,800°F. The turbine blades are made from a nickel alloy that is more heat resistant and much stronger than earlier blades. The jumbo jets also have an increased emphasis on safety. Many systems have multiple backups to provide a wider margin of safety. For example, there are four main landing-gear legs on a 747 instead of two. A middle spar was added to the wings in the event one is damaged, and redundant hydraulic systems operate the critical system of elevators, stabilizers, and flaps that control the motion of the plane.

Lasers   Light waves from a laser have the same frequency and thus create a beam with one characteristic color. More importantly, the light waves travel in phase forming a narrow beam that can easily be directed and focused. CO_2 lasers can be used to drill holes in materials that range from ceramics to composite materials to rubber. Lasers can be used medically to weld detached retinas, seal leaky blood vessels, vaporize brain tumors, remove warts and cysts, and perform delicate inner-ear surgery. Laser light can also be used inside the body through fiber-optic endoscopes to burn fatty deposits out of clogged arteries, pulverize kidney stones, and open blocked Fallopian tubes. Lasers are also used to make three-dimensional pictures called holograms. One part of a laser beam is focused directly onto photographic film, while the other part bounces off an object and onto the film. The waves from the two beams interfere with each other in complex patterns that are recorded on the film. When the film is developed, the patterns reflect a slightly different image in slightly different directions, thus giving the three-dimensional image. Hologram images are used to develop tamper-proof seals, bar-code readers, and sensitive sensors. Other applications of lasers include fiber-optic cable communications, compact disc recording, and laser radar.

Fiber-Optic Communications   An optical fiber is a transparent glass thread that is thinner than a human hair, but it can carry more information than either radio waves or electric waves in copper telephone wires. In addition, fiber-optic communication signals do not produce electromagnetic waves that cause "cross-talk" noise on communication lines. The TAT-8, the first transoceanic fiber-optic cable, was laid in 1988 across the Atlantic. It contains four fibers that can handle up to 40,000 calls at once. An undersea fiber-optic cable linking the United States and Japan was activated in 1989. AT&T has just added its second transatlantic fiber-optic cable that will carry 80,000 simultaneous calls. Other applications for optical fibers include motion sensing in gyroscopes, linking industrial lasers to machining tools, and threading light into the human body for examinations and laser surgery. Current research is aimed at designing optical computers that would theoretically be thousands of times faster than the best modern supercomputers.

Genetically Engineered Products   Genetically engineered products are produced by splicing a gene that produces a valuable substance from one organism into another organism that will multiply itself and the foreign gene along with it. Once the new organism has been created, a system has to be designed to produce and process the product in large quantities at a reasonable cost. The first commercial product of genetic engineering was human insulin, which appeared commercially under the trade name Humulin. The human insulin molecule is composed of two parts called A and B chains. The original Humulin process used two versions (one containing a gene producing A chains and the other containing the B-chain gene) of a common bacterium, *Escherichia coli* (*E. coli*). Each version was grown in a large tank, and the chains extracted and purified. In a third vessel the chains were combined, purified, and crystallized into human insulin. Genetic engineering is also being used to develop new products such as tomatoes that resist softening and thus have a longer shelf life. Current research is investigating the use of genetically altered microbes to clean up toxic waste, to degrade pesticides, and to turn organic waste into useful products.

Grand Challenges Remaining

Office of Science and Technology Policy

In spite of the many significant engineering breakthroughs that have occurred during the past few decades, there are many challenging problems facing engineers. The Office of Science and Technology Policy in Washington, D.C., recently published a research and development strategy for high-performance computing, where high-performance computing refers to the full range of supercomputing activities and the new generation of large-scale parallel architectures. As part of this strategy, the grand challenges   Office of Science and Technology Policy identified a number of grand challenges—fundamental problems in science or engineering, having broad applications, whose solutions would benefit from high-performance computing. We will use some of these grand challenges as a general theme for the text by highlighting them in the chapter opening applications and by using them in many of the examples and applications. The following paragraphs [2, 3] briefly present several of the grand challenges and the types of benefits that will come with their solutions.

Prediction of Weather, Climate, and Global Change   In order to make long-range predictions about the coupled biosphere system formed by the atmosphere and the ocean, we must understand its behavior with a great amount of detail. This detail includes understanding CO_2 dynamics in the atmosphere and ocean, ozone depletion, and climatological changes due to the releases of chemicals or energy. This complex interaction also includes solar interactions. A major eruption from a solar storm near a "coronal hole" (a venting point for the solar wind) can eject vast amounts of hot gases from the sun's surface toward the earth's surface at speeds over a million miles per hour. This ejection of hot gases bombards the earth with X-rays and can interfere with communication and cause power fluctuations in power lines. Learning to predict changes in weather, climate, and global change involves collecting large amounts of data for study and developing new mathematical models that can represent the interdependency of many variables.

Speech Recognition and Understanding   Automatic speech understanding by computers could revolutionize our communication systems, but many problems are involved. Teaching a computer to understand words from a small vocabulary spoken by the same person is currently possible. However, it is very difficult to develop systems that are speaker independent and that understand words from large vocabularies and from different languages. Subtle changes in one's voice, caused by a cold or stress, can affect the performance of speech recognition systems. Even assuming that the computer can recognize the words, it is not simple to determine their meaning. Many words are context dependent and thus cannot be analyzed separately. Intonation such as raising one's voice can change a statement into a question. While there are still many difficult problems left to address in automatic speech recognition and understanding, exciting applications are everywhere. Imagine a telephone system that translates the language of the speaker into the language of the listener.

Machine Vision   Human vision and object recognition represent an incredibly powerful and complex sensory process. Adapting this process to machines represents a formidable challenge as well as significant advancements in many areas. Advanced sensors are needed to collect high-resolution information about the images being collected. Since a high-resolution image contains a great deal of information, memory devices are needed that can quickly and efficiently store the information. Signal processing and artificial intelligence are needed to extract the desired information from the images and translate that information into the form needed. For example, to identify an object and to track it from one image to another is a very complex problem. Machine vision for computers and robots requires technological advances in many areas, including image signal processing, texture and color modeling, geometric processing, object modeling, and advanced sensors.

Vehicle Performance   Substantial improvements in the performance of vehicles (cars, planes, ships, trains) requires more complex physical modeling in the areas of fluid dynamic behavior for three-dimensional flow fields and flow inside engine turbomachinery and ducts. Turbulence in fluid flows impacts the stability and control, thermal characteristics, and fuel performance of aerospace vehicles, and modeling this flow is necessary for the analysis of new configurations. The analysis of the aeroelastic behavior of vehicles also impacts new designs. The efficiency of combustion systems is also related because attaining significant improvements in combustion efficiency requires understanding the relationships between the flows of the various substances and the quantum chemistry that causes the substances to react. Vehicle performance is also being addressed through the use of onboard computers and microprocessors. Transportation systems are currently being studied in which cars have computers with small video screens mounted on the dash. The driver enters the destination location, and the video screen shows the street names and path from the current location to the desired location. A communication network keeps the car's computer aware of any traffic jams so that it can automatically reroute the path if necessary. Other transportation research addresses totally automated driving, with computers and networks handling all the control and information interchange.

Superconductivity   Normal metals, such as copper, have a resistance to electrical flow that leads to power losses. As these normal metals are cooled, the resistance decreases to a constant value. Therefore, the cooler the metals, the more efficient they are for conducting electricity. Using the work of the Dutch physicist H. Kamerlingh Onnes, who succeeded in liquefying helium, other experimenters showed that some metals have zero resistance at very low temperatures. These metals are called superconductors. Some of the materials that have been identified as superconductors include aluminum, lead, niobium, niobium tin, and niobium-germanium. Recent research has discovered materials such as an oxide compound of barium, lanthanum, and copper that is superconducting at higher temperatures. Superconductivity has the potential for spectacularly energy-efficient power transmission and the design of ultrasensitive instrumentation. However, the materials currently supporting high-temperature superconductivity are difficult to form, stabilize, and use.

Enhanced Oil and Gas Recovery   Dependence on oil and gas dictates that technology be used to locate the estimated 300 billion barrels of oil reserves in the United States. Current techniques for identifying structures likely to contain oil and gas use seismic techniques that can evaluate structures down to 20,000 feet below the surface. These techniques use a group of sensors (called a sensor array) that is located near the area to be tested. A ground shock signal is sent into the earth and is then reflected by the different geologic layer boundaries and is received by the sensors.

Using sophisticated signal processing, the boundary layers can be mapped and some estimate can be made as to the materials in the various layers, such as sandstone, shale, and water. The ground shock signals can be generated in several ways—a hole can be drilled, and an explosive charge can be exploded in the hole; a ground shock can be generated by an explosive charge on the surface; or a special truck that uses a hydraulic hammer can be used to pound the earth several times per second. Continued research is needed to improve the resolution of the information and to find economical methods of production and recovery that are economical and ecologically sound.

Nuclear Fusion   During nuclear fission, a large unstable nucleus breaks into two components and releases a large amount of energy plus additional neutrons, which can then be used to cause more fissions that release more neutrons and more energy. Nuclear fusion is the process of combining two nuclei to form a heavier one; under certain conditions, the process releases a large amount of energy in the process. The energy released from fusion is what keeps the sun and other stars hot. The most likely way of accomplishing a self-sustaining fusion reaction on earth is to compress a high-temperature plasma using magnetic fields. The development of controlled nuclear fusion would clearly revolutionize energy development. However, the behavior of fully ionized gases at very high temperatures under the influence of strong magnetic fields is not well understood. The behavior must also be analyzed in terms of the complex three-dimensional geometry. In addition, fusion reactors will be extremely radioactive as a result of their high flux of neutrons, and they could be sources of thermal pollution.

The grand challenges listed here are only a few of the many interesting problems waiting to be solved by engineers and scientists. The solutions to problems of this magnitude will be the result of organized approaches that combine ideas and technologies through logical processes. The use of computers and engineering application programs such as MATLAB will be a key element in the solution process. Therefore, in the next section we present a problem-solving process for developing computer solutions to engineering problems.

1.2   AN ENGINEERING PROBLEM-SOLVING METHODOLOGY

Problem solving is a key part of not only engineering courses, but also courses in mathematics, physics, chemistry, and computer science. Therefore, a consistent approach to solving problems is important. It is also helpful if the approach is general enough to work for all these different areas, so that we do not have to learn a technique for mathematics problems, and a different technique for physics problems, and so on. The problem-solving technique that we present works for engineering problems and can be tailored to solve problems in other areas as well; however, it does assume that we are using MATLAB to help solve the problem. (In the next chapter we begin to use MATLAB. For now, think of MATLAB as a software tool that can be used to perform many types of numerical computations for us.)

The process or methodology for problem solving [2] that we will use throughout this text has five steps:

5-step process

1. State the problem clearly.
2. Describe the input and output information.
3. Work the problem by hand (or with a calculator) for a simple set of data.
4. Develop a MATLAB solution.
5. Test the solution using a variety of data sets.

We now discuss each of these steps using the example of computing the distance between two points in a plane.

 1. PROBLEM STATEMENT
The first step is to state the problem clearly. It is extremely important to give a clear, concise problem statement to avoid any misunderstandings. For this example, the problem statement is the following:

Compute the straight-line distance between two points in a plane.

 2. INPUT/OUTPUT DESCRIPTION
The second step is to describe carefully the information that is given to solve the problem and then to identify the values to be computed. These items represent the input and the output for the problem and collectively can be called input/output, or I/O. For many problems, a diagram that shows the input and output is useful. Sometimes this type of diagram is called a "black box" because we aren't defining at this point all the steps to determine the output, but we are showing the information that is used to compute the output. For this example, we could use the diagram in Figure 1.1.

 3. HAND EXAMPLE
The third step is to work the problem by hand or with a calculator, using a simple set of data. This is a very important step and should not be skipped even for simple problems. This is the step in which you work out the details of

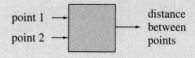

Fig. 1.1   I/O diagram.

the problem solution. If you can't take a simple set of numbers and compute the output (either by hand or with a calculator), then you aren't ready to move on to the next step; you should reread the problem and perhaps consult reference material. Once you can work the problem for a simple set of data, you are then ready to develop an algorithm, or a step-by-step outline of the solution. This outline of the solution is then converted to MATLAB commands so that we can use the computer to do all the computations. The hand example for this specific example is shown below:

Let the points p1 and p2 have the following coordinates:

$$p1 = (1,5), \; p2 = (4,7)$$

We want to compute the distance between the two points, which is the hypotenuse of a right triangle, as shown in Figure 1.2. Using the Pythagorean theorem, we can compute the distance d with the following equation:

$$
\begin{aligned}
d &= \sqrt{s1^2 + s2^2} \\
&= \sqrt{(4 - 1)^2 + (7 - 5)^2} \\
&= \sqrt{13} \\
&= 3.61
\end{aligned}
$$

 4. MATLAB SOLUTION

In the next chapter we discuss MATLAB commands. However, from the following solution you can see that the commands are very similar to the equations that were used in the hand example. The percent signs are used to precede comments that explain the MATLAB commands.

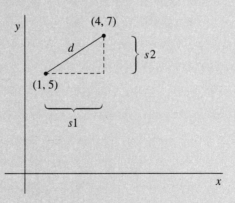

Fig. 1.2 Distance between two points.

```
%
%              This program computes and prints the
%              straight-line distance between two points.
%
p1 = [1,5];                      % initialize point1
p2 = [4,7];                      % initialize point2
d = sqrt(sum((p2-p1).^2))        % calculate distance
```

 5. TESTING

The final step in our problem-solving process is testing the solution. We should first test the solution with the data from the hand example, since we have already computed the solution. When the MATLAB statements in this solution are executed, the computer displays the following output:

```
d =
    3.6056
```

This output matches the value that we calculated by hand. If the MATLAB solution did not match the hand solution, then we should review both solutions to find the error. Once the solution works for the hand example, we should also test it with several sets of data to be sure that the solution works for other valid sets of data.

1.3   ANALYZING CLIMATOLOGY DATA

We now return to the grand challenge discussed at the beginning of the chapter. Since we have not yet learned how to use MATLAB commands, we cannot develop a program to analyze climatology data. However, we can discuss the preliminary types of analyses that must precede the development of a computer solution. The first step in attempting to develop an equation or a model to predict the weather is to study the past history of the weather. Fortunately, a number of national agencies are interested in collecting and storing weather information [2]. The National Oceanic and Atmospheric Administration (NOAA) is a research-oriented organization that studies the oceans and the atmosphere. It also funds environmental research in data analysis, modeling, and experimental work relative to global changes. The National Environmental Satellite, Data, and Information Service collects and distributes information relative to the weather. The National Climatic Data Center collects and compiles climatology information from National Weather Service offices across the country. It is also the National Weather Service offices that interact with state and local weather forecasters to keep the general public aware of current weather information.

JAN 1991
DENVER, CO
NAT'L WEA SER OFC
10230 SMITH ROAD

ISSN 0198-7690

LOCAL
CLIMATOLOGICAL DATA
Monthly Summary

STAPLETON INTERNATIONAL AP

LATITUDE 39° 45'N   LONGITUDE 104° 52'W   ELEVATION (GROUND) 5282 FEET   TIME ZONE MOUNTAIN   23062

DENVER, CO   JAN 1991

DATE	TEMPERATURE °F						DEGREE DAYS BASE 65°F		WEATHER TYPES 1 FOG 2 HEAVY FOG 3 THUNDERSTORM 4 ICE PELLETS 5 HAIL 6 GLAZE 7 DUSTSTORM 8 SMOKE, HAZE 9 BLOWING SNOW	SNOW ICE PELLETS OR ICE ON GROUND AT 0500 INCHES	PRECIPITATION		AVERAGE STATION PRESSURE IN INCHES ELEV. 5332 FEET ABOVE M.S.L	WIND (M.P.H.)				PEAK GUST		FASTEST 1-MIN		SUNSHINE		SKY COVER (TENTHS)		
	MAXIMUM	MINIMUM	AVERAGE	DEPARTURE FROM NORMAL	AVERAGE DEW POINT	HEATING (SEASON BEGINS WITH JUL)	COOLING (SEASON BEGINS WITH JAN)			WATER EQUIVALENT (INCHES)	SNOW, ICE PELLETS (INCHES)		RESULTANT DIR.	RESULTANT SPEED	AVERAGE SPEED	SPEED	DIRECTION	SPEED	DIRECTION	MINUTES	PERCENT OF TOTAL POSSIBLE	SUNRISE TO SUNSET	MIDNIGHT TO MIDNIGHT
1	2	3	4	5	6	7A	7B	8	9	10	11	12	13	14	15	16	17	18	19	20	21	22	23
01	59*	26	43	13	14	22	0		T	0.00	0.0	24.750	19	3.9	6.3	15	S	14	20	552	98	3	4
02	35	12	24	-6	14	41	0	2	8	0	0.01	0.1	24.865	04	1.9	5.0	16	N	14	36	467	83	7	6
03	21	14	18	-12	16	47	0	2		T	0.05	0.7	24.860	03	1.9	3.4	14	N	10	36	1	0	10	10
04	28	16	22	-8	17	43	0	1	8	T	0.00	0.0	24.560	03	1.1	2.9	10	S	6	14	127	22	9	7
05	42	15	29	0	17	36	0			T	0.00	0.0	24.640	08	4.3	6.3	21	E	13	07	317	56	6	4
06	44	10	27	-2	14	38	0	1		T	0.00	0.0	24.800	04	2.1	3.8	16	N	12	01	494	87	0	0
07	55	8	32	3	8	33	0	1		T	0.00	0.0	24.690	15	2.6	5.3	15	SE	12	21	477	84	7	4
08	49	21	35	6	13	30	0			0	0.00	0.0	24.675	22	0.9	7.3	29	NW	17	29	439	77	4	3
09	37	15	26	-3	16	39	0			0	0.00	0.0	24.700	22	0.4	4.0	12	N	9	36	214	37	10	8
10	50	14	32	3	15	33	0			0	0.00	0.0	24.630	12	3.1	5.1	21	E	14	07	489	85	1	2
11	48	19	34	5	11	31	0			0	0.00	0.0	24.790	24	1.6	6.7	24	NW	15	32	485	84	1	1
12	55	26	41	12	15	24	0			0	0.00	0.0	24.705	24	7.1	9.9	31	W	22	29	437	76	4	3
13	55	32	44*	15	17	21	0			T		T	24.610	26	3.7	6.9	40	W	29	29	198	34	10	8
14	46	23	35	6	22	30	0			T		T	24.640	04	0.7	6.0	23	NE	14	05	325	56	6	5
15	43	22	33	4	16	32	0			0	0.06	0.8	24.540	09	1.0	6.2	18	N	13	33	20	3	9	8
16	39	20	30	1	23	35	0	2		2	0.10	2.1	24.770	33	1.3	4.5	21	N	14	01	281	48	7	6
17	42	15	29	0	16	36	0			2	0.00	0.0	24.780	18	2.2	3.8	14	S	8	19	302	52	9	5
18	51	21	36	7	15	29	0			T	0.00	0.0	24.680	18	7.0	7.2	20	S	15	19	582	99	1	0
19	50	18	34	5	20	31	0	1		T	0.23	4.6	24.550	04	4.5	10.0	31	N	23	06	83	14	9	7
20	31	6	19	-10	9	46	0	1		4	0.03	0.3	24.840	16	3.1	4.0	15	SE	7	18	550	94	1	4
21	32	6	19	-10	8	46	0			3	0.00	0.0	24.760	17	3.2	4.4	17	S	10	19	537	91	1	2
22	36	17	27	-2	10	38	0			2	0.00	0.0	24.550	17	5.0	6.5	17	S	14	19	376	64	5	3
23	27	9	18	-12	14	47	0			2	0.07	0.7	24.660	14	1.2	7.9	23	N	16	01	174	29	9	6
24	37	8	23	-7	9	42	0			3	0.02	0.2	24.570	11	1.0	4.7	23	N	18	36	309	52	8	6
25	24	-1	12	-18	6	53	0	1		4	0.07	1.9	24.770	05	2.2	4.3	20	N	15	01	537	90	0	3
26	41	-4*	19	-11	5	46	0			4	0.00	0.0	24.560	22	1.7	4.9	25	W	17	28	552	92	3	1
27	43	19	31	1	11	34	0			4	0.00	0.0	24.390	29	0.9	8.4	38	NW	28	31	460	77	8	5
28	45	7	26	-4	10	39	0		9	3	0.08	2.0	24.380	02	3.4	10.9	37	N	30	36	342	57	7	8
29	17	0	9*	-21	-1	56	0			5	0.04	0.7	24.570	06	2.4	5.1	20	NE	12	01	562	93	1	4
30	47	8	28	-3	6	37	0			5	0.00	0.0	24.660	26	5.3	10.0	33	W	17	29	591	98	1	2
31	53	20	37	6	12	28	0			4	0.00	0.0	24.830	17	4.7	5.7	17	SW	10	19	120	20	7	8

| | SUM 1282 | SUM 442 | | | | TOTAL 1143 | TOTAL 0 | NUMBER OF DAYS | | TOTAL 0.76 | TOTAL 14.1 | 24.670 | 17 | 0.8 | 6.0 | FOR THE MONTH: 40 | W | 30 | 36 | TOTAL 11400 | % 164 | SUM 142 | SUM |
| | AVG. 41.4 | AVG. 14.3 | AVG. 27.9 | DEP. -1.6 | AVG. 12.7 | DEP. 42 | DEP. 0 | PRECIPITATION ≥ .01 INCH. 11 | | DEP. 0.25 | | | | DATE:13 | | DATE: 28 | POSSIBLE 18088 | MONTH 63 | AVG. 5.3 | AVG. 4.6 |

NUMBER OF DAYS				SEASON TO DATE TOTAL 3444	TOTAL 0	SNOW, ICE PELLETS ≥ 1.0 INCH 4	GREATEST IN 24 HOURS AND DATES				GREATEST DEPTH ON GROUND OF SNOW, ICE PELLETS OR ICE AND DATE		
MAXIMUM TEMP.		MINIMUM TEMP.		DEP.	DEP.	THUNDERSTORMS 0	PRECIPITATION		SNOW, ICE PELLETS				
≥ 90°	≤ 32°	≤ 32°	≤ 0°			HEAVY FOG 3	0.26	19-20	4.9	19-20	5	30+
0	7	31	3			CLEAR 12 PARTLY CLOUDY 8 CLOUDY 11						

* EXTREME FOR THE MONTH - LAST OCCURRENCE IF MORE THAN ONE.
T TRACE AMOUNT.
+ ALSO ON EARLIER DATE(S).
HEAVY FOG: VISIBILITY 1/4 MILE OR LESS.
BLANK ENTRIES DENOTE MISSING OR UNREPORTED DATA.

DATA IN COLS 6 AND 12-15 ARE BASED ON 21 OR MORE OBSERVATIONS AT HOURLY INTERVALS. RESULTANT WIND IS THE VECTOR SUM OF WIND SPEEDS AND DIRECTIONS DIVIDED BY THE NUMBER OF OBSERVATIONS. COLS 16 & 17: PEAK GUST - HIGHEST INSTANTANEOUS WIND SPEED. ONE OF TWO WIND SPEEDS IS GIVEN UNDER COLS 18 & 19: FASTEST MILE - HIGHEST RECORDED SPEED FOR WHICH A MILE OF WIND PASSES STATION (DIRECTION IN COMPASS POINTS). FASTEST OBSERVED ONE MINUTE WIND - HIGHEST ONE MINUTE SPEED (DIRECTION IN TENS OF DEGREES). ERRORS WILL BE CORRECTED IN SUBSEQUENT PUBLICATIONS.

I CERTIFY THAT THIS IS AN OFFICIAL PUBLICATION OF THE NATIONAL OCEANIC AND ATMOSPHERIC ADMINISTRATION, AND IS COMPILED FROM RECORDS ON FILE AT THE NATIONAL CLIMATIC DATA CENTER

noaa   NATIONAL OCEANIC AND ATMOSPHERIC ADMINISTRATION   NATIONAL ENVIRONMENTAL SATELLITE, DATA AND INFORMATION SERVICE   NATIONAL CLIMATIC DATA CENTER ASHEVILLE NORTH CAROLINA   Kenneth D. Hadeen   DIRECTOR NATIONAL CLIMATIC DATA CENTER

Fig. 1.3   Local climatological data.

The National Climatic Data Center in North Carolina is responsible for maintaining climatological data from National Weather Service offices. These data are available in many forms, including local climatology data by month, data by state, and data for the world. It also maintains historical climatology data beginning with 1931. Figure 1.3 contains a month summary of local climatology data that was collected by the National Weather Service office at Stapleton International Airport in Denver, Colorado, for January 1991. The summary contains 23 different pieces of weather information collected for each day, including maximum and minimum temperatures, amount of precipitation, peak wind gust, and minutes of sunshine. These data are then analyzed to generate the monthly summary information at the bottom of the form, which includes average temperature, total rainfall, total snowfall, and the number of days that were partly cloudy.

To analyze this weather data for one month, one of the first things that we might do is plot some of the different pieces of data in order to see if we could observe any visible trends in the data. For example, we should be able to observe if the maximum temperature seems to stay the same, increases, decreases, or fluctuates around a common point. Figure 1.4 contains a graph of the maximum temperatures for January 1991. We can see that the temperature has some wide fluctuations or

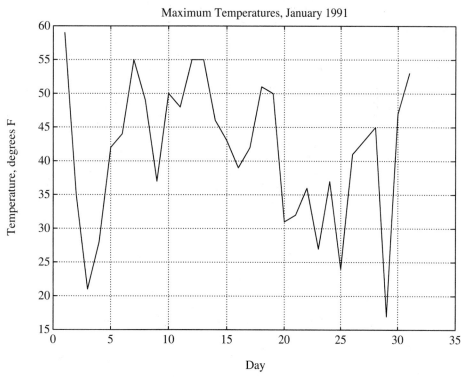

Fig. 1.4   Maximum temperatures for January 1991.

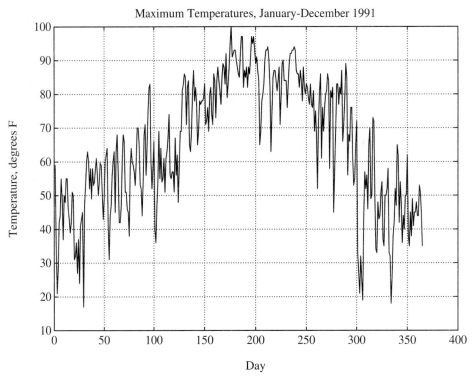

Fig. 1.5   Maximum temperatures for 1991.

variations, but there is no steady increase or decrease. We could also analyze temperature data for a year in the same way; Figure 1.5 contains the maximum daily temperatures from January to December 1991. To observe gradual warming trends, it would be important to look at the temperatures over many years.

We often are interested in analyzing several different sets of data at the same time to see if there are relationships. For example, we would expect that the maximum temperatures and the average temperatures over a period of time would be related. That is, we expect that days with higher maximum temperatures will also have higher average temperatures. However, we also expect that on some days, the maximum temperature and the average temperature are close together, while on other days, the maximum temperature and the average temperature are not close together. Figure 1.6 contains plots of the maximum temperatures and the average temperatures for the month of January 1991 at Stapleton International Airport, and illustrates this relationship between the maximum temperatures and the average temperatures.

Graphs give us a quick intuitive feel for data trends, but we need more analytical methods for using past history to predict the future. Analytical methods commonly used to model data include linear regression, cubic splines, and polynomial

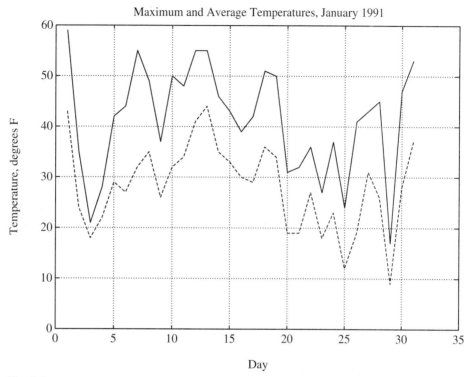

Fig. 1.6   Maximum and average temperatures for January 1991.

models. MATLAB provides easy access to both the graphics and the analytical methods to analyze data and search for trends. We will discuss both these capabilities of MATLAB later in this text.

SUMMARY

The description of the top 10 engineering achievements and the discussion of the grand challenges will hopefully spark your interest in learning more about engineering disciplines. Since engineering is built around problem solving, it is important to begin your investigation with a solid methodology for solving problems, and then to learn how to use the computer to help solve the problems encountered. In this chapter we presented the following five-step process for developing problem solutions:

1. State the problem clearly.
2. Describe the input and output information.
3. Work the problem by hand (or with a calculator) for a simple set of data.
4. Develop a MATLAB solution.
5. Test the solution using a variety of data sets.

This five-step process will be illustrated throughout the text as we develop solutions to problems from various applications.

PROBLEMS

1. Write a short report on one of the outstanding engineering achievements listed below:

Moon Landing	Composite Materials
Application Satellites	Jumbo Jets
Microprocessors	Lasers
CAD/CAM	Fiber-Optic Communications
CAT Scans	Genetically Engineered Products

   Use at least two references. A good starting point for finding references is the set of suggested readings that follow this problem set.

2. Write a short report on one of the grand challenges listed below:

   Predication of Weather, Climate, and Global Change
   Speech Recognition and Understanding
   Machine Vision
   Vehicle Performance
   Superconductivity
   Enhanced Oil and Gas Recovery
   Nuclear Fusion

   Use at least two references. A good starting point for finding references is the set of suggested readings that follow this problem set.

3. Write a short report discussing the collection of climatological data. A good source of information would be the national weather service office near you. Your university library may contain additional sources of climatological data in its government records. Climatological information can also be ordered from the National Climatic Data Center in Asheville, North Carolina.

SUGGESTED READINGS

For further reading on the top engineering achievements of the past 25 years and on the grand challenges, we recommend the following articles from *Scientific American*. Several of these references are from the National Academy of Engineering's 1989 brochure entitled "10 Outstanding Achievements 1964–1989."

BARTON, JOHN H.   "Patenting Life." *Scientific American*, March 1991, pp. 40–46.
BERNS, MICHAEL W.   "Laser Surgery." *Scientific American*, June 1991, pp. 84–90.
CAVA, ROBERT J.   "Superconductors beyond 1-2-3." *Scientific American*, August 1990, pp. 42–49.

CHOU, TSU-WEI, ROY L. MCCULLOUGH, and R. BYRON PIPES. "Composites." *Scientific American,* October 1986, pp. 192–203.

CONN, ROBERT W., et al. "The International Thermonuclear Experimental Reactor." *Scientific American,* April 1992, pp. 103–110.

CORCORAN, ELIZABETH. "Calculating Reality." *Scientific American,* January 1991, pp. 100–109.

CURL, ROBERT. F., and RICHARD E. SMALLEY. "Fullerenes." *Scientific American,* October 1991, pp. 54–63.

DESURVIRE, EMMANUELF. "Lightwave Communications: The Fifth Generation." *Scientific American,* January 1992, pp. 114–121.

DREXHAGE, MARTIN G., and CORNELIUS T. MOYNIHAN. "Infrared Optical Fibers." *Scientific American,* November 1988, pp. 110–116.

GASSER, CHARLES S., and ROBERT T. FRALEY. "Transgenic Crops." *Scientific American,* June 1992, pp. 62–69.

GREENBERG, DONALD P. "Computers and Architecture." *Scientific American,* February 1991, pp. 104–109.

HESS, WILMOT, et al. "The Exploration of the Moon." *Scientific American,* October 1969, pp. 55–72.

JEWELL, JACK L., JAMES P. HARBISON, and AXEL SCHERER. "Microlasers." *Scientific American,* November 1991, pp. 86–94.

JOHNSTON, ARCH C., and LISA R. KANTER. "Earthquakes in Stable Continental Crust." *Scientific American,* March 1990, pp. 68–75.

MAHOWALD, MISHA A., and CARVER MEAD. "The Silicon Retina." *Scientific American,* May 1991, pp. 76–82.

MATTHEWS, DENNIS L., and MORDECAI D. ROSEN. "Soft-X-Ray Lasers." *Scientific American,* December 1988, pp. 86–91.

RICHELSON, JEFFREY T. "The Future of Space Reconnaissance." *Scientific American,* January 1991, pp. 38–44.

ROSS, PHILIP E. "Eloquent Remains." *Scientific American,* May 1992, pp. 114–125.

STEINBERG, MORRIS A. "Materials for Aerospace." *Scientific American,* October 1986, pp. 67–72.

VELDKAMP, WILFRID B., and THOMAS J. MCHUGH. "Binary Optics." *Scientific American,* May 1992, pp. 92–97.

WALLICH, PAUL. "Silicon Babies." *Scientific American,* December 1991, pp. 124–134.

Courtesy of National Aeronautics and Space Administration

GRAND CHALLENGE: Vehicle Performance

Wind tunnels are test chambers built to generate precise wind speeds. Accurate scale models of new aircraft can be mounted on force-measuring supports in the test chamber, and measurements of the forces on the model can then be made at many different wind speeds and angles of the model relative to the wind direction. Some wind tunnels can operate at hypersonic velocities, generating wind speeds of thousands of miles per hour. The size of wind tunnel test sections vary from a few inches across to sizes large enough to accommodate a business jet. At the completion of a wind tunnel test series, many sets of data have been collected that can be used to determine the lift, drag, and other aerodynamic performance characteristics of a new aircraft at its various operating speeds and positions.

An Introduction to MATLAB

Introduction

The computer is an essential tool in engineering problem solving. In this chapter we present a brief introduction to computing systems, including both computer hardware and computer software. We then turn to MATLAB, which is the premier software package for interactive numeric computation, data analysis, and graphics. After a brief introduction to MATLAB, we discuss the ways of representing data in MATLAB and then use an example to illustrate the use of MATLAB to generate a plot of x-y data. With this capability, you will be able to use MATLAB to prepare graphs for your next assignments in your engineering, science, and mathematics courses.

2.1 COMPUTING SYSTEMS

Before we begin discussing MATLAB, a brief discussion on computing [2] is useful especially for those who have not worked with computers prior to this. A computer is a machine that is designed to perform operations that are specified with a set of instructions called a program. Computer hardware refers to the computer equipment, such as the keyboard, the mouse, the terminal, the hard disk, and the printer. Computer software refers to the programs that describe the steps that we want the computer to perform.

Computer Hardware

All computers have a common internal organization, as shown in Figure 2.1. The processor is the part of the computer that controls all the other parts. It accepts input values (from a device such as a keyboard) and stores them in the memory. It also interprets the instructions in a computer program. If we want to add two values, the processor will retrieve the values from memory and send them to the arithmetic logic unit, or ALU. The ALU performs the addition, and the processor then stores the result in memory. The processing unit and the ALU use a small amount of memory, called the internal memory, in their processing; most data are stored in external memory or secondary memory using hard disk drives or floppy disk drives that are attached to the processor. The processor and ALU together are called the central processing unit, or CPU. A microprocessor is a CPU that is contained in a single integrated circuit chip that contains thousands of components in an area smaller than your fingernail.

    The computed values can be displayed on the terminal screen or printed to paper. Dot matrix printers use a matrix (or grid) of pins to produce the shape of a

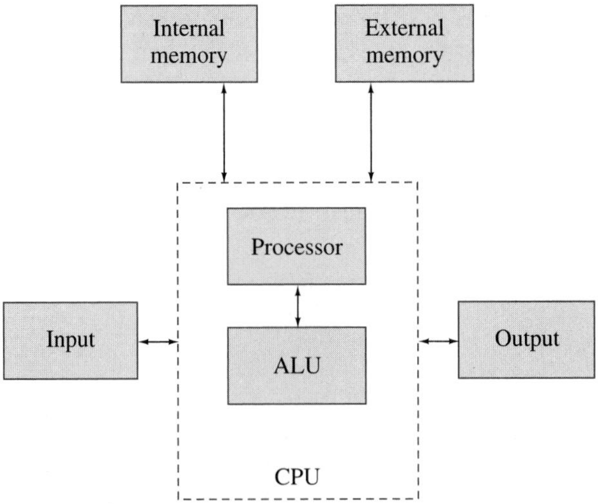

ALU

CPU

printers

Fig. 2.1   Internal organization of a computer.

character on paper, while a laser printer uses a light beam to transfer images to paper. The computer can also write information to diskettes, which store the information magnetically. A printed copy of information is called a hard copy, and a magnetic copy of information is called an electronic copy.

personal
computers

Computers come in all sizes, shapes, and forms. Personal computers (PCs) are small, inexpensive computers that are commonly used in offices, homes, and laboratories; PCs are also referred to as microcomputers. Minicomputers and mainframes are more powerful computers that are often used in businesses and research laboratories. A workstation is a minicomputer or mainframe computer that is small enough to fit on a desktop. Supercomputers are the fastest of all computers and are capable of solving very complex problems that cannot be feasibly solved on other computers. Mainframes and supercomputers require special facilities and a specialized staff to run and maintain the computer systems.

The type of computer needed to solve a particular problem depends on the problem requirements. If the computer is part of a home security system, a microprocessor is sufficient; if the computer is running a flight simulator, a mainframe is probably needed. Computer networks allow computers to communicate with one another so that they can share resources and information.

Computer Software

Computer software contains the instructions or commands that we want the computer to perform. There are several important categories of software; we discuss these separately.

Operating Systems   Some software, such as the operating system, typically comes with the computer hardware when it is purchased. The operating system provides an interface between you (the user) and the hardware by providing a convenient and efficient environment in which you can select and execute the software on your system. Operating systems also contain a group of programs called utilities that allow you to perform functions such as printing files, copying files from one diskette to another, and listing the files that you have saved on a diskette. While these utilities are common to most operating systems, the commands themselves vary from computer to computer. For example, to list the files in the current directory using DOS (a disk operating system used mainly with PCs), the command is dir; to list the files with Unix (a powerful operating system used frequently with workstations), the command is ls. Some operating systems simplify the interface with the operating system; examples of user-friendly systems are the Macintosh environment and the Windows environment.

word processors

Software Tools   Software tools are programs that have been written to perform common operations. For example, word processing programs such as Microsoft Word and Word Perfect help you enter and format text. Word processors allow you to move sentences and paragraphs and often include the capabilities to check your spelling and grammar. Very sophisticated word processors allow you to produce well-designed pages that combine elaborate charts and graphics with text and head-lines; these word processing programs use a technology called desktop publishing, which combines a very powerful word processing program with a high-quality printer to produce professional-looking documents.

spreadsheets

Spreadsheet programs are software tools that allow you to easily work with data that can be displayed in a grid of rows and columns. Spreadsheets were initially used for financial and accounting applications, but many science and engineering problems can easily be solved using spreadsheets. Most spreadsheet packages include plotting capabilities, so they can be especially useful in analyzing and display-ing information. Lotus 1-2-3, Quattro, and Excel are popular spreadsheet packages.

database
management

Another popular group of software tools are database management programs, such as dBase III. These programs allow you to store a large amount of data, easily retrieve pieces of the data, and then format them into reports. Databases are used by large organizations such as banks, hospitals, hotels, and airlines. Scientific databases are also used to analyze large amounts of data. Meteorology data are an example of scientific data that require large databases for storage and analysis.

graphics

Graphics packages are software tools that allow you to create a variety of graphs with your data. These packages allow you to generate *x-y* graphs, bar graphs, three-dimensional graphs, and contour plots. Graphics packages usually include the software to interface to a variety of graphics printers. Computer-aided design pack-ages such as AutoCAD, AutoSketch, and MathCAD also have graphics capabilities.

mathematical
computation

There are also some very powerful mathematical computational software tools, such as MATLAB and Mathematica. Not only do these tools have very power-ful mathematical commands, but they also provide extensive capabilities for generat-ing graphs. This combination of computational power and visualization power makes them particularly useful tools for engineers.

Computer Languages   Machine language is the language that is understood by the computer hardware. Since computer designs are based on two-state technology (devices with two states such as open or closed circuits, on or off switches, positive or negative charges), machine language is written using two symbols that are usually
machine language
binary language
represented using the digits 0 and 1. Therefore, machine language is also a binary language, and the instructions are written as sequences of 0's and 1's called binary strings. Machine language (also called a low-level language) is closely tied to the de-sign of the computer hardware, and thus the machine language for a Sun computer is different from the machine language for a VAX computer.

high-level
language

High-level languages are computer languages that have English-like commands and instructions, and include languages such as C, Fortran, Pascal, and Basic. Writ-ing programs in high-level languages is certainly easier than writing programs in machine language. However, in a high-level language, your program must include a great deal of detail on the steps that you want to perform. In addition, a high-level language contains a large number of commands and an extensive set of syntax (or grammar) rules for using the commands.

If a problem can be easily solved using a software tool, then we suggest using the software tool because the problem can be solved more quickly and easily. However, if the problem cannot be easily solved using a software tool, then a computer language gives you the flexibility to write the necessary program.

2.2   GENERAL MATLAB INFORMATION

matrix laboratory

The MATLAB software was originally developed to be a "matrix laboratory." Today's MATLAB has capabilities far beyond the original MATLAB, and is an interactive system and programming language for general scientific and technical computation. Its basic element is a matrix (which we discuss in detail in the next section). Because the MATLAB commands are similar to the way that we express engineering steps in mathematics, writing computer solutions in MATLAB is much quicker than writing computer solutions using a high-level language such as C or Fortran. In this section we explain the differences between the student version and the professional version of MATLAB, and we give you some initial workspace information.

Student Edition

The student edition is identical to the professional version 3.5 of MATLAB except for four features:

- Each vector or matrix is limited to 1,024 elements.
- The metafile and graphics postprocessor feature for graphics hardcopy is not available.
- A math coprocessor is not required but will be used if it is available.
- A Signals and Systems Toolbox is included for student use.

If your copy of this text also includes the software for the student edition, be sure to complete the registration card. As a registered student user, you are entitled to replacement of defective disks at no charge. You also qualify for a discount on upgrades to professional versions of MATLAB.

If MATLAB is not already installed on your computer, refer to either Appendix C ("MATLAB on MS-DOS PCs") or Appendix D ("MATLAB on Macintosh Computers") for instructions on installing MATLAB. These appendices also contain information specific to MATLAB on MS-DOS PCs or to MATLAB on Macintosh computers.

Workspace Information

matlab
>>
demo
quit
exit
save

To begin MATLAB, select the MATLAB program from a menu in your operating system or by entering matlab with the keyboard. You should see the MATLAB prompt (>>), which tells you that MATLAB is waiting for you to enter a command. If you are a first-time user of MATLAB, use the command demo to obtain a list of demonstrations that you can run to illustrate some of the capabilities of MATLAB. The commands to exit MATLAB are quit or exit. However, if you want to save the variables in the workspace before you quit, use the command save. This

load
computer

clc
clg
clear

ˆc

casesen off
casesen

who
whos

ans
size

%

help

automatically saves all the variables in a file called matlab.mat. When you enter MATLAB again, you can restore these variables with the command load. The computer command prints a character string that specifies the type of computer on which MATLAB is running.

MATLAB uses two windows: A command window is used to enter commands and data and to print results, and a graphics window is used to generate plots. Both windows are cleared when you begin a MATLAB session. If you want to clear the command window during a work session, use the command clc. To clear the graphics window, use the command clg. You can also clear the workspace (which contains all the variables that you have defined) using the command clear.

It is also important to know how to abort a command in MATLAB. For example, there may be times when your commands cause the computer to print seemingly endless lists of numbers or when the computer seems to go into an endless loop. In these cases, hold down the control key and then press the letter c to generate a local abort within MATLAB. The control-c sequence is sometimes written as ˆc; however, this can be confusing because the sequence does not include the ˆ character.

MATLAB is a case-sensitive language, which means that we can assign one value to the variable named TIME, another value to the variable named time, and another value to the variable named Time. If you do not want the commands to be case sensitive, you can use the command casesen off. Case sensitivity can also be turned on and off with the command casesen, which changes (or toggles) the case-sensitivity setting each time that the command is executed.

Several commands are useful for checking the status of the variables and matrices that you have defined in a program. The command who lists the variables that you have defined. The command whos lists the variables that you have defined, along with their sizes and whether they have nonzero imaginary parts. In addition to listing the variables that you have used, who and whos will list a variable named ans. This variable is used to store the value of an expression that is computed but not given a name. We will discuss ans more in Chapter 3. The size command is used to request the size of a specific matrix, as in size(A).

MATLAB commands are typically entered on separate lines, although multiple statements can be entered on the same line if they are separated by semicolons. Text following a percent sign (%) is used to give explanations to someone reading the commands and is ignored by MATLAB. In our examples, we use these comments to initially describe the purpose of a MATLAB program and then to document the individual statements. This is useful when someone needs to understand or modify our programs, and it is also useful when we need to modify a program that we wrote and have not used recently.

The help facility is an important resource in MATLAB. When you execute the command help, a list of help topics is generated. You can then select the topic for which you would like further information. Using this resource allows you to have access to all of MATLAB's capabilities without constantly carrying a text or manual around with you.

M-Files

In addition to executing commands entered through the keyboard, MATLAB is also capable of executing sequences of commands that are stored in files with names that have an extension of .m, as in program1.m or labtest.m. These files are called M-files because of the filename extension. Most of the examples that we use are short enough to be entered through the keyboard. However, when you have long sequences of commands, or when you want to save a sequence of commands to use later, you should use an M-file. An M-file, also called a script file, is a file of ASCII

script file

characters and is generated using an editor or word processor. (ASCII stands for American Standard Code for Information Interchange and is a common code used to store information.)

To invoke the script, enter the name of the M-file, without the extension. Thus, to execute the commands in the M-file labtest.m, we would enter the command labtest. Any variables used in the script are available when the script is finished. Normally, while an M-file is executing, the commands are not displayed on

echo

the screen. The command echo causes M-files to be viewed as they execute.

M-files can also be used to write your own MATLAB functions. In Chapter 3 we discuss the functions that are automatically available in MATLAB and how to write your own additional functions.

what
type

The what command shows a directory listing of the M-files that are available in the current directory on your disk. The type command will list the contents of a specified filename; if the filename used with this command does not have an extension, the command will assume that the file is an M-file.

2.3   MATRICES, VECTORS, AND SCALARS

When solving engineering problems, it is important to be able to visualize the data related to the problem. Sometimes the data is just a single number, such as the radius of a circle. Other times the data may be a coordinate on a plane that can be represented as a pair of numbers, with one number representing the x-coordinate and the other number representing the y-coordinate. In another problem we might have a set of four x-y-z coordinates that represent the four vertices of a pyramid with a triangular base in a three-dimensional space. We can represent all these examples us-

matrix

ing a special type of data structure called a matrix, where a matrix is a set of numbers arranged in a rectangular grid of rows and columns. Thus, a single point can be considered to be a matrix with one row and one column, an x-y coordinate can be considered to be a matrix with one row and two columns, and a set of four x-y-z coordinates can be considered to be a matrix with four rows and three columns. Examples of these matrices are shown below:

$$A = [3.5]$$
$$B = [1.5 \quad 3.1]$$

$$C = \begin{bmatrix} -1 & 0 & 0 \\ 1 & 1 & 0 \\ 1 & -1 & 0 \\ 0 & 0 & 2 \end{bmatrix}$$

Note that the values within a matrix are written within brackets. In mathematical notation, matrices are usually given names with uppercase letters.

scalar

vector

When a matrix has one row and one column, we can also refer to the number as a scalar. Similarly, when a matrix has one row or one column, we can refer to it as a row vector or a column vector.

subscripts

When we use a matrix, we need a way to refer to individual elements or numbers in the matrix. A simple method for specifying an element in the matrix uses the row and column numbers. For example, if we refer to the value in row 4 and column 3 in the matrix C in the previous example, there is no ambiguity—we are referring to the value 2. We use the row and column numbers as subscripts, and thus $C_{4,3} = 2$. To refer to the entire matrix, we use the name without subscripts, as in C, or we use brackets around an individual element reference that uses letters instead of numbers for the subscripts, as in $[C_{i,j}]$. In formal mathematical notation, the matrix name is usually an uppercase letter, with the subscripted references using a lowercase letter, as in C, $c_{4,3}$ and $[c_{i,j}]$. However, since MATLAB is case sensitive, C and c represent different matrices. Therefore, in your MATLAB programs, you will need to consistently use all uppercase or all lowercase letters in references to a specific matrix.

The size of a matrix is specified by the number of rows and columns. Thus, using our previous examples, C is a matrix with four rows and three columns, or a 4 × 3 matrix. If a matrix contains m rows and n columns, then it contains a total of m · n numbers. If a matrix has the same number of rows as columns, it is called a square matrix.

Practice!

Answer the following questions about this matrix:

$$G = \begin{bmatrix} 0.6 & 1.5 & 2.3 & -0.5 \\ 8.2 & 0.5 & -0.1 & -2.0 \\ 5.7 & 8.2 & 9.0 & 1.5 \\ 0.5 & 0.5 & 2.4 & 0.5 \\ 1.2 & -2.3 & -4.5 & 0.5 \end{bmatrix}$$

1. What is the size of G?

2. Is G a square matrix?

3. Give the references for all locations that contain the value 0.5.

4. Give the references for all locations that contain negative values.

Initializing Matrices

We are now ready to define some matrices using MATLAB. We present four methods for initializing matrices. The first method explicitly lists the values, the second method reads the data from a data file, the third method uses the colon operator, and the fourth method accepts input values from the keyboard.

Explicit Lists   The simplest way to define a matrix is to define it using a list of numbers in brackets as shown in the following examples that define the matrices A, B, and C that we used in our previous example:

```
A = [3.5];
B = [1.5,3.1];
C = [-1,0,0;  1,1,0;  1,-1,  0;  0,0,2];
```

These statements are examples of the assignment statement in MATLAB that allows us to define a matrix by explicitly giving the values. The name of the matrix must start with a letter and can contain up to 19 characters that are digits, letters, or the underscore character, and appears on the left side of an equal sign. The right side of the equal sign contains the data enclosed in brackets in row order. Semicolons separate the rows, and the values in the rows can be separated by commas or blanks. A value can contain a plus or minus sign, and a decimal point, but it cannot contain a comma, as in 32,010. When we define a matrix, MATLAB will print the value of the matrix on the next line unless we suppress the printing with a semicolon after the definition. Try entering the three previous statements without the final semicolon. In our examples, we will generally include the semicolon to suppress printing. However, when you are first learning to define matrices, it is helpful to see the matrix. Therefore, you may want to omit the semicolon after a matrix definition until you are confident that you know how to properly define matrices.

suppress printing

A matrix can also be defined by listing each row on a separate line, as in the following set of MATLAB commands:

```
C = [-1,0,0
      1,1,0
      1,-1,0
      0,0,2];
```

ellipsis

If there are too many numbers in a row of the matrix to fit on one line, you can end one line with an ellipsis consisting of three or more periods followed by the carriage return, and then continue the statement on the next line. For example if we want to define the row vector F to have 10 values, we could use either of the following statements:

```
F = [1,52,64,197,42,-42,55,82,22,109];
F = [1,  52,  64,  197,  42,  -42,  ...
         55,  82,  22,  109];
```

MATLAB also allows you to define a matrix using another matrix that has already been defined. For example, consider the following statements:

```
B = [1.5, 3.1];
S = [3.0 B];
```

These commands are equivalent to the following:

```
S = [3.0 1.5 3.1];
```

We can also change values in a matrix or add additional values using a subscript reference in parentheses. Thus, the following command:

```
S(2) = -1.0;
```

changes the second value in the matrix S from 1.5 to -1.0.

You can also extend a matrix by defining new elements. If we execute the following command:

```
S(4) = 5.5;
```

then the matrix S will have four values instead of three. If we execute the following command:

```
S(8) = 9.5;
```

then the matrix S will have eight values, and the values of S(5), S(6), and S(7) are automatically set to zero, since no values were given for them.

Practice!

Give the sizes of the following matrices. Check your answers using MATLAB.

1. `A = [1,0,0,0,0,1];`

2. `B = [ 2; 4; 6; 10];`

3. `C = [5 3 5; 6 2 -3];`

4. `D = [     3 4`
   `       5 7`
   `       9 10 ];`

5. `E = [ 3 5 10 0; 0 0 ...`
   `            0 3; 3 9 9 8];`

```
6.  T = [ 4 24 9];
    Q = [ T 0 T];

7.  X = [3 6];
    Y = [ D; X];

8.  R = [ C; X, 5 ];

9.  V = [ C(2,1); B];

10. A(2,1) = -3;
```

MAT files
ASCII files

Data Files   Matrices can also be defined from information that has been stored in a data file. MATLAB can interface to two different types of data files—MAT-files and ASCII files. A MAT-file contains data stored in a memory-efficient binary format, while an ASCII file contains information stored in ASCII characters. MAT-files are preferable for data that are going to be generated and used by MATLAB programs. ASCII files are necessary if the data are to be shared (imported or exported) with programs other than MATLAB programs.

file extension

MAT-files are generated by a MATLAB program using the save command, which contains the filename and the matrices to be stored in the file. The .mat extension is automatically added to the filename. For example, the following command:

```
save data1 x y;
```

will save the matrices x and y in a file named data1.mat. To restore these matrices in a MATLAB program, use the command:

```
load data1;
```

(In the previous discussion on the MATLAB workspace, we discussed the form of the load and save commands, which save and restore all the variables in the workspace.)

An ASCII data file that is going to be used with a MATLAB program must contain only numeric information, and each row of the file must contain the same number of data values. The file can be generated using a word processing program or an editor. It can also be generated by running a program written in a computer language, such as a Fortran program. It can also be generated by a MATLAB program using the following form of the save command:

```
save data1.dat z /ascii;
```

Each row of the matrix z will be written to a separate line in the data file. The .mat extension is not added to an ASCII file. However, as we illustrated in this example, we recommend that ASCII filenames include the extension .dat so that it is easy to distinguish them from MAT-files and M-files.

Suppose that an ASCII file named data1.dat contained a set of time and distance values, with a time and its corresponding distance value on each line of the data file. Thus, the first few lines in the data file might have the following form:

```
0.00   0.00
0.01   0.1255
0.02   0.2507
```

The load command followed by the filename will read the information into a matrix with the same name as the data file. For example, consider this statement:

```
load data1.dat;
```

The data values will automatically be stored in the matrix data1, which has two columns.

See Appendix B for information on reading or writing MAT-files from Fortran, C, or Pascal programs.

Colon Operator   The colon operator can be used to create vectors from a matrix. This is useful in a number of applications. For example, as we will see in the next section, if we want to generate an *x-y* plot of data, it is convenient to have the x data in a vector and the y data in another vector. When a colon is used in a matrix reference in place of a specific subscript, the colon represents all the rows or all the columns. For example, using the data1 matrix that was read from a data file in the previous discussion, the following commands will store the first column of data1 in the column vector x and the second column of data1 in the column vector y:

```
x = data1 (:,1);
y = data1 (:,2);
```

The colon operator can also be used to generate new matrices. If a colon is used to separate two integers, the colon operator generates all the integers between the two specified integers. For example, the following notation generates a vector named H that contains the numbers from one to eight:

```
H = 1:8;
```

If colons are used to separate three numbers, then the colon operation generates values between the first and third numbers, using the second number as the increment. For example, the following notation generates a row vector named TIME that contains the numbers from 0.0 to 5.0 in increments of 0.5:

TIME = 0.0:0.5:5.0;

The increment can also be negative, as shown in the following example, which generates the numbers 10, 9, 8, . . . 0 in the row vector named VALUES:

VALUES = 10:-1:0;

The colon operator can also be used to select a submatrix from another matrix. For example, assume that C is the following matrix:

$$C = \begin{bmatrix} -1 & 0 & 0 \\ 1 & 1 & 0 \\ 1 & -1 & 0 \\ 0 & 0 & 2 \end{bmatrix}$$

If we then execute the following commands:

C_PARTIAL_1 = C(:,2:3);
C_PARTIAL_2 = C(3:4,1:2);

we have defined the following matrices:

$$C\_PARTIAL\_1 = \begin{bmatrix} 0 & 0 \\ 1 & 0 \\ -1 & 0 \\ 0 & 2 \end{bmatrix} \quad C\_PARTIAL\_2 = \begin{bmatrix} 1 & -1 \\ 0 & 0 \end{bmatrix}$$

If the colon notation defines a matrix with invalid subscripts, as in C(5:6,:), then an error message is displayed.

empty matrix
In MATLAB it is valid to have an empty matrix. For example the following statements will each generate an empty matrix:

a = [];
b = 4:-1:5;

Note that an empty matrix is different from a matrix that contains only zeros.
Finally, the use of the expression C(:) is equivalent to one long column matrix that contains the first column of C, followed by the second column of C, and so on.

We will discuss the colon operator more in later sections because it is a very powerful operator that can be used in many ways.

Practice!

Give the sizes and contents of the following matrices. Use the following matrix G where referenced:

$$G = \begin{bmatrix} 0.6 & 1.5 & 2.3 & -0.5 \\ 8.2 & 0.5 & -0.1 & -2.0 \\ 5.7 & 8.2 & 9.0 & 1.5 \\ 0.5 & 0.5 & 2.4 & 0.5 \\ 1.2 & -2.3 & -4.5 & 0.5 \end{bmatrix}$$

Check your answers using MATLAB.

1. A = G(:, 2);

2. B = G(4,:);

3. C = [10:15];

4. D = [4:9; 1:6];

5. E = [-5,5];

6. F = [0.0:0.1:1.0];

7. T1 = G(4:5,1:3);

8. T2 = G(1:2:5,:);

input

User Input   The values for a matrix can also be entered through the keyboard using the `input` command which displays a text string and then waits for input. The value (or values) entered by the user are then stored in the matrix specified. If the user strikes the return key without entering input values, an empty matrix is returned. If the command does not end with a semicolon, the values entered for the matrix are printed.

Consider the following command:

```
z = input('Enter values for z');
```

When this command is executed, the text string 'Enter values for *z*' is displayed on the terminal screen. The user can then enter an expression such as [5.1 6.3 −18.0] which then specifies values for z. Since this `input` command ends with a semicolon, the values of z are not printed when the command is completed.

Printing Matrices

There are several ways to print the contents of a matrix. The simplest way is to enter the name of the matrix. The name of the matrix will be repeated, and the values of the matrix will be printed starting with the next line. There are also several commands that can be used to print matrices with more control over the form of the output.

Display Format   When elements of a matrix are printed, integers are always printed as integers. Values with decimal fractions are printed using a default format (called a short format) that shows five significant decimal digits. MATLAB allows you to specify other formats that show more significant digits. For example, to specify that we want values to be displayed in a decimal format with 15 significant digits, we use the command `format long`. The format can be returned to a decimal format with five significant digits using the command `format short`.

> format long
> format short

When a value is very large or very small, decimal notation does not work satisfactorily. For example, a value that is used frequently in chemistry is Avogadro's constant, whose value to four significant places is 602,300,000,000,000,000,000,000. Obviously we need a more manageable notation for very large values like Avogadro's constant or for very small values like 0.0000000031. Scientific notation expresses a value as a number between 1 and 10 multiplied by a power of 10. In scientific notation, Avogadro's constant becomes 6.023×10^{23}. This form is also referred to as a mantissa (6.023) and an exponent (23). In MATLAB, values in scientific notation are printed with the letter e to separate the mantissa and the exponent, as in $6.023e+23$. If we want MATLAB to print values in scientific notation with five significant digits, we use the command `format short e`. To specify scientific notation with 15 significant digits, we use the command `format long e`. We can also enter values in a matrix using scientific notation, but it is important to omit any blanks because MATLAB will interpret $6.023\ e+23$ as two values (6.023 and $e+23$), while $6.023e+23$ will be interpreted as one value.

> format short e
> format long e

Another format command is `format +`. When a matrix is printed with this format, the only characters printed are plus and minus signs. If a value is positive, a plus sign will be printed; if a value is zero, a space will be skipped; if a value is negative, a negative sign will be printed. This format allows us to view a large matrix in terms of its signs when otherwise we would not be able to see it easily because the number of values in a row are too many to fit on a single line.

> format +

For long and short formats, a common scale factor is applied to the entire matrix if the elements become very large. This scale factor is printed along with the scaled values.

Finally, the command `format compact` suppresses many of the line-feeds that appear between matrix displays and allows more lines of information to be seen together on the screen. In our example output, we will assume that this command

> format
> compact

format loose   has been executed. The command format loose will return to the less compact display mode.

disp   **Printing Text or Matrices**   The disp command can be used to display text enclosed in single quotation marks. It can also be used to print the contents of a matrix without printing the matrix name. Thus, if a scalar temp contained a temperature value in degrees Fahrenheit, we could print the value on one line plus the units on the next line using these commands:

disp(temp); disp('degrees F')

If the value of temp is 78, then the output will be the following:

    78
degrees F

fprintf   **Formatted Output**   The fprintf command gives you even more control over the output than you have with the disp command. In addition to printing both text and matrix values, you can also specify the format to be used in printing the values. You can also include a skip to a new line within the fprintf command. The general form of this command is the following:

fprintf(format,matrices)

%e
%f
%g
\n
where the format contains the text and format specifications (enclosed in single quotation marks) to be printed along with the matrices listed. Within the text, the specifiers %e, %f, and %g are used to show where the matrix values are printed. If %e is used, the values will be printed in an exponential notation; if %f is used, the values will be printed in a fixed point or decimal notation; if %g is used, the values will use either %e or %f, depending on which is shorter. If the string \n appears in the format, the line specified up to that point is printed, and the rest of the information will be printed on the next line. The format should always end with \n so that the output will end on a line by itself.

A simple example of the fprintf command is shown below:

fprintf('The temperature is %f degrees F \n',temp)

The output would be the following:

The temperature is 78.000000 degrees F

If we modify the command to this form:

fprintf('The temperature is \n %f degrees F \n',temp)

then the output is the following:

```
The temperature is
 78.000000 degrees F
```

The format specifiers %f, %e, and %g can also contain information to specify the number of decimal places to print and the number of positions to allot for the corresponding value. Consider this command:

```
fprintf('The temperature is %4.1f degrees F \n',temp)
```

The output will contain the value of temp printed with 4 positions, one of which will be a decimal position, as shown in the following:

```
The temperature is 78.0 degrees F
```

The fprintf statement allows you to have a great deal of control over the form of the output. We will use it frequently in our examples to help you become familiar with it.

X-Y Plots

Suppose that we want to plot the values in a matrix instead of printing them. MATLAB makes it very easy to plot information. In Chapter 7 we discuss the different types of graphs and the options that are available; in this section we show you how to generate a simple *x-y* plot from data stored in two vectors. Then, even without knowing any other statements, you can immediately begin using MATLAB to generate the plots that you need for your homework assignments or laboratory reports.

Assume that we want to plot the following temperature data that we collected during a physics experiment:

Time, s	Temperature, degrees F
0	54.2
1	58.5
2	63.8
3	64.2
4	67.3
5	71.5
6	88.3
7	90.1
8	90.6
9	89.5
10	90.4

Also assume that we have stored the time values in a vector called x, and that we have stored the temperature values in a vector called y. (The data could have been entered explicitly into the vectors, or read from a data file into a matrix and stored in the vectors using the colon operator, or entered with the input command.) To plot these points, we simply use the plot command, where x and y are either both row vectors or are both column vectors:

```
plot(x,y)
```

The plot in Figure 2.2 is automatically generated. If you are using MATLAB on a Macintosh, your graphics window may contain a plot with a y-axis range of 50 to 100, instead of the range of 50 to 95 as shown in Figure 2.2. This variation is due to automatic scaling for a small graphics window. If you enlarge the graphics window, the y-axis will rescale to the one shown in Figure 2.2. If you print the plot in the graphics window, the printed plot will match the one shown in Figure 2.2, even though the plot in the graphics window may not have the same y-axis range.

Good engineering practice requires that we include units and a brief description with every plot. Therefore, we can improve the graph with the following com-

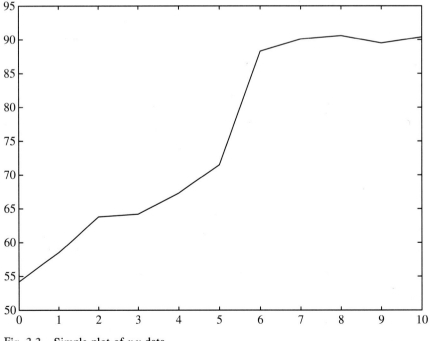

Fig. 2.2   Simple plot of x-y data.

mands that add a title, x and y labels, and a background grid:

```
plot(x,y),...
title('Laboratory Experiment 1'),...
xlabel('Time, s'),...
ylabel('Temperature, degrees F'),...
grid
```

The ellipses after the first four commands were used so that MATLAB would execute these five commands together. Otherwise, the first command would generate the plot; then, since the second command modified the plot, it would be redrawn on the screen. This would happen four times unless we group the commands together as shown. These commands generate the plot in Figure 2.3.

If you display two plots in the same program, or if you display a plot and then continue with more computations, MATLAB will generate and display the first plot and then return immediately to the rest of the commands in the program. Since the

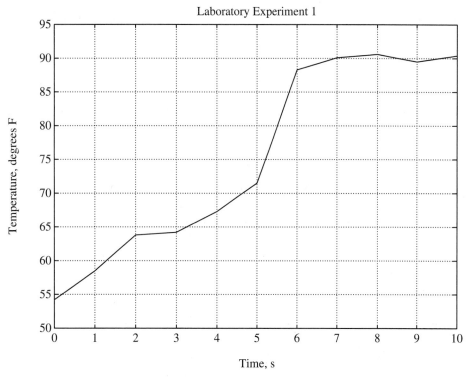

Fig. 2.3   Enhanced plot of *x-y* data.

plot is replaced by the command window when MATLAB returns to finish the computations, you may want to use the pause command to halt the program temporarily to give you a chance to study the plot. Execution will continue when any key is pressed. If you want to pause for a specified number of seconds, use the pause(n) command, which will pause for n seconds before continuing.

If you want to learn more now about the options for generating x-y plots, refer to Chapter 7.

PROBLEM SOLVING APPLIED: WIND TUNNEL DATA ANALYSIS

A wind tunnel is a test chamber built to generate different wind speeds, or Mach numbers (which is the wind speed divided by the speed of sound). Accurate scale models of aircraft can be mounted on force-measuring supports in the test chamber, and then measurements of the forces on the model can be made at many different wind speeds and angles of the model relative to the wind direction. At the end of an extended wind tunnel test, many sets of data have been collected and can be used to determine the lift, drag, and other aerodynamic performance characteristics of the new aircraft at its various operating speeds and positions.

We will use this application several times in our problems throughout this text. In this section, we assume that the data collected from the wind tunnel test have been stored in an ASCII file named wind1.dat. We would like to view a plot of the data to be sure that the sensors on the scale model appear to have been working properly. We assume that each line of the file contains a flight path angle in degrees and a corresponding coefficient of lift. For this example we will use the following data[2]:

Flight Path Angle (degrees)	Coefficient of Lift
−4	−0.202
−2	−0.050
0	0.108
2	0.264
4	0.421
6	0.573
8	0.727
10	0.880
12	1.027
14	1.150
15	1.195
16	1.225

Flight Path Angle (degrees)	Coefficient of Lift
17	1.244
18	1.250
19	1.245
20	1.221
21	1.177

We are using a small set of data for this example, but, even for this data, you can see the advantage of using a data file. If we did not use a data file, then each time that we wanted to use the data we would need to use the MATLAB commands to explicitly define the matrices. It is not only time-consuming to enter this much data by hand; it is also easy to make errors.

Even though reading and plotting these data are simple to do in MATLAB, we will use the five-step process to illustrate the fact that it too is a simple process that allows us to structure our thoughts in developing a problem solution.

 1. PROBLEM STATEMENT
Generate a plot of the flight path angle and coefficient of lift data.

 2. INPUT/OUTPUT DESCRIPTION
Where possible, we use a diagram for I/O, as shown in Figure 2.4. In this example we are reading information from a data file and then using MATLAB to plot it. The diagram contains an icon or symbol for a diskette to represent the data file that is the input. Note that we put the name of the file under the symbol. The output is a plot of the data, so we use another icon to represent a plot. The plot in the icon is not supposed to represent this specific data—it is just a symbol to represent a plot.

 3. HAND EXAMPLE
Even with a plot, we should skim at least part of the data and determine roughly what we expect to see in the plot. If we get a plot that is very different

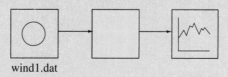
wind1.dat

Fig. 2.4   I/O diagram.

from what we expected, we then know to recheck the MATLAB commands and to recheck the data. In this example, a look at this data tells us that the coefficient of lift is increasing initially from a value of −0.2, and should reach a maximum of 1.25 at 18 degrees.

◆ 4. MATLAB SOLUTION

```
%
%        This program reads and plots the information
%        in a data file containing flight path angles
%        and coefficients of lift.
%
load wind1.dat;            % read data from file to matrix
x = wind1(:,1);            % copy first column into x vector
```

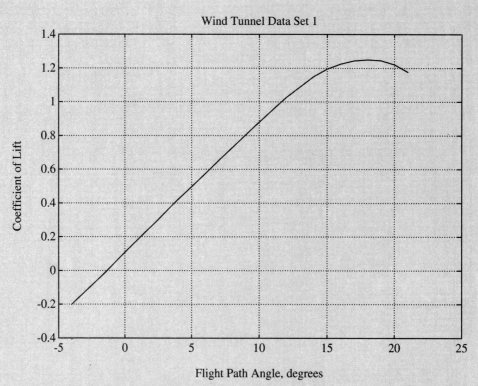

Fig. 2.5 Plot of wind tunnel data.

```
y = wind1(:,2);              % copy second column into y vector
plot(x,y),...                % plot x and y
title('Wind Tunnel Data Set 1'),...
xlabel('Flight Path Angle, degrees'),...
ylabel('Coefficient of Lift'),...
grid
```

◆ 5. TESTING
The test of this MATLAB solution generates the graph in Figure 2.5.

SUMMARY

In this chapter we introduced you to the MATLAB environment. The primary data structure in MATLAB is a matrix, which can be a single point (a scalar), a list of values (a vector), or a rectangular grid of values with rows and columns. Values can be entered into a matrix using an explicit listing of the values. Values can also be loaded into a matrix from a MAT-files or ASCII files. Values can also be entered into a matrix using a colon operator that allows us to specify a starting value, an increment, and an ending value for generating the sequence of values. After a number of examples, we then demonstrated how to generate a simple x-y plot of data values. The chapter closed with an application that plotted data from a wind tunnel test.

MATLAB SUMMARY

This MATLAB summary lists all the special symbols, commands, and functions that were defined in this chapter. A brief description is also included for each one.

Special Characters

[ ]	forms matrices
( )	forms subscripts
,	separates subscripts and matrix elements
;	separate matrix or commands, suppresses printing
>>	prompts user for next command
...	continues command to a new line
%	indicates comments, formats
:	generates matrices
\n	indicates new line
^c	generates a local abort

Commands and Functions

ans	default variable for storing results
casesen off	sets case sensitivity off
casesen	toggles case sensitivity
clc	clears command screen
clear	clears workspace
clg	clears graph screen
computer	prints type of computer
demo	runs demonstrations
disp	displays matrix or text
echo	enables echoing with an M-file
exit	terminates MATLAB
format +	sets format to plus and minus signs only
format compact	sets format to compact form
format long	sets format to long decimal
format long e	sets format to long exponential
format loose	sets format to noncompact form
format short	sets format to short decimal
format short e	sets format to short exponential
fprintf	prints formatted information
grid	draws grid lines on a plot
help	invokes help facility
input	accepts keyboard input
load	loads matrices from a file
matlab	initiates MATLAB
pause	temporarily halts a program
plot	generates a linear x-y plot
quit	terminates MATLAB
save	saves variables in a file
size	prints row and column dimensions
title	adds a title to a plot
type	lists contents of a file
what	lists M-files on disk
who	lists variables in memory
whos	lists variables in memory plus sizes
xlabel	adds x-axis label to a plot
ylabel	adds y-axis label to a plot

PROBLEMS

In Problems 1 to 10, use the colon operator to generate the vectors specified, and then print the corresponding tables. You do not need to do any arithmetic computations in MATLAB to do the following problems, but you may need to compute the increment for the colon operator using a calculator. Include a table heading and column headings as shown here:

```
Angle Conversion
Degrees and Radians
```

You may find it convenient to use the transpose of a vector A which is denoted by A'. The transpose of a row vector is a column vector, and the transpose of a column vector is a row vector. More details on the transpose of a matrix are included on page 158.

1. Generate a table of conversions from degrees to radians. The first line should contain the values for 0°, the second line should contain the values for 10°, and so on. The last line should contain the values for 360°. (Recall that π radians = 180°.)

2. Generate a table of conversions from radians to degrees. Start the radian column at 0.0, and increment by $\frac{\pi}{10}$, until the radian amount is 2π. (Recall that π radians = 180°.)

3. Generate a table of conversions from inches to centimeters. Start the inches column at 0.0, and increment by 0.5 in. The last line should contain the value 20.0 in. (Recall that 1 in. = 2.54 cm.)

4. Generate a table of conversions from centimeters to inches. Start the centimeters column at 0, and increment by 2 cm. The last line should contain the value 50 cm. (Recall that 1 in. = 2.54 cm.)

5. Generate a table of conversions from mi/hr to ft/s. Start the mi/hr column at 0, and increment by 5 mi/hr. The last line should contain the value 65 mi/hr. (Recall that 1 mi = 5,280 ft.)

6. Generate a table of conversions from ft/s to mi/hr. Start the ft/s column at 0, and increment by 5 ft/s. The last line should contain a value entered by the user. (Recall that 1 mi = 5,280 ft.)

The following currency conversions apply to problems 7 to 10:

$$\$1 = 5.3 \text{ francs (Fr)}$$
$$1 \text{ yen (Y)} = \$ .0079$$
$$1.57 \text{ deutsche mark (DM)} = \$1$$

7.   Generate a table of conversions from francs to dollars. Start the francs column at 5 Fr, and increment by 5 Fr. Print 25 lines in the table.

8.   Generate a table of conversions from deutsche marks to francs. Start the deutsche marks column at 1 DM and increment by 2 DM. Allow the user to enter the number of lines of output for the table.

9.   Generate a table of conversions from yen to deutsche marks. Start the yen column at 100 Y, and print 25 lines, with the final line containing the value 10,000 Y.

10.   Generate a table of conversions from dollars to francs, deutsche marks, and yen. Start the column with $1, and increment by $1. Print 50 lines in the table.

Problems 11 to 15 refer to an ASCII data file named `temps.dat` contained on the data disk that accompanies this text. The data file contains 100 lines of data. Each line of data represents information collected during the test of a new jet engine. The first value on the line contains time, with 0.0 representing the time in seconds when the engine was first started. At each time, temperature measurements were made at four different locations on the engine. Thus, each line contains a time, and four temperatures (in degrees Fahrenheit), which we will refer to as T_1, T_2, T_3, and T_4.

11.   Generate a plot with time on the x-axis and T_1 on the y-axis. Label the plot appropriately.

12.   Generate a plot with time on the x-axis and T_2 on the y-axis. Label the plot appropriately.

13.   Generate a plot with time on the x-axis and T_3 on the y-axis. Label the plot appropriately.

14.   Generate a plot with T_1 on the x-axis and T_3 on the y-axis. Label the plot appropriately.

15.   Generate a plot with T_4 on the x-axis and time on the y-axis.

PART II

Courtesy of National Aeronautics and Space Administration

Fundamental Engineering Computations

In this part we present material that is fundamental to basic computations in all engineering disciplines. For example, we present the MATLAB functions for trigonometric computations, exponentials, and logarithms. We also include a discussion on complex numbers that are especially important for the Electrical Engineering and Mechanical Engineering disciplines. In addition to computing values, we also need to be able to ask questions and to generate loops in our MATLAB programs. Therefore, a chapter on control flow presents the structures for asking questions and then performing different operations based on the answers, and for building repetitive loops. A chapter on statistical measurements and their interpretation provides important insight into analyzing and understanding engineering data collected from experiments. We also present functions for generating random numbers which can be used to develop computer simulations or to generate random sequences. MATLAB also contains additional operations and functions that relate specifically to matrices; these are presented in a separate chapter. Finally, since understanding engineering concepts is aided by visual representations, it is important to be able to easily display information in a variety of ways. MATLAB has an extremely powerful set of graphic commands that are presented in the last chapter in Part II.

3

Courtesy of Phillips Petroleum Company

GRAND CHALLENGE: Enhanced Oil and Gas Recovery

Current techniques for identifying structures likely to contain oil and gas use techniques that can evaluate structures down to 20,000 feet below the ground surface. The techniques use a group of sensors (called a sensor array) that is located near the area to be tested. A ground shock signal is sent into the earth and is then reflected by the different geological layer boundaries and is received by the sensors. Using sophisticated signal processing, the boundary layers can be mapped and some estimate can be made as to the materials in the various layers, such as sandstone, shale, oil, and water. Sonar signals are used in the exploration for oil under the ocean floor. The study of sonar (SOund Navigation And Ranging) includes the generation, transmission, and reception of sound energy in water. Oceanological applications of sonar, in addition to oil exploration, include ocean floor mapping, biological signal measurements, fish finding, acoustic beacon navigation, and submarine navigation.

Scalar and Array Computations

Introduction

The operations of adding, subtracting, multiplying, and dividing are the most fundamental operations used by engineers and scientists. We also perform other routine operations, such as computing the square root of a value, the logarithm of a value, or the tangent of an angle. These operations might be performed on a single value (a scalar), applied to a list of values (a vector), or applied to a set of values stored in a matrix with both rows and columns. In this chapter we learn how to use MATLAB to perform all these common operations and functions. In addition, we learn how to use complex numbers with MATLAB. One application problem discusses echoes in communication channels, and the other application problem relates to sonar signals.

3.1   SPECIAL VALUES AND SPECIAL MATRICES

functions

MATLAB includes a number of predefined constants, special values, and special matrices that are available to our programs. Most of these special values and special matrices are generated by MATLAB using functions. A MATLAB function typically uses inputs called arguments to compute a matrix, although some functions do not require any input arguments. In this section, we give some examples of MATLAB functions, and in Section 3.4 we review a large number of additional MATLAB functions.

Special Values

The scalar values that are available to use in MATLAB programs are described in the following list:

π pi   The value of π is automatically stored in this variable.

$\sqrt{-1}$ i,j   These variables are initially set to the value $\sqrt{-1}$. See Section 3.5 for a complete discussion on using complex numbers.

∞ Inf   This variable is MATLAB's representation for infinity, which typically occurs as a result of a division by zero. A warning message will be printed, and if you display the result of the division, the value will be ∞.

Not-a-number   NaN   This value stands for Not-a-Number and typically occurs when an expression is undefined, as in the division of zero by zero.

         clock   This function returns the current time in a six-element row vector containing year, month, day, hour, minute, and seconds.

         date   This function returns the current date in a character string format, such as 20-Jun-92.

         eps   This function contains the floating-point precision for the computer being used. This epsilon precision is the difference between 1.0 and the next larger decimal value.

         ans   This variable is used to store values computed by an expression that is computed but not stored in a variable name.

Special Matrices

MATLAB contains a group of functions that generate special matrices. Some of these matrices have specific application to the numerical techniques discussed in Part III, so they will be discussed in the appropriate sections there. Other special matrices have more general utility and are discussed now.

Magic Squares   A magic square of order n is an $n \times n$ matrix constructed from the integers from 1 through n^2. The integers are ordered such that all the row sums and all the column sums are equal to the same number. For example, the magic square of order 3 is the following matrix:

$$\begin{bmatrix} 8 & 1 & 6 \\ 3 & 5 & 7 \\ 4 & 9 & 2 \end{bmatrix}$$

The function to generate magic squares has one argument—the order of the magic square. The magic square shown above was generated with the expression `magic(3)`.

magic

Matrix of Zeros   The `zeros` function generates a matrix containing all zeros. If the argument to the function is a scalar, as in `zeros(6)`, the function will generate a square matrix using the argument as both the number of rows and the number of columns. If the function has two scalar arguments, as in `zeros(m, n)`, the function will generate a matrix with m rows and n columns. If the argument for the zero function is a matrix A, then the function will generate a matrix the same size as A containing all zeros. The following statements illustrate these various cases:

zeros

```
A = zeros(3);
B = zeros(3,2);
C = [ 1 2 3 ; 4 2 5];
D = zeros(C);
```

The matrices generated are the following:

$$A = \begin{bmatrix} 0 & 0 & 0 \\ 0 & 0 & 0 \\ 0 & 0 & 0 \end{bmatrix} \quad B = \begin{bmatrix} 0 & 0 \\ 0 & 0 \\ 0 & 0 \end{bmatrix}$$

$$C = \begin{bmatrix} 1 & 2 & 3 \\ 4 & 2 & 5 \end{bmatrix} \quad D = \begin{bmatrix} 0 & 0 & 0 \\ 0 & 0 & 0 \end{bmatrix}$$

Matrix of Ones   The `ones` function generates a matrix containing all ones, just as the `zeros` function generates a matrix containing all zeros. If the argument to the function is a scalar, as in `ones(6)`, the function will generate a square matrix using the argument as both the number of rows and the number of columns. If the function has two scalar arguments, as in `ones(m, n)`, the function will generate a matrix with m rows and n columns. If the argument for the ones function is a matrix A, then the function will generate a matrix the same size as A containing all ones. The following statements illustrate these various cases:

ones

```
A = ones(3);
B = ones(3,2);
C = [1 2 3 ; 4 2 5];
D = ones(C);
```

The matrices generated are the following:

$$A = \begin{bmatrix} 1 & 1 & 1 \\ 1 & 1 & 1 \\ 1 & 1 & 1 \end{bmatrix} \qquad B = \begin{bmatrix} 1 & 1 \\ 1 & 1 \\ 1 & 1 \end{bmatrix}$$

$$C = \begin{bmatrix} 1 & 2 & 3 \\ 4 & 2 & 5 \end{bmatrix} \qquad D = \begin{bmatrix} 1 & 1 & 1 \\ 1 & 1 & 1 \end{bmatrix}$$

Identity Matrix   An identity matrix is a matrix with ones on the main diagonal and zeros elsewhere. For example, the following matrix is an identity matrix with four rows and four columns:

$$\begin{bmatrix} 1 & 0 & 0 & 0 \\ 0 & 1 & 0 & 0 \\ 0 & 0 & 1 & 0 \\ 0 & 0 & 0 & 1 \end{bmatrix}$$

Note that the main diagonal is the diagonal containing elements in which the row number is the same as the column number. Therefore, the subscripts for elements on the main diagonal are (1,1), (2,2), (3,3), and so on.

eye   In MATLAB, identity matrices can be generated using the eye function. The arguments of the eye function are similar to those for the `zeros` function and the `ones` function. If the function has one scalar argument, as in eye (6), the function will generate a square matrix using the argument as both the number of rows and the number of columns. If the function has two scalar arguments, as in eye (m, n), the function will generate an identity matrix with m rows and n columns. If the argument for the eye function is a matrix A, then the function will generate an identity matrix the same size as A. While most applications use a square identity matrix, the definition can be extended to nonsquare matrices. The following statements illustrate these various cases:

```
A = eye (3) ;
B = eye (3, 2) ;
C = [1 2 3 ;   4   2   5] ;
D = eye (C) ;
```

The matrices generated are the following:

$$A = \begin{bmatrix} 1 & 0 & 0 \\ 0 & 1 & 0 \\ 0 & 0 & 1 \end{bmatrix} \qquad B = \begin{bmatrix} 1 & 0 \\ 0 & 1 \\ 0 & 0 \end{bmatrix}$$

$$C = \begin{bmatrix} 1 & 2 & 3 \\ 4 & 2 & 5 \end{bmatrix} \qquad D = \begin{bmatrix} 1 & 0 & 0 \\ 0 & 1 & 0 \end{bmatrix}$$

We recommend that you do not name an identity matrix i because this can cause problems if you also use complex numbers in the same program.

Pascal's Triangle   MATLAB will generate a square matrix containing the entries from Pascal's triangle. Recall that Pascal's triangle gives the coefficients of a binomial expansion of the form $(a + b)^n$. The triangle with coefficients for a fourth-degree expansion has the following form:

$$
\begin{array}{ccccccccc}
& & & & 1 & & & & \\
& & & 1 & & 1 & & & \\
& & 1 & & 2 & & 1 & & \\
& 1 & & 3 & & 3 & & 1 & \\
1 & & 4 & & 6 & & 4 & & 1
\end{array}
$$

pascal   If we use the command `pascal (5)`, the following matrix is generated:

$$
\begin{bmatrix}
1 & 1 & 1 & 1 & 1 \\
1 & 2 & 3 & 4 & 5 \\
1 & 3 & 6 & 10 & 15 \\
1 & 4 & 10 & 20 & 35 \\
1 & 5 & 15 & 35 & 70
\end{bmatrix}
$$

If you now look at the diagonals generated between values in the first column and values in the first row, you will see the elements from Pascal's triangle. For example, the values 1 2 1 are in positions (3,1), (2,2), and (1,3). The values 1 3 3 1 are in the diagonal just below that diagonal.

3.2   SCALAR OPERATIONS

arithmetic expressions

Arithmetic computations are specified using expressions. An expression can be as simple as a constant, or it may contain matrices and constants combined with arithmetic operations. In this section, we discuss operations involving only scalars. In the next section, we extend the operations to include element-by-element operations between scalars and matrices or between two matrices.

The arithmetic operations between two scalars are shown in Table 3.1. An expression can be evaluated and stored in a specified variable, as in the following statement, which specifies that the values in a and b are to be added, and the sum stored in a variable x:

```
x = a + b;
```

This assignment statement should be interpreted as specifying that the value in a is added to the value in b, and the sum is stored in x. If we interpret assignment statements in this way, then we are not disturbed by the following valid MATLAB statement:

```
count = count + 1;
```

TABLE 3.1   Arithmetic Operations between Two Scalars

Operation	Algebraic Form	MATLAB
Addition	$a + b$	a + b
Subtraction	$a - b$	a - b
Multiplication	$a \times b$	a*b
Right division	$\dfrac{a}{b}$	a/b
Left division	$\dfrac{b}{a}$	a\b
Exponentiation	a^b	a^b

scalar arithmetic

Clearly this statement is not a valid algebraic statement, but in MATLAB it specifies that 1 is to be added to the value in count, and the result stored back in count. Therefore, it is equivalent to specifying that the value in count should be incremented by 1.

It is important to recognize that a variable can store only one value at a time. For example, suppose that the following statements were executed one after another:

```
time = 0.0;
time = 5.0;
```

The value 0.0 is stored in the variable time when the first statement is executed and is then replaced by the value 5.0 when the second statement is executed.

ans

When you enter an expression without specifying a variable to store the result, the result or answer is automatically stored in a variable named ans. Each time a new value is stored in ans, the previous value is lost.

Precedence of Arithmetic Operations

Since several operations can be combined in a single arithmetic expression, it is important to know the order in which operations are performed. Table 3.2 contains the precedence of arithmetic operations in MATLAB. Note that this precedence also follows the standard algebraic precedence.

TABLE 3.2   Precedence of Arithmetic Operations

Precedence	Operation
1	parentheses
2	exponentiation, left to right
3	multiplication and division, left to right
4	addition and subtraction, left to right

precedence of operations

Assume that we want to compute the area of a trapezoid, and also assume that the scalar base contains the length of the base and that height_1 and height_2 contain the two heights. The area of a trapezoid can be computed using the following statement:

```
area = 0.5*base*(height_1 + height_2);
```

Suppose that we omitted the parentheses in the expression, as in:

```
area = 0.5*base*height_1 + height_2;
```

This statement would be executed as if it were this statement:

```
area = (0.5*base*height_1) + height_2;
```

Note that although the incorrect answer has been computed, there are no error messages printed to alert us to the error. Therefore, it is important to be very careful when converting equations into MATLAB statements. Adding extra parentheses is an easy way to be sure that computations are done in the order that you want. If an expression is long, break it into multiple statements. For example, consider the following equation:

$$f = \frac{x^3 - 2x^2 + x - 6.3}{x^2 + 0.05005x - 3.14}$$

The value of f could be computed using the following statements, where x is a scalar:

```
numerator = x^3 - 2*x^2 + x - 6.3;
denominator = x^2 + 0.05005*x - 3.14;
f = numerator/denominator;
```

It is better to use several statements that are easy to understand than to use one statement that requires careful thought to figure out the order of operations.

Practice!

Give MATLAB commands to compute the following values. Assume that the variables in the equations are scalars and have been assigned values.

1. Coefficient of friction between tires and pavement:

$$\text{friction} = \frac{v^2}{30s}$$

2. Correction factor in pressure calculation:

$$\text{factor} = 1 + \frac{b}{v} + \frac{c}{v^2}$$

3. Slope between two points:

$$\text{slope} = \frac{y_2 - y_1}{x_2 - x_1}$$

4. Resistance of a parallel circuit:

$$\text{resistance} = \frac{1}{\dfrac{1}{r_1} + \dfrac{1}{r_2} + \dfrac{1}{r_3}}$$

5. Pressure loss from pipe friction:

$$\text{loss} = f \cdot p \cdot \frac{1}{d} \cdot \frac{v^2}{2}$$

Computational Limitations

The variables stored in a computer have a wide range of values that they can assume. For most computers, the range of values extends from 10^{-308} to 10^{308}, which should be enough to accommodate most computations. However, it is possible for the result of an expression to be outside of this range. For example, suppose that we execute the following commands:

```
x = 2.5e200;
y = 1.0e200;
z = x*y;
```

If we assume the range of values is from 10^{-308} to 10^{308}, then the values of x and y are within the allowable range. However, the value of z is $2.5e400$, and this value exceeds the range. This error is called exponent overflow because the exponent of the result of an arithmetic operation is too large to store in the computer's memory. In MATLAB, the result of an exponent overflow is ∞.

exponent over-
flow

Exponent underflow is a similar error caused by the exponent of the result of an arithmetic operation being too small to store in the computer's memory. Using the same allowable range, we obtain an exponent underflow with the following commands:

exponent un-
derflow

```
x = 2.5e-200;
y = 1.0e200;
z = x/y;
```

Again, the values of x and y are within the allowable range, but the value of z should be $2.5e-400$. Since the exponent is less than the minimum, we have caused an exponent underflow error to occur. In MATLAB, the result of an exponent underflow is zero.

division by zero   We know that division by zero is an invalid operation. If an expression results in a division by zero in MATLAB, the result of the division is ∞. MATLAB will print a warning message and subsequent computations continue. The operations that follow use ∞ as the result of the division.

3.3   ARRAY OPERATIONS

element-by-element operations   An array operation is performed element by element. For example, suppose that A is a row vector with five elements, and B is a row vector with five elements. One way to generate a new row vector C with values that are the products of corresponding values in A and B is the following:

```
C(1)  =  A(1)*B(1);
C(2)  =  A(2)*B(2);
C(3)  =  A(3)*B(3);
C(4)  =  A(4)*B(4);
C(5)  =  A(5)*B(5);
```

These commands are essentially scalar commands because each command multiplies a single value by another single value and stores the product in a third value. To indicate that we want to perform an element-by-element multiplication between two matrices of the same size, we use the operation preceded by a period. Thus, the five statements above can be replaced by the following:

```
C  =  A.*B;
```

Omitting the period before the asterisk is a serious omission because the statement then specifies a matrix operation, not an element-by-element operation. Matrix operations are discussed in Chapter 6.

For addition and subtraction, array operations and matrix operations are the same, so we do not need to distinguish between them. However, array operations for multiplication, division, and exponentiation are different from matrix operations for multiplication, division, and exponentiation, so we need to be sure to use the period when we want to specify an array operation. These rules are summarized in Table 3.3.

Element-by-element operations, or array operations, apply not only to operations between two matrices of the same size, but also to operations between a scalar and nonscalar. However, multiplication of a matrix by a scalar and left division of a matrix by a scalar can be written either way. Thus, the two statements in each set of

array
operations

TABLE 3.3   Element-by-Element Operations

Operation	Algebraic Form	MATLAB
Addition	$a + b$	a + b
Subtraction	$a - b$	a - b
Multiplication	$a \times b$	a.*b
Right division	$\dfrac{a}{b}$	a./b
Left division	$\dfrac{b}{a}$	a.\b
Exponentiation	a^b	a.^b

statements below are equivalent for a nonscalar matrix A:

```
B = 3*A;
B = 3.*A;
```

```
C = A/5;
C = A./5;
```

The resulting matrices B and C will be the same size as A.

To illustrate the array operations for vectors, consider the following two row vectors:

$$A = [2 \quad 5 \quad 6]$$
$$B = [2 \quad 3 \quad 5]$$

array product

If we compute the array product of A and B using the following statement:

```
C = A.*B;
```

then C will contain the following values:

$$C = [4 \quad 15 \quad 30]$$

array division

MATLAB has two division operators—a forward division that uses the symbol / and a backward division that uses the symbol \. The forward array division command:

```
C = A./B;
```

will generate a new vector in which each element of A is divided by the corresponding element of B. Thus, C will contain the following values:

$$C = [1 \quad 1.667 \quad 1.2]$$

The backward array division command:

```
C = A. \B;
```

will generate a new vector in which each element is the corresponding element of B divided by the corresponding element of A. Thus, C will contain the following values:

$$C = [1 \quad 0.6 \quad 0.833]$$

array exponentiation
Array exponentiation is also an element-wise operation. For example, use the same values for A and B:

$$A = [2 \quad 5 \quad 6]$$
$$B = [2 \quad 3 \quad 5]$$

Consider these commands:

```
C = A. ^2;
D = A. ^B;
```

The vectors C and D are the following;

$$C = [4 \quad 25 \quad 36]$$
$$D = [4 \quad 125 \quad 7776]$$

We can also use a scalar base to a vector exponent, as in the following:

```
C = 3.0. ^A;
```

which generates a vector with the following values:

$$C = [9 \quad 243 \quad 729]$$

This vector could also have been computed with the following statement:

```
C = (3). ^A;
```

However, the following statement is incorrect:

```
C = 3. ^A;
```

MATLAB assumes that the period is part of the constant 3, and then attempts to do a matrix exponentiation, which will be discussed in Chapter 6. If a space were inserted before the period, as in the following:

```
C = 3 . ^A;
```

then the statement would perform an element-wise operation as desired. This example points out the fact that we need to be careful when specifying array operations. If you are not sure that you have written the correct expression, always test it with simple examples like the ones we have used.

The previous examples used vectors, but the same rules apply to matrices with rows and columns, as shown by the following statements:

```
d = [1:5;  -1:-1:-5];
z = ones(d);
s = d - z;
p = d.*s;
sq = d.^3;
```

The values of these matrices are shown below:

$$d = \begin{bmatrix} 1 & 2 & 3 & 4 & 5 \\ -1 & -2 & -3 & -4 & -5 \end{bmatrix} \quad z = \begin{bmatrix} 1 & 1 & 1 & 1 & 1 \\ 1 & 1 & 1 & 1 & 1 \end{bmatrix}$$

$$s = \begin{bmatrix} 0 & 1 & 2 & 3 & 4 \\ -2 & -3 & -4 & -5 & -6 \end{bmatrix} \quad p = \begin{bmatrix} 0 & 2 & 6 & 12 & 20 \\ 2 & 6 & 12 & 20 & 30 \end{bmatrix}$$

$$sq = \begin{bmatrix} 1 & 8 & 27 & 64 & 125 \\ -1 & -8 & -27 & -64 & -125 \end{bmatrix}$$

Practice!

Give the values in the vector C after executing the following statements, where A and B contain the values shown. Check your answers using MATLAB.

$$A = [2 \quad -1 \quad 5 \quad 0] \qquad B = [3 \quad 2 \quad -1 \quad 4]$$

1. C = A - B;

2. C = B + A - 3;

3. C = 2*A + A.^B;

4. C = B./A;

5. C = B.\A;

6. C = A.^B;

7. C = (2).^B + A;

8. C = 2*B/3.0.*A;

◆ PROBLEM SOLVING APPLIED: ECHOES
◈ IN COMMUNICATION SIGNALS

Interesting research is currently being done to develop computer systems that respond to verbal commands. The design of such a system assumes that the microphone picking up the speech commands has a clear representation of the speech. Unfortunately, sensors such as microphones introduce distortions, called noise. Systems with two-way communications also often have echoes that are inadvertantly introduced by the instrumentation. Therefore, a speech recognition system must be able to perform some processing on the speech signal to remove some of the distortion and undesirable components, such as echoes, before attempting to recognize the words. In order to test a program that has been designed to remove echoes, we need to be able to generate a digital signal and add echoes to it. We can then evaluate the performance of the program that is supposed to remove the echoes. In this section, we define digital signals[4], and we then develop a MATLAB program to add echoes to a digital signal.

Digital Signals

sensors

A/D converter

A signal is a function (usually with respect to time) that represents information. This information or data are collected with a sensor. Some of the more common types of sensors are microphones, which measure acoustic or sound data (such as speech); seismometers, which measure earth motion; photocells, which measure light intensity; thermistors, which measure temperature; and oscilloscopes, which measure voltage. Sensors are usually connected to another piece of instrumentation called an analog-to-digital (A/D) converter, which samples the signal periodically and records the time and the signal values so that they can be stored in a data file. The original signal is usually a continuous (or analog) function; the sequence of values collected from the original signal is called a digital signal. Figure 3.1 contains an example of a continuous signal, and Figure 3.2 contains a set of data points that represent a digital signal collected from the continuous signal. The digital signal is composed of a set of x-y coordinates and thus could easily be stored in a data file, and then read into matrices in a MATLAB program. When we plot a digital signal, we usually connect the points with line segments instead of plotting just the points.

Generating Echoes in a Signal

An echo of a signal is represented by a scaled (or attenuated) version of the original signal that occurs later in time. For example, Figure 3.3 contains an original signal $s(t)$ in the top sketch. The second sketch contains an echo of the original signal that has been attenuated by approximately 0.5 and shifted (or delayed) 2 seconds in time. The third sketch contains an echo of the original signal that has been scaled by approximately 0.3 and delayed 5 seconds in time; this is a

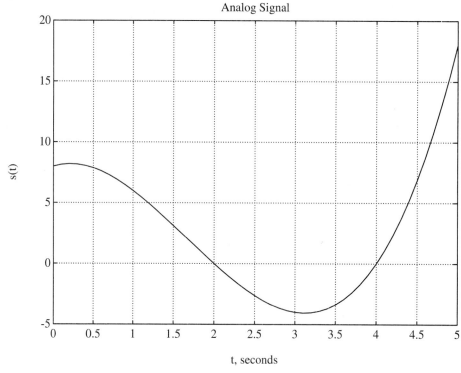

Fig. 3.1   Continuous or analog signal.

rolled echo   rolled echo because echo values are the negatives of the expected echo. The fourth sketch contains the original signal plus the two echos added to it.

Assume that a digital signal was collected over a period of 10 seconds, with a sampling time interval of 0.1 seconds. The following set of coordinates were collected in the first second, and all the signal values after these values were zeros:

Time, s	Signal Value
0.0	0.0
0.1	0.5
0.2	1.0
0.3	−0.5
0.4	0.75
0.5	0.0
0.6	−0.2
0.7	−0.1
0.8	0.0
0.9	0.0
1.0	0.0

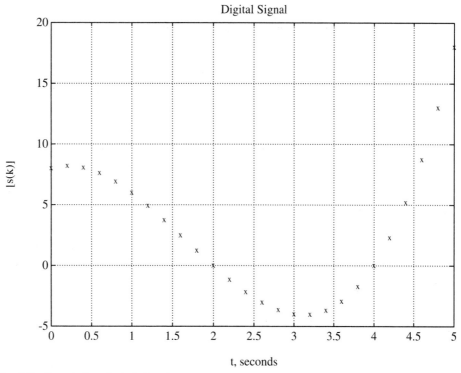

Fig. 3.2   Sampled or digital signal.

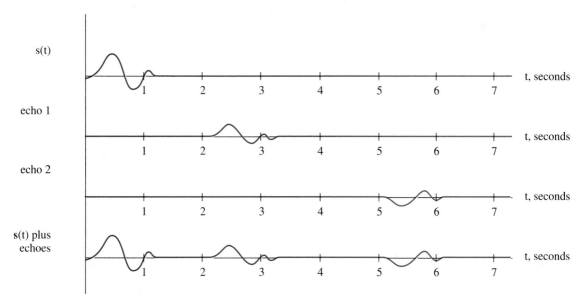

Fig. 3.3   Original signal and echoes.

Write a MATLAB program to generate a signal that contains this original signal with three echoes added to it. The first echo is scaled by 0.5 and delayed 2 seconds; the second echo is a rolled echo with a time delay of 4 seconds and a scaling of 0.3; the third echo is delayed 7.5 seconds and scaled by 0.1. Plot the original signal and the new signal with the echoes added to it. Then store the time signal and the signal with echoes in a MAT-file named echo.mat.

 1. PROBLEM STATEMENT
Given an original signal, generate a new signal containing the original signal plus three specified echoes added to it.

 2. INPUT/OUTPUT DESCRIPTION
The dotted box in Figure 3.4 contains a detailed picture of the process of generating the echoes from the input signal $[s_n]$. This signal is delayed and multiplied by a scale factor (represented by the triangle) to generate each echo. Then the original signal and all the echoes are added together in a new signal $[g_n]$, which is plotted and stored in a data file named echo.mat.

 3. HAND EXAMPLE
For a hand example, we use the first three values of the original signal:

Time, s	Signal Value
0.0	0.0
0.1	0.5
0.2	1.0

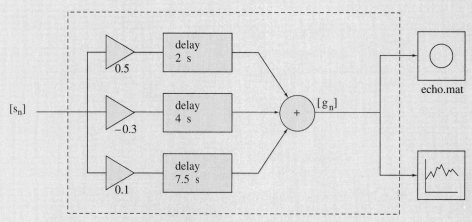

Fig. 3.4 I/O diagram.

The specified echoes then have the following non-zero values:

Time, s	Echo 1
2.0	$(0.5) \cdot (0.0) = 0.0$
2.1	$(0.5) \cdot (0.5) = 0.25$
2.2	$(0.5) \cdot (1.0) = 0.5$

Time, s	Echo 2
4.0	$(-0.3) \cdot (0.0) = 0.0$
4.1	$(-0.3) \cdot (0.5) = -0.15$
4.2	$(-0.3) \cdot (1.0) = -0.3$

Time, s	Echo 3
7.5	$(0.1) \cdot (0.0) = 0.0$
7.6	$(0.1) \cdot (0.5) = 0.05$
7.7	$(0.1) \cdot (1.0) = 0.1$

The sum of the original signal plus the three echoes is shown in Figure 3.5.

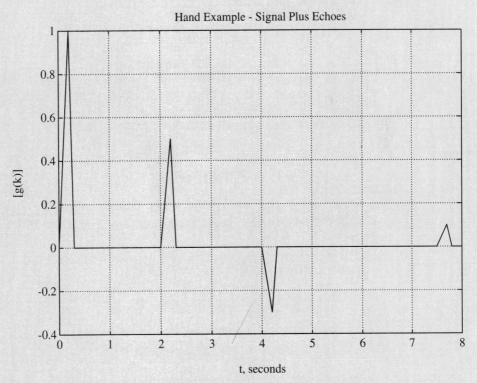

Fig. 3.5   Signal plus echoes from the hand example.

4. MATLAB SOLUTION

We are going to develop the solution so that the echoes will be generated correctly for any signal; that is, we are not going to assume that only the first few values of the signal are nonzero.

To compute the subscripts for the echo, we use the fact that the sampling interval is 0.1 seconds. Therefore to generate an echo with a delay of 2 seconds, we need to delay the echo 20 samples. The value in s(1) would create an echo in echo_1(21), the value in s(2) would create an echo in echo_1(22), and so on.

```
%
%            This program generates three echoes from an
%            original signal. The sum of the original
%            signal and the three echoes is stored in a
%            data file and then plotted.
%
t = 0.0:0.1:10;
s = zeros(t);
s(1:8) = [0 0.5 1 -0.5 0.75 0 -0.2 -0.1];
%
%      Generate three echoes by scaling and
%      delaying original signal.
%
echo_1 = zeros(t);
echo_1(21:101) = 0.5*s(1:81);
echo_2 = zeros(t);
echo_2(41:101) = -0.3*s(1:61);
echo_3 = zeros(t);
echo_3(76:101) = 0.1*s(1:26);
%
%      Add echoes to original signal giving new signal.
%      Save time signal and new signal.
%
g = s + echo_1 + echo_2 + echo_3;
save echo t g;
%
%      Plot new signal.
%
plot(t,g),...
title('Signal with Echoes'),...
xlabel('t, seconds'),...
ylabel('[g(k)]'),...
grid
```

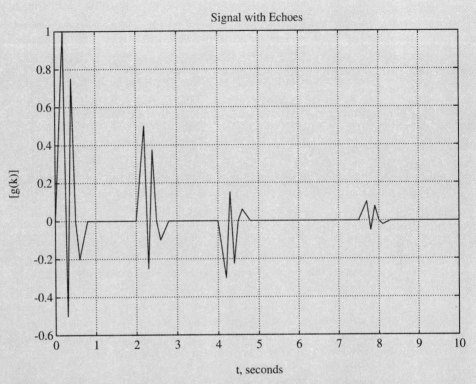

Fig. 3.6 Signal plus three echoes.

 5. TESTING

The plot in Figure 3.6 contains the original signal with the echoes added to it. Additional problems related to this application are included in the end-of-chapter problems.

3.4   COMMON FUNCTIONS

Arithmetic expressions often require computations other than addition, subtraction, multiplication, division, and exponentiation. For example, many expressions require the use of logarithms, exponentials, and trigonometric functions. MATLAB allows us to use function references to perform these types of computations instead of requiring us to compute them using the basic arithmetic operations. For example, if we want to compute the sine of an angle and store the result in b, we can use the following command:

```
b = sin(angle);
```

The sin function assumes that the argument is in radians. If the argument contains a value in degrees, we can convert the degrees to radians within the function reference:

```
b = sin(angle*pi/180);
```

We could also have done the conversion in a separate statement:

```
angle_radians = angle*pi/180;
b = sin(angle_radians);
```

These statements are valid if angle is a scalar or if angle is a matrix. If angle is a matrix, then the function will be applied element by element to the values in the matrix.

Now that you have seen several examples of functions, we review the rules regarding them. A function is a reference that represents a matrix. The arguments or parameters of the function are contained in parentheses following the name of the function. A function may contain no arguments, one argument, or many arguments, depending on its definition. For example, pi is a function that has no argument; when we use the function reference pi, the value for π automatically replaces the function reference. If a function contains more than one argument, it is very important to give the arguments in the correct order. Some functions also require that the arguments be in specific units. For example, the trigonometric functions assume that the arguments are in radians. In MATLAB, some functions use the number of arguments to determine the output of the function. For example, the zeros function can have one or two arguments, which then determine the output.

A function reference cannot be used on the left side of an equal sign, since it represents a value and not a variable. Functions can appear on the right side of an equal sign and in expressions. A function reference can also be part of the argument of another function reference. For example, the following statement computes the logarithm of the absolute value of x:

```
log_x = log(abs(x));
```

When one function is used to compute the argument of another function, be sure to enclose the argument of each function in its own set of parentheses. This nesting of functions is also called composition of functions. Function names must be in lowercase unless the case sensitivity is turned off.

We now discuss several categories of functions that are commonly used in engineering computations. Other functions will be presented throughout the remaining chapters as we discuss relevant subjects. Tables of common functions are included on the inside front cover.

Elementary Math Functions   The elementary math functions include functions to perform a number of common computations such as computing the absolute value and the square root. In addition, we also include a group of functions used to perform rounding. We now list these functions with a brief description:

abs(x)	This function computes the absolute value of x.
sqrt(x)	This function computes the square root of x.
round(x)	This function rounds x to the nearest integer.
fix(x)	This function rounds x to the nearest integer toward zero.
floor(x)	This function rounds x to the nearest integer toward $-\infty$.
ceil(x)	This function rounds x to the nearest integer toward ∞.
sign(x)	This function returns a value of -1 if x is less than zero, a value of zero if x equals zero, and a value of 1 otherwise.
rem(x,y)	This function returns the remainder of $\frac{x}{y}$. For example, rem(25,4) is 1, and rem(100,21) is 16.
exp(x)	This function returns the value of e^x, where e is the base for natural logarithms, or approximately 2.718282.
log(x)	This function returns ln x, the natural logarithm of x to the base e.
log10(x)	This function returns $\log_{10}$ x, the common logarithm of x to the base 10.

Practice!

Evaluate the following expressions, and then check your answer by entering the expressions in MATLAB.

1. round(-2.6)

2. fix(-2.6)

3. floor(-2.6)

4. ceil(-2.6)

5. sign(-2.6)

6. abs(round(-2.6))

7. sqrt(floor(10.7))

8. rem(15,2)

9. floor(ceil(10.8))

10. log10(100) + log10(0.001)

11. abs(-5:5)

12. round([0:0.3:2,1:0.75:4])

Trigonometric Functions   The trigonometric functions assume that angles are represented in radians. To convert radians to degrees or degrees to radians, use the following conversions, which use the fact that $180° = \pi$ radians:

```
angle_degrees = angle_radians*(180/pi);
angle_radians = angle_degrees*(pi/180);
```

We now list the trigonometric functions with a brief description:

sin(x)	This function computes the sine of x, where x is in radians.
cos(x)	This function computes the cosine of x, where x is in radians.
tan(x)	This function computes the tangent of x, where x is in radians.
asin(x)	This function computes the arcsine or inverse sine of x, where x must be between -1 and 1. The function returns an angle in radians between $\dfrac{-\pi}{2}$ and $\dfrac{\pi}{2}$.
acos(x)	This function computes the arccosine or inverse cosine of x, where x must be between -1 and 1. The function returns an angle in radians between 0 and π.
atan(x)	This function computes the arctangent or inverse tangent of x. The function returns an angle in radians between $\dfrac{-\pi}{2}$ and $\dfrac{\pi}{2}$.
atan2(x,y)	This function computes the arctangent or inverse tangent of the value $\dfrac{y}{x}$. The function returns an angle in radians that will be between $-\pi$ and π, depending on the signs of x and y.

The other trigonometric functions can be computed using the following equations:

$$\sec x = \frac{1}{\cos x}$$

$$\csc x = \frac{1}{\sin x}$$

$$\cot x = \frac{1}{\tan x}$$

Practice!

Give MATLAB commands for computing the following values. Assume that the variables in the equations are scalars and have been assigned values.

1. Uniformly accelerated motion:

$$\text{motion} = \sqrt{vi^2 + 2 \cdot a \cdot x}$$

2. Electrical oscillation frequency:

$$\text{frequency} = \frac{1}{\sqrt{\dfrac{2\pi \cdot c}{L}}}$$

3. Range for a projectile:

$$\text{range} = 2\, vi^2 \cdot \frac{\sin(b) \cdot \cos(b)}{g}$$

4. Length contraction:

$$\text{length} = k\sqrt{1 - \left(\frac{v}{c}\right)^2}$$

5. Volume of a fillet ring:

$$\text{volume} = 2\pi x^2\left(\left(1 - \frac{\pi}{4}\right) \cdot y - \left(0.8333 - \frac{\pi}{4}\right) \cdot x\right)$$

6. Distance of the center of gravity from a reference plane in a hollow cylinder sector:

$$\text{center} = \frac{38.1972 \cdot (r^3 - s^3)\sin a}{(r^2 - s^2) \cdot a}$$

Hyperbolic Functions   The hyperbolic functions are functions of the natural exponential function e^x; the inverse hyperbolic functions are functions of the natural logarithm function ln x. These functions are useful in applications such as the design of some types of digital filters. MATLAB includes several hyperbolic functions, as shown in these brief descriptions:

sinh(x)   This function computes the hyperbolic sine of x, which is equal to $\dfrac{e^x - e^{-x}}{2}$.

cosh(x)   This function computes the hyperbolic cosine of x, which is equal to $\dfrac{e^x + e^{-x}}{2}$.

tanh(x)   This function computes the hyperbolic tangent of x, which is equal to $\dfrac{\sinh x}{\cosh x}$.

asinh(x)   This function computes the inverse hyperbolic sine of x, which is equal to $\ln(x + \sqrt{x^2 + 1})$.

acosh(x)   This function computes the inverse hyperbolic cosine of x, which is equal to $\ln(x + \sqrt{x^2 - 1})$ for x greater than or equal to 1.

atanh(x)   This function computes the inverse hyperbolic tangent of x, which is equal to $\dfrac{1}{2}\ln\left(\dfrac{1 + x}{1 - x}\right)$ for $|x| < 1$.

Other hyperbolic and inverse hyperbolic functions can be computed using the following equations:

$$\coth x = \frac{\cosh x}{\sinh x} \quad \text{for } x \neq 0$$

$$\operatorname{sech} x = \frac{1}{\cosh x}$$

$$\operatorname{csch} x = \frac{1}{\sinh x}$$

$$\operatorname{acoth} x = \frac{1}{2}\ln\left(\frac{x + 1}{x - 1}\right) \quad \text{for } |x| > 1$$

$$\operatorname{asech} x = \ln\left(\frac{1 + \sqrt{1 - x^2}}{x}\right) \quad \text{for } 0 < x \leq 1$$

$$\operatorname{acsch} x = \ln\left(\frac{1}{x} + \frac{\sqrt{1 + x^2}}{|x|}\right) \quad \text{for } x \neq 0$$

Practice!

Give MATLAB expressions for calculating the following values, given the value of x. (Assume that the value of x is in the proper range of values for the calculations.)

1. coth x

2. sec x

3. csc x

4. acoth x

5. asech x

6. acsch x

M-File Functions

As you use MATLAB to perform more and more computations, you will find calculations that you wish were included as functions. In these cases, you can write the function in an M-file, and then your program can refer to the function in the same way that it refers to a MATLAB function. The function file has very specific rules that must be followed when writing it. Before we list the rules, we consider a simple example of an M-file function that has been stored in a file named `circum.m`:

```
function c = circum(r)
%   CIRCUM       Circumference of a circle with radius r.
%                For matrices, CIRCUM(r) returns a matrix
%                containing the circumferences of circles
%                with radii equal to the values in the
%                original vector.
c = pi*2*r;
```

The following statements are all valid references to this function:

```
r = [0 1.4 pi];
circum(r);

a = 5.6;
disp(circum(a))

c = [1.2 3; 5 2.3];
circum(c);
```

We now summarize the rules for writing an M-file function. Refer to the example function as you read each rule.

function

1. The function must begin with a line containing the word `function`, followed by the output argument, an equal sign, and the name of the function. The input arguments to the function follow the function name and are enclosed in parentheses. This line defines the input arguments and the output arguments, and distinguishes the function file from a script file.

2. The first few lines should be comments because they will be displayed if help is requested for the function name, as in `help circum`.

3. The only information returned from the function is contained in the output arguments, which are, of course, matrices. Always check to be sure that the function includes a statement that assigns a value to the output argument.

4. The same matrix names can be used in both a function and the program that references it. No confusion occurs as to which matrix is referenced, because the function and the program are completely separate. However, any values computed in the function, other than the output arguments, are not accessible from the program.

5. A function that is going to return more than one value should show all values to be returned as a vector in the function statement, as in this example, which will return three values:

```
function [dist, vel, accel] = motion(x)
```

All three values need to be computed within the function.

6. A function that has multiple input arguments must list the arguments in the function statement, as shown in this example, which has two input arguments:

```
function error = mse(w, d)
```

nargin
nargout

7. The special variables `nargin` and `nargout` can be used to determine the number of input arguments passed to the function and the number of output arguments used in the reference to the function. These variables are accessible only within the function.

Recall that the `what` command lists all the M-files and MAT files that are available in the current workspace. The `type` command followed by a filename will list a file on the screen. If an extension is not included with the filename, the `type` command automatically assumes that the extension is `.m`.

 PROBLEM SOLVING APPLIED: SONAR SIGNALS

The study of sonar (*so*und *n*avigation *a*nd *r*anging) includes the generation, transmission, and reception of sound energy in water. Oceanological applications include ocean topography, geological mapping, and biological signal measurements; industrial sonar applications include fish finding, oil and mineral exploration, and acoustic beacon navigation; naval sonar applications include submarine navigation and submarine tracking. An active sonar system transmits a signal that is usually a sinusoid with a known frequency. The reflections or echoes of the signal are then received and analyzed to provide information about the surroundings. A passive sonar system does not transmit signals but collects signals from sensors and analyzes them based on their frequency content [4].

In this section, we first discuss sinusoids, since they are the basic signal used in sonar systems. We then develop a MATLAB program for generating a sonar signal.

Sinusoid Generation

We are familiar with the sine function, which is a function of an angle in radians:

$$g(\theta) = \sin \theta$$

A sinusoid is a sine function that is written as a function of time:

$$g(t) = \sin (2\pi ft)$$

where f is the frequency of the sinusoid in cycles per second, or Hertz (Hz). (Note that the units of $2\pi ft$ are radians $\times \dfrac{\text{cycles}}{\text{second}} \times$ seconds, or radians.)

If the frequency of a sinusoid is 5 Hz, we have:

$$g(t) = \sin (2\pi\, 5t)$$
$$= \sin (10\pi t)$$

From the plot of this function in Figure 3.7, we see that there are five cycles of the sinusoid in one second, thus verifying that the frequency of this sinusoid is 5 cycles per second, or 5 Hz. The period P of a sinusoid is the period of time that corresponds to one cycle; thus, the period of this sinusoid is 0.2 seconds, or $\frac{1}{5}$. This relationship can be written with the following equations;

$$f = \frac{1}{P}; \qquad P = \frac{1}{f}$$

where f is the frequency in Hz and P is the period in seconds.

If a sinusoid is multiplied by a scalar A, the equation can be written in this form:

$$g(t) = A \sin (2\pi ft)$$

The scalar is also called the amplitude of the sinusoid. A sinusoid with a phase shift ϕ in radians can be written in this form:

$$g(t) = A \sin (2\pi ft + \phi)$$

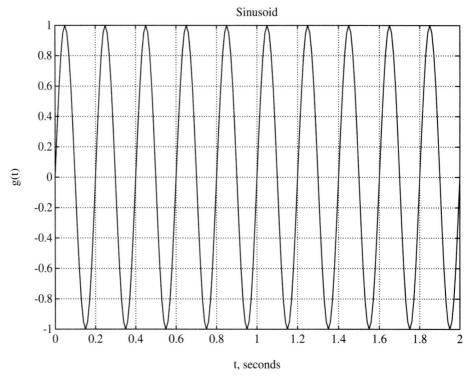

Fig. 3.7   Plot of a 5 Hz sinusoid.

If the phase shift is equal to $\dfrac{\pi}{2}$ radians, we then have the following:

$$A \sin \left(2\pi ft + \frac{\pi}{2} \right) = A \cos \left(2\pi ft \right)$$

Therefore, a sinusoid is a function of time that can be written in terms of either a sine function or a cosine function, and may or may not include an additional phase shift.

Generating a Sonar Signal

An active sonar system transmits a signal of a specified frequency. The reflections or echoes of that signal are then received and analyzed to provide information about the surroundings. A passive sonar system does not transmit signals but collects signals from sensors and analyzes them based on their frequency content. One of the types of signals used in sonar systems is the sinusoid. These sinusoids can be represented by the following equation[4,5]:

$$s(t) = \begin{cases} \sqrt{\dfrac{2E}{PD}} \cos{(2\pi f_c t)}, & 0 \le t \le PD \\ 0 & \text{elsewhere,} \end{cases}$$

where E is the transmitted energy,
    PD is the pulse duration in seconds,
    f_c is the frequency in Hz.

Durations of a sonar signal can range from a fraction of a millisecond to several seconds, and frequencies can range from a few hundred Hz to tens of kHz (kilo-Hertz) depending on the system and its desired operating range.

Write a MATLAB program that allows the user to enter values for E, PD, and f_c to generate a sonar signal. Store the signal values in a MAT-file named sonar.mat. The sampling of the signal should cover the pulse duration and be sampled such that there are 10 samples for every period of $x(t)$. In addition, add a period of 200 points of silence after the pulse.

1. PROBLEM STATEMENT
   Write a program to generate a sonar signal that contains 10 samples from each period of a specified sinusoid, covering a specified time duration.

2. INPUT/OUTPUT DESCRIPTION
   The values of E (transmitted energy in joules), PD (pulse duration in seconds), and f_c (frequency in Hz) are the input values. The output is a data file named sonar.mat, which contains time and signal values for the sonar pulse duration, as shown in Figure 3.8. We also plot the sonar signal.

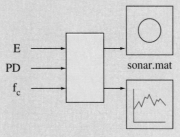

Fig. 3.8 I/O diagram.

3. HAND EXAMPLE

For a hand example, we use the following values:

$$E = 500 \text{ joules}$$
$$PD = 5 \text{ milliseconds (ms)}$$
$$f_c = 3.5 \text{ kHz}$$

The period of the sinusoid is $\dfrac{1}{3500}$, or approximately 0.3 ms. Thus, to have 10 samples per period, the sampling interval needs to be approximately 0.03 ms. The pulse duration is 0.5 ms, and thus we need 167 samples of the following signal:

$$s(t) = \sqrt{\frac{2E}{PD}} \cos (2\pi f_c t)$$

$$= \sqrt{\frac{1000}{0.005}} \cos (2\pi (3500)t)$$

$$= 447.2 \cos (2\pi\, 3500t).$$

The first few values for the sonar signal are computed below to one decimal place of accuracy:

t (ms)	s(t)
0.00	447.2
0.03	353.4
0.06	111.2
0.09	−177.6
0.12	−391.9
0.15	−441.7
0.18	−306.1
0.21	−42.1
0.24	239.6
0.27	420.8
0.30	425.3
0.33	251.4

We would then add 200 points of silence by adding data values with the correct times and corresponding signal values of zero.

 4. MATLAB SOLUTION

The most interesting part of this program is computing the time values for the period of silence. We want to add 200 new times, and these times need to continue the sequence generated for the sonar pulse. Therefore, we generate a scalar named `silence_start`, which contains the starting time for the silence period. We then generate the ending time for the silence period and use the colon operator to compute the new times in a vector `silence`. We then append the new vector to the time vector, and append the same number of zeros to the sonar signal vector.

```
%
%             This program generates a sonar signal
%             using values obtained from the user for
%             energy, pulse duration, and frequency.
%             An additional 200 points of silence are
%             added to the signal before it is stored.
%
energy = input('Enter energy in joules ');
duration = input('Enter pulse duration in seconds ');
fc = input('Enter frequency in Hz ');
%
%      Generate sonar signal.
%
A = sqrt(2*energy/duration);
period = 1/fc;
t_incr = period/10;
t = [0:t_incr:duration];
s = A*cos(2*pi*fc*t);
%
%      Generate and add 200 points of silence.
%
last = length(t);
silence_start = t(last) + t_incr;
silence_end = t(last) + 200*t_incr;
silence = [silence_start:t_incr:silence_end];
t = [t silence];
s = [s zeros(silence)];
%
%      Save and plot sonar signal.
%
save sonar t s
plot(t,s),...
title('Sonar Signal'),...
xlabel('t, seconds'),...
ylabel('[s(k)]'),...
grid
```

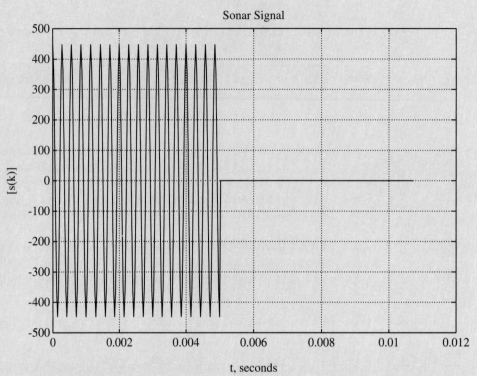

Fig. 3.9   Sonar signal followed by silence.

 5. TESTING

Figure 3.9 contains the plot generated by this program with the following input:

```
Enter energy in joules 500
Enter pulse duration in seconds 0.005
Enter frequency in Hz 3500
```

Since the period of the 3.5 kHz sinusoid is approximately 0.0003 second, and the pulse has a duration of 0.005 second, the pulse should contain approximately 17 periods; this is confirmed by the plot. Additional problems related to this application are included in the end-of-chapter problems.

3.5   COMPLEX NUMBERS

roots of an equation

The solutions to many engineering problems involve finding the roots of an equation of the following form:

$$y = f(x)$$

where the roots are the values of x for which y is equal to zero. Examples of applications in which we need to find roots of equations include designing the control system for a robot arm, designing springs and shock absorbers for an automobile, analyzing the response of a motor, and analyzing the stability of an electric circuit.

For a quadratic equation (polynomial of degree 2), we use factoring or the quadratic formula to determine the two roots of the equation. For example, consider the following quadratic equation:

$$f(x) = x^2 + 3x + 2.$$

By factoring this equation, we have:

$$f(x) = (x + 1) \cdot (x + 2)$$

real roots

We now set this equation with the factored terms to zero; it becomes clear that the roots are -1 and -2. If we use the quadratic formula, we also obtain the same values for the roots.

Now, consider this quadratic equation:

$$f(x) = x^2 + 3x + 3$$

Using the quadratic formula, we compute the following roots:

complex roots

$$x_1 = \frac{-3 + \sqrt{-3}}{2} = -1.5 + 0.87 \sqrt{-1},$$

$$x_2 = \frac{-3 - \sqrt{-3}}{2} = -1.5 - 0.87 \sqrt{-1}.$$

In order for these roots to have meaning, we need to define $\sqrt{-1}$. We do this by expanding our concept of a number system to the system of complex numbers, where a complex number has the form $a + ib$ with $i = \sqrt{-1}$. This form can also represent real values if b is zero.

Figure 3.10 contains plots of both quadratic polynomials that we have just discussed. As we would expect, the polynomial for the first equation crosses the x-axis in two points, yielding two real roots. The polynomial for the second equation (dotted line) does not cross the x-axis and thus has two complex roots.

Consider the following general form for a polynomial of degree N:

$$a_0 x^N + a_1 x^{N-1} + a_2 x^{N-2} + \cdots + a_{N-2} x^2 + a_{N-1} x + a_N = 0$$

A polynomial of degree N will always have exactly N roots, some of which may be multiple roots or complex roots. In Chapter 10 we present MATLAB commands for computing the roots of polynomials. In this section, we discuss operations with complex numbers and MATLAB functions that use complex numbers.

Arithmetic Operations with Complex Numbers

Complex numbers are an important part of many problem solutions in engineering. Table 3.4 reviews the results of arithmetic operations between two complex numbers.

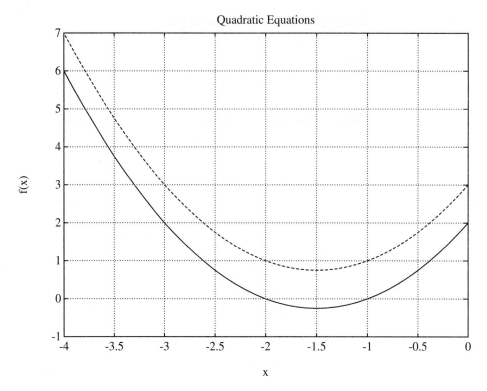

Fig. 3.10 Plots of two quadratic equations.

One of the advantages of using MATLAB for engineering computations is its ease in handling complex numbers. MATLAB commands recognize complex numbers using i to represent $\sqrt{-1}$. (MATLAB also recognizes the use of j to represent $\sqrt{-1}$. This notation is commonly used in electrical engineering.) The following

TABLE 3.4 Arithmetic Operations with Complex Numbers

Operation	Result
$c_1 + c_2$	$(a_1 + a_2) + i(b_1 + b_2)$
$c_1 - c_2$	$(a_1 - a_2) + i(b_1 - b_2)$
$c_1 \cdot c_2$	$(a_1 a_2 - b_1 b_2) + i(a_1 b_2 + a_2 b_1)$
$\dfrac{c_1}{c_2}$	$\left(\dfrac{a_1 a_2 + b_1 b_2}{a_2^2 + b_2^2}\right) + i\left(\dfrac{a_2 b_1 - b_2 a_1}{a_2^2 + b_2^2}\right)$
$|c_1|$	$\sqrt{a_1^2 + b_1^2}$ (magnitude or absolute value of c_1)
c_1^*	$a_1 - ib_1$ (conjugate of c_1)

(Assume that $c_1 = a_1 + ib_1$ and $c_2 = a_2 + ib_2$.)

command defines a complex variable x:

```
x = 1 - 0.5*i;
```

When we perform operations between two complex numbers, MATLAB automatically performs the necessary computations, as outlined in Table 3.4. If an operation is performed between a real number and a complex number, MATLAB assumes that the imaginary part of the real number is zero. Be careful not to use the name i or j for other variables in a program in which you also use complex numbers; the new values will replace the value of $\sqrt{-1}$ and could cause many problems.

MATLAB includes several functions that are specific to complex numbers.

real(x)   This function computes the real portion of the complex number x.

imag(x)   This function computes the imaginary portion of the complex number x.

conj(x)   This function computes the complex conjugate of the complex number x. Thus, if x = $a + ib$, then conj(x) = $a - ib$.

abs(x)   This function computes the magnitude of the complex number x.

angle(x) This function computes the angle using the value of atan2(imag(x),real(x)), and thus the angle value is between $-\pi$ and π.

These functions make it very easy to convert between polar and rectangular form, as we will see in the next section.

Polar and Rectangular Coordinates

complex number system

We can view the complex number system as a plane with real and imaginary axes. Real numbers (those with no imaginary part) represent the x axis, imaginary numbers (those with no real part) represent the y axis, and numbers with both a real part and an imaginary part represent the rest of the plane. Thus, the real number system (the one with which we are most familiar) is a subset of the complex number system.

rectangular notation

When we represent a complex number with a real part and an imaginary part, as in $2 + i3$, we are using a rectangular notation. In Figure 3.11, we see that a complex number could also be described with an angle θ and a radius r relative to

polar notation

the origin. This form is called polar notation, and the point $2 + i3$ can be represented in polar notation with an angle of .98 radian and a radius of 3.6. From Figure 3.11, it is easy to determine the following relationships for converting between rectangular coordinates and polar coordinates:

conversions

Rectangular to polar

$$r = \sqrt{a^2 + b^2}$$

$$\theta = \tan^{-1}\frac{b}{a}$$

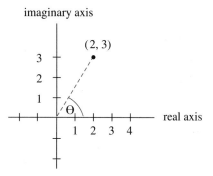

Fig. 3.11   Complex plane.

Polar to rectangular

$$a = r \cos \theta$$
$$b = r \sin \theta$$

Most calculators perform the transformations between rectangular and polar coordinates, but you should also be sure that you can do the transformations without the calculator. When you are converting rectangular to polar, be especially careful determining the quadrant for the angle θ. Some calculators and computers automatically return a value in the first quadrant, instead of analyzing the values of a and b to determine the correct quadrant. (In MATLAB, the atan(x) function returns values in the range $-\pi/2$ to $\pi/2$, while the atan2(y, x) returns the inverse tangent of y/x in the range $-\pi$ to π. The angle(x) function uses a complex x and returns the value of atan2(imag(x), real(x))).

angle quadrant

magnitude, phase

If x is a complex number, then the magnitude and phase can be computed with the following commands:

```
r = abs(x);
theta = angle(x);
```

To compute the complex number using a specified magnitude and angle, we use this command:

```
y = r*exp(i*theta);
```

This relationship is discussed in more detail shortly.

real, imaginary

We can compute the real and imaginary portion of a complex number with these statements:

```
a = real(x);
b = imag(x);
```

To compute the complex number with a specified real and imaginary portion, we use the command:

```
y = a + i*b;
```

If you are interested in polar plots, you may want to read Section 7.2 at this point.

Practice!

Convert the complex values in problems 1 to 4 to polar form. Then check your answers using MATLAB functions.

1. $3 - i2$

2. $-i$

3. -2

4. $0.5 + i$

Convert the complex exponential values in problems 5 to 8 to rectangular form. Check your answers using MATLAB functions.

5. e^i

6. $e^{i\pi.75}$

7. $0.5e^{i2.3}$

8. $3.5e^{i3\pi}$

Euler's Formula

Maclaurin series

To derive some important properties for complex numbers, we need the following Maclaurin series representations, which are usually discussed in a calculus course:

$$\sin x = x - \frac{x^3}{3!} + \frac{x^5}{5!} - \cdots \tag{3.1}$$

$$\cos x = 1 - \frac{x^2}{2!} + \frac{x^4}{4!} - \cdots \tag{3.2}$$

$$e^x = 1 + x + \frac{x^2}{2!} + \frac{x^3}{3!} + \cdots \tag{3.3}$$

Now, let x be the imaginary value ib. Then, from Equation (3.3), we have:

$$e^{ib} = 1 + ib + \frac{(ib)^2}{2!} + \frac{(ib)^3}{3!} + \frac{(ib)^4}{4!} + \frac{(ib)^5}{5!} + \cdots$$

$$= 1 + ib - \frac{b^2}{2!} - i\frac{b^3}{3!} + \frac{b^4}{4!} + i\frac{b^5}{5!} + \cdots . \tag{3.4}$$

We now separate the infinite sum in Equation (3.4) into two parts, giving:

$$e^{ib} = \left(1 - \frac{b^2}{2!} + \frac{b^4}{4!} - \cdots\right) + i\left(b - \frac{b^3}{3!} + \frac{b^5}{5!} - \cdots\right) \quad (3.5)$$

Finally, we substitute the infinite sums from Equation (3.1) and Equation (3.2) in Equation 3.5 to obtain:

$$e^{ib} = \cos b + i \sin b. \quad (3.6)$$

Euler's formula   Equation (3.6) is a very important formula called Euler's formula. We frequently use it and these two additional formulas that can be derived from it:

$$\sin \theta = \frac{e^{i\theta} - e^{-i\theta}}{2i} \quad (3.7)$$

$$\cos \theta = \frac{e^{i\theta} + e^{-i\theta}}{2} \quad (3.8)$$

Using Euler's formula, we can express a complex number in a rectangular co-ordinate form or in polar form. This relationship is derived as shown:

$$a + ib = (r \cos \theta) + i(r \sin \theta)$$
$$= r(\cos \theta + i \sin \theta). \quad (3.9)$$

We then use Equation (3.7) and Equation (3.8) in Equation (3.9), giving:

exponential form
$$a + ib = re^{i\theta} \quad (3.10)$$

where $r = \sqrt{a^2 + b^2}$

$$\theta = \tan^{-1} \frac{b}{a}$$
$$a = r \cos \theta$$
$$b = r \sin \theta$$

Thus, we can represent a complex number in either rectangular form $(a + ib)$ or in exponential form $(re^{i\theta})$.

SUMMARY

In this chapter we explored the various mathematical operations and MATLAB functions for creating matrices and for calculating new matrices from existing matrices. The computations between matrices were performed on an element-by-element basis and are called array operations. We also learned how to write user-defined functions using M-files. Functions and mathematical computations involving complex numbers were also discussed. Engineering applications relating to echoes in communication signals and relating to sonar signals were presented, and related problems were solved.

MATLAB SUMMARY

This MATLAB summary lists all the special symbols, commands, and functions that were defined in this chapter. A brief description is also included for each one.

Special Characters

+	scalar and array addition
−	scalar and array subtraction
*	scalar multiplication
.*	array multiplication
/	scalar right division
./	array right division
\	scalar left division
.\	array left division
^	scalar exponentiation
.^	array exponentiation

Commands and Functions

abs	computes absolute value or magnitude
acos	computes arccosine
acosh	computes inverse hyperbolic cosine
angle	computes angle of complex value
ans	stores expression values
asin	computes arcsine
asinh	computes inverse hyperbolic sine
atan	computes 2-quadrant arctangent
atan2	computes 4-quadrant arctangent
atanh	computes inverse hyperbolic tangent
ceil	rounds toward ∞
clock	represents the current time
conj	computes complex conjugate
cos	computes cosine of angle
cosh	computes hyperbolic cosine
date	represents the current date
eps	represents floating-point precision
exp	computes value with base e
eye	generates identity matrix
function	generates user-defined function
fix	rounds toward zero

floor	rounds toward $-\infty$
i	represents the value $\sqrt{-1}$
imag	computes imaginary part of complex value
Inf	represents the value ∞
j	represents the value $\sqrt{-1}$
log	computes natural logarithm
log10	computes common logarithm
magic	generates magic square
NaN	represents the value Not-a-Number
nargin	determines input arguments in a function
nargout	determines output arguments in a function
ones	generates matrix of ones
pascal	generates Pascal's triangle in a matrix
pi	represents the value π
real	computes real part of complex value
rem	computes remainder from division
round	rounds to nearest integer
sign	generates $-1,0,1$ based on sign
sin	computes sine of angle
sinh	computes hyperbolic sine
sqrt	computes square root
tan	computes tangent of angle
tanh	computes hyperbolic tangent
zeros	generates matrix of zeros

PROBLEMS

Problems 1 to 8 relate to the engineering applications presented in this chapter. Problems 9 to 23 relate to new engineering applications.

Echoes   These problems relate to signals containing echoes. In addition to generating signals with echoes, some of the problems also look at various techniques for modifying a signal containing an echo. These modifications would be useful in algorithms that attempt to identify the location of the echoes.

1. Generate and plot 500 points of a signal that contains the original signal presented in "Problem Solving Applied" plus an echo scaled by 0.3 and delayed 3 seconds.

2. Generate and plot 500 points of a signal that contains the original signal presented in "Problem Solving Applied" plus a rolled echo scaled by 0.6 and delayed 5.5 seconds.

3. Generate and plot 500 points of a signal that contains the original signal presented in "Problem Solving Applied" plus two echoes. The first echo is delayed 0.5 seconds and is scaled by 0.7; the second echo is delayed 4 seconds and is scaled by 0.1.

4. Generate and plot 500 points of the signal that contains only the echoes described in problem 3.

5. Using the signal with echoes generated in "Problem Solving Applied" on echoes generate and plot a new signal that contains the absolute value of the signal with echoes.

6. Using the signal with echoes generated in "Problem Solving Applied" on echoes generate and plot a new signal that contains the square of the values of the signal with echoes.

Sonar Signals   These problems use sonar signals such as the one developed in "Problem Solving Applied" on sonar signals.

7. Generate and plot a sonar signal that contains the sonar signal specified by the parameters in "Problem Solving Applied" on sonar signals followed the 200 points of silence, followed by 2 milliseconds of the sonar signal.

8. Generate and plot a sonar signal that contains the sonar signal specified by the parameters in "Problem Solving Applied" on sonar signals followed by 1 millisecond of silence.

Temperature Conversions   The following equations give relationships between temperatures in degrees Fahrenheit (T_F), degrees Celsius (T_C), degrees Kelvin (T_K), and degrees Rankin (T_R):

$$T_F = T_R - 459.67°R$$

$$T_F = \frac{9}{5} T_C + 32°F$$

$$T_R = \frac{9}{5} T_K$$

9. Write a program to generate a table with the conversions from Fahrenheit to Celsius for values from 0° F to 100° F. Print a line in the table for each 5-degree change.

10. Write a program to generate a table with the conversions from Fahrenheit to Kelvin and Rankin for values from 0° F to 100° F. Print a line in the table for each 5-degree change.

Bacteria Growth   The growth of bacteria in a colony can be modeled with the following exponential equation[2]:

$$y_{new} = y_{old} e^{1.386\, t}$$

where y_{new} is the new number of bacteria in the colony, y_{old} is the initial number of bacteria in the colony, and t is the elapsed time in hours.

11. Use the bacteria growth equation to predict the number of bacteria in the colony after 6 hours if the initial population is one. Print a table that shows the number of bacteria every hour up through 6 hours.

12. Modify the program developed in problem 11 so that the user enters the elapsed time in hours.

13. Modify the program developed in problem 11 so that the user enters the elapsed time in minutes even though the equation still requires a time in hours.

14. Modify the program developed in problem 11 so that the user enters the initial population.

15. Modify the program developed in problem 11 so that the user enters two time values, and the program computes the amount of growth between the two times.

Carbon Dating   Carbon dating is a method for estimating the age of organic substances such as shells, seeds, and wooden artifacts[2]. The technique compares the amount of carbon 14, a radioactive carbon, contained in the remains of the substance with the amount of carbon 14 that would have been in the object's environment at the time it was alive. The following equation gives the estimated age in years of an artifact:

$$\text{age} = \frac{-\log_e(\text{carbon 14 proportion remaining})}{0.0001216}$$

The half life of carbon 14 is approximately 5700 years. If we use 0.5 for the proportion remaining, the age is computed to be 5700.22 years.

16. Write a program that allows the user to enter the proportion of carbon 14 remaining (which will be a value between 0.0 and 1.0). Print the age of the artifact.

17. Modify the program written in problem 16 so that it truncates the age to the nearest year.

18. Modify the program written in problem 16 so that it rounds the age to the nearest century.

Frequency Measurements   The frequency of a sinusoid can be given in cycles per seconds (Hz) or in radians per seconds, where 1 cycle per second is equal to 2π radians per second.

19. Write a program that allows the user to enter a frequency in cycles per second and then converts the value to radians per second.

20. Write a program that allows the user to enter a frequency in radians per second and then converts the value to Hz.

21. A normalized frequency is defined to be equal to ωT, where ω is a frequency in radians per second and T is in units of seconds. Thus, a normalized frequency will be in units of radians. Write a program that allows a user to enter a frequency in radians per seconds and a value of T. Compute and print the corresponding normalized frequency.

22. Modify the program in problem 21 such that the user enters the frequency units in Hz.

23. Modify the program in problem 21 such that the normalized frequency is always a value between 0 and 2π by subtracting multiples of 2π. Thus, if the normalized frequency is equal to 2.5π, then the desired value is 0.5π; if the normalized frequency is equal to 7π, then the desired value is π. (Hint: review the rem function.)

Courtesy of Japan National Tourist Organization

GRAND CHALLENGE: Superconductivity

Fast trains (also called "bullet trains") can travel at speeds approaching 300 miles per hour, suspended inches above a rail guideway by a repellent force generated between a superconductive magnet on the vehicle and a coil on the ground. Electronic sensors detect an approaching train and adjust the electromagnetic field as it passes over, giving it a "jolt" to get to the next set of sensors. When there is not a train over the sensors, the energy shuts off. Wheels are retracted inside the vehicle when it is moving, but the "landing wheels" are lowered for extra stability when the train slows down to stop. New materials that are both lighter and stronger must be developed to further improve transportation systems such as magnetic levitation trains. These new composite materials go through rigorous testing to determine properties such as their strength under stress and their temperature distribution properties. The temperature distribution properties become very important when these new materials will be near engine parts or high temperature components.

Control Flow

Introduction

The MATLAB programs that we have written so far have included only sequential steps. One operation was performed after another until we had completed the desired computations. There are many problem solutions in which we need a selection command that will allow us to execute one set of statements if a specified condition is true and another set of statements if the specified condition is false. We also need commands that allow us to build loops, which are groups of statements that are repeated. Both types of statements are called control statements because they allow us to control the order in which statements are executed.

4.1   IF STATEMENTS

if

A selection statement is a conditional control statement that allows us to ask a question or test a condition to determine which steps are to be performed next. The most common form of selection structure is an `if` statement. Since the `if` statement uses relational and logical operations, we discuss these first and then come back to the `if` statement.

Relational Operators

MATLAB has six relational operators for comparing two matrices of equal size, as shown in the following list:

Relational Operator	Interpretation
<	less than
<=	less than or equal
>	greater than
>=	greater than or equal
==	equal
~=	not equal

logical expression
0-1 matrix

Matrices or matrix expressions are used on both sides of a relational operator to yield another matrix of the same size. Each entry in the resulting matrix contains a 1 if the comparison is true when applied to the values in the corresponding position of the matrices; otherwise, the entry in the resulting matrix contains a 0. An expression that contains a relational operator is a logical expression because the result is a matrix containing ones and zeros that can be interpreted as true values and false values, respectively; the resulting matrix is also called a 0-1 matrix.

Consider the following logical expression:

a < b

If a and b are scalars, then the value of this expression is 1 (for true) if a is less than b; otherwise, the expression is 0 (for false). Let a and b be vectors with the following values:

$$a = [2 \quad 4 \quad 6]$$
$$b = [3 \quad 5 \quad 1]$$

and
or
not

Then, the value of a < b is the vector [1   1   0], while the value of a ~= b is [1   1   1].

We can also combine two logical expressions using the logical operators and, or, and not. These logical operators are represented by the following symbols:

Logical Operator	Symbol
and	&
or	$\vert$
not	$\sim$

Logical operators allow us to compare 0-1 matrices such as those computed by relational operators, as shown in the following logical expression:

a<b & b<c

This operation is valid only if the two resultant matrices (represented by a<b and b<c) are the same size. Then an entry in the matrix represented by this logical expression is 1 if the values in the corresponding entries in a, b, and c, are such that a<b<c; otherwise, the entry is zero.

When two logical expressions are joined by or, entries in the resulting 0-1 matrix are 1 (true) if either or both expressions are true; it is 1 (false) only when both expressions are false. When two logical expressions are joined by and, the entire expression is true only if both expressions are true. Table 4.1 lists all possible combinations for logical operators between logical expressions A and B.

Logical operators are used with complete logical expressions. For example, a>b&b>c is a valid logical expression, but a>b&c is not an equivalent expression.

Logical expressions can also be preceded by the logical operator not. This operator changes the value of the expression to the opposite value; hence, if a>b is true, then $\sim$(a>b) is false.

A logical expression may contain several logical operators as in the following:

$\sim$(b==c $\vert$ b==5.5)

hierarchy of
logical operators

The hierarchy, from highest to lowest, is not, and, and or. Of course, parentheses can be used to change the hierarchy. In the example above, the expressions b==c and b==5.5 are evaluated first. Suppose b = 3 and c = 5. Then neither expression is true, so the expression b==c $\vert$ b==5.5 is false. We then apply the not operator, which changes the value of the expression to true. Suppose that we did not

TABLE 4.1 Logical Operators

A	B	$\sim$A	A $\vert$ B	A & B
false	false	true	false	false
false	true	true	true	false
true	false	false	true	false
true	true	false	true	true

have the parentheses around the logical expression, as in the following:

~b==c | b==5.5

In this case, the expression ~b==c would be evaluated along with b==5.5. For the values given for b and c, the value of each relational expression is false, and thus the value of the entire logical expression is false. You might wonder how we can evaluate ~b when the value in b is a number. In MATLAB, any values that are nonzero are considered to be true; values of zero are false. As a result, we have to be very careful using relational and logical operators to be sure that the steps being performed are the ones that we want to perform.

Practice!

Determine if the following expressions in problems 1 to 8 are true or false. Then check your answers using MATLAB. Remember that to check your answer, all you need to do is enter the expression. Assume that the following variables have the indicated values:

$$a = 5.5 \qquad b = 1.5 \qquad k = -3$$

1. a < 10.0

2. a + b >= 6.5

3. k ~= 0

4. b - k > a

5. ~(a == 3*b)

6. -k <= k + 6

7. a < 10 & a > 5

8. abs(k) > 3 | k < b - a

Simple If Statement

The if statement can be implemented as a simple if statement, or it can include else clauses or elseif clauses. We discuss the simple if statement first; its general form is the following:

```
if   logical expression
        statement group A
end
```

If the logical expression is true, we execute the statements between the if statement and the end statement. If the logical expression is false, we jump immediately to the statement following the end statement. It is important to indent the statements within an if structure in order to see the control flow easily.

An example of the if statement is shown below:

```
if  a < 50
    count = count + 1;
    sum = sum + a;
end
```

Assume that a is a scalar. If a < 50, then count is incremented by 1 and a is added to sum; otherwise, these two statements are skipped. If a is not a scalar, then count is incremented by 1 and a is added to sum only if every element in a is less than 50.

nested if statements

If statements can also be nested; the following structure includes an if statement within an if statement:

```
if  logical expression 1
    statement group A
    if  logical expression 2
        statement group B
    end
    statement group C
end
statement group D
```

If logical expression 1 is true, we then always execute statement groups A and C. If logical expression 2 is also true, we then also execute statement group B before executing statement group C. If logical expression 1 is false we immediately skip to statement group D. The indenting is very important in understanding the structure of nested if statements.

Here is an example of nested if statements that extends the previous example:

```
if  a < 50
    count = count + 1;
    sum = sum + a;
    if  b > a
        b = 0;
    end
end
```

Again, first assume that a and b are scalars. Then, if a < 50 we increment count by 1 and add a to sum. In addition, if b > a, then we also set b to zero. If a is not less than 50, then we skip immediately to the statement following the second end statement. If a is not a scalar, then the condition a < 50 is only true if every element of a is less than 50. If neither a nor b is a scalar, then b is greater than a only if for every corresponding pair of elements from a and b, b is greater than a. If a or b is a scalar, then the other matrix is compared to the scalar element-wise.

Relational and Logical Functions

MATLAB contains a set of relational and logical functions that are very useful with if statements. We now discuss each of these functions.

any (x)	For each column of x, this function returns a value of 1 (or true) if any elements of the matrix x are nonzero; otherwise, the function returns a value of 0 (or false).
all (x)	For each column of x, this function returns a value of 1 (or true) if all the elements of the matrix x are nonzero; otherwise, the function returns a value of 0 (or false).
find (x)	This function returns a vector containing the indices of the nonzero elements of a vector x. If x is a matrix, then the indices are selected from x (:), which is a long column vector formed from the columns of x.
exist ('A')	This function returns a value of 1 (true) if A exists as a variable in the workspace, a value of 2 if A or A.m is a file, and 0 (false) if A does not exist. Note that the variable name must be in quotes.
isnan (x)	This function returns a matrix with ones where the elements of x are NaNs and zeros where they are not.
finite (x)	This function returns a matrix with ones where the elements of x are finite and zeros where they are infinite or NaNs.
isempty (x)	This function returns 1 if x is an empty matrix and 0 otherwise.
isstr (x)	This function returns 1 if x is a string and 0 otherwise.
strcmp (y1, y2)	This function compares strings y1 and y2 and returns 1 if the two are identical and 0 otherwise. The comparison is case sensitive, and any leading and trailing blanks are included.

Assume that A is a matrix with three rows and three columns of values. Consider the following statement:

```
if all(A) == 0
    disp ('A contains all zeros')
end
```

The string 'A contains all zeros' is printed only if all nine values in A are zero.

Practice!

Determine the value of the following expressions. Then check your answers using MATLAB. Remember that to check your answer, all you need to do is enter the expression. Assume that the matrix b has the indicated values:

$$b = \begin{bmatrix} 1 & 0 & 4 \\ 0 & 0 & 3 \\ 8 & 7 & 0 \end{bmatrix}$$

1. any (b)

2. find (b)

3. all (any (b))

4. any (all (b))

5. finite (b (: , 3))

6. any (b (1: 2, 1: 3))

Else Clause

The else clause allows us to execute one set of statements if a logical expression is true and a different set if the logical expression is false. The general form of an if statement combined with an else clause (sometimes called an if-else statement) is shown below. If the logical expression is true, then statement group A is executed. If the logical expression is false, then statement group B is executed. The statement groups can also contain if or if-else statements to provide a nested structure.

```
if   logical expression
     statement group A
else
     statement group B
end
```

To illustrate, we use a cable car that goes between two towers. Assume that the scalar d contains the distance of the cable car from the nearest tower. If the cable car is within 30 feet of the tower, the velocity is computed using this equation:

$$velocity = 0.425 + 0.00175 \; d^2$$

If the cable car is farther than 30 feet from the tower, use the following equation:

$$velocity = 0.625 + 0.12 \; d - 0.00025 \; d^2$$

We can compute the correct velocity with this statement:

```
if d <= 30
   velocity = 0.425 + 0.00175*d^2;
else
   velocity = 0.625 + 0.12*d - 0.00025*d^2;
end
```

Elseif Clause

When we nest several levels of if-else statements, it may be difficult to determine which logical expressions must be true (or false) to execute each set of statements. In this case, the elseif clause is often used to clarify the program logic:

```
if   logical expression 1
     statement group A
elseif logical expression 2
     statement group B
elseif logical expression 3
     statement group C
end
```

We have shown two elseif clauses, although more may be used. If logical expression 1 is true, then only statement group A is executed. If logical expression 1 is false and logical expression 2 is true, then only statement group B is executed. If logical expressions 1 and 2 are false and logical expression 3 is true, then only statement group C is executed. If more than one logical expression is true, the first logical expression that is true determines which statement group is executed. If none of the logical expressions is true, then none of the statements within the if structure is executed.

The else clause can be combined with the elseif clause, as shown in this general form:

```
if   logical expression 1
      statement group A
elseif logical expression 2
      statement group B
elseif logical expression 3
      statement group C
else
      statement group D
end
```

case structure

If none of the three logical expressions is true, then statement group D is executed. The if-elseif structure is also sometimes called a case structure because a number of cases are tested. Each case is defined by its corresponding logical expression.

Practice!

In problems 1 to 7, give MATLAB statements that perform the steps indicated. Assume that the variables are scalars.

1. If time is greater than 50.0, increment time by 1.0.

2. When the square root of poly is less than 0.001, print the value of poly.

3. If the difference between volt_1 and volt_2 is larger than 2.0, print the values of volt_1 and volt_2.

4. If the value of den is less than 0.003, set result to zero; otherwise, set result equal to num divided by den.

5. If the natural logarithm of x is greater than or equal to 10, set time equal to zero and increment count.

6. If dist is less than 50.0 and time is greater than 10.0, increment time by 2; otherwise, increment time by 5.

7. If dist is greater than or equal to 100.0, increment time by 10. If dist is between 50 and 100, increment time by 1. Otherwise, increment time by 0.5.

4.2   FOR LOOPS

MATLAB contains two statements for generating loops, the for statement and the while statement. In this section we discuss the for statement, and in Section 4.3 we discuss the while loop.

The for statement has the following general structure:

```
for index  =  expression
     statement group A
end
```

The expression is a matrix (which could also be a scalar or a vector), and the statements in statement group A are repeated as many times as there are columns in the expression matrix. Each time through the loop, the index has the value of one of the elements in the expression matrix. Before we discuss the rules for using a for loop, we look at a simple example.

Suppose that we have a vector containing a group of distance values that represent the distance of a cable car from the nearest tower. We want to generate a corresponding vector containing corresponding velocities. If the cable car is within 30 feet of the tower, we use this equation:

$$\text{velocity} = 0.425 + 0.00175\ d^2$$

If the cable car is further than 30 feet from the tower, we use the following equation:

$$\text{velocity} = 0.625 + 0.12\ d - 0.00025\ d^2$$

Since the choice of velocity equation depends on the value of d, we must make each selection separately for d(1), d(2), and so on. However, we do not want to write a separate statement for computing velocity(1), velocity(2), and so on. Therefore, we use a loop, with the index used as a subscript.

In the first solution we assume that there are 25 elements in the vector d.

```
for k = 1:25
   if d(k) <= 30
      velocity = 0.425 + 0.00175*d(k)^2;
   else
      velocity = 0.625 + 0.12*d(k) - 0.00025*d(k)^2;
   end
end
```

In the following solution we assume that the size of the vector d is not known. Therefore, we use the length function to determine the number of times that we want to execute the loop.

```
for k = 1: length (d)
    if d(k) <= 30
        velocity = 0.425 + 0.00175*d(k)^2;
    else
        velocity = 0.625 + 0.12*d(k) - 0.00025*d(k)^2;
    end
end
```

The rules for a for loop are the following:

1. The index of the for loop must be a variable.
2. If the expression matrix is the empty matrix, the loop will not be executed. Control will pass to the statement following the end statement.
3. If the expression matrix is a scalar, then the loop will be executed one time, with the index containing the value of the scalar.
4. If the expression matrix is a row vector, then each time through the loop, the index will contain the next value in the vector.
5. If the expression matrix is a matrix, then each time through the loop, the index will contain the next column in the matrix.
6. Upon completion of the for loop, the index contains the last value used.
7. If the colon operator is used to define the expression matrix using the following format:

```
for k = initial: increment: limit
```

then the number of times that the loop will be executed can be computed using the following equation:

$$\text{floor}\left(\frac{\text{limit - initial}}{\text{increment}}\right) + 1$$

If this value is negative, the loop is not executed. Thus, if a for statement contained the following information:

```
for k = 5: 4: 83
```

then it would be executed the following number of times:

$$\text{floor}\left(\frac{83 - 5}{4}\right) + 1 = \text{floor}\left(\frac{78}{4}\right) + 1 = 20$$

The value of k would be 5, then 9, then 13, and so on until the final value of 81. The loop would not be executed with the value of 85, because it is greater than the limit, 83.

Practice!

Determine the number of times that the for loop defined by the following statements will be executed. Then to test your answer, modify and enter the following statement, which corresponds to the number of times that the for loop will be executed in the first problem:

length(3:20)

1. for k = 3:20

2. for count = -2:14

3. for k = -2:-1:-10

4. for time = 10:-1:0

5. for time = 10:5

6. for index = 52:-12

break

The break statement can be used to exit a loop before it has been completed. This statement is useful if an error condition is detected within the loop.

 PROBLEM SOLVING APPLIED: OPTICAL FIBERS

light pipe

If light is directed into one end of a long rod of glass or plastic, the light is totally reflected by the walls, bouncing back and forth until it emerges at the far end of the rod [2,6]. This interesting optical phenomenon can be used to transmit light, and even images, from one place to another. If we bend a "light pipe," the light will follow the shape of the pipe and emerge only at the end, as shown in the diagram in Figure 4.1.

image conduit

An optical fiber is a very thin glass fiber. Bundles of very thin optical fibers can be used to make light pipes. If the fiber ends are polished and the spatial arrangement is the same at both ends (a coherent bundle), the fiber bundle can be used to transmit an image, and the bundle is called an image conduit. If the fibers do not have the same arrangement at both ends (an incoherent bundle), light is transmitted instead of an image, and it is called a light guide. Because the optical fibers are so flexible, light guides and image conduits are used in instruments designed to permit visual observation of objects or areas that would otherwise be inaccessible. For example, an endoscope is an instrument used by physicians to examine the interior of a patient's body with only a very small incision. The

light guide

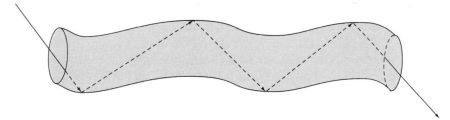

Fig. 4.1 Optical fiber.

endoscope uses an optical fiber to transmit the light needed inside the patient's body. Optical fibers can also be used to transmit laser light, which can be used to unclog obstructed arteries, break up kidney stones, and clear cataracts.

This phenomenon of total internal reflection can be predicted using Snell's law and the indices of refraction for the materials being considered for the light pipe. A light pipe is actually composed of two materials—the material forming the rod or pipe itself and the material that surrounds the pipe. Normally, the material forming the rod is denser than the surrounding medium. When light passes from one material into another material with a different density, the light is bent, or "refracted," at the interface of the two materials. The amount of refraction depends on the indices of refraction (see Table 4.2) of the materials and the angle of

indices of refraction

TABLE 4.2   Index of Refraction for Common Materials [6]

Material	Index of Refraction
Gases (at atmospheric pressure and 0°C)	
Hydrogen	1.0001
Air	1.0003
Carbon dioxide	1.0005
Liquids (at 20°C)	
Water	1.333
Ethyl Alcohol	1.362
Glycerine	1.473
Solids (at room temperature)	
Ice	1.31
Polystyrene	1.59
Crown glass	1.50–1.62
Flint glass	1.57–1.75
Diamond	2.417
Acrylic (polymethylmethacrylate)	1.49

critical angle

incidence of the light. If the light striking the interface comes from within the denser material, it may reflect off the interface, rather than pass through it. The angle of incidence where the light will be reflected from the surface, rather than cross it, is called the critical angle θ_c. Since the critical angle depends on the indices of refraction of the two materials, we can compute this angle and determine if light entering the pipe at a particular angle will stay within the pipe. Assume n_2 is the index of refraction of the surrounding medium, and n_1 is the index of refraction of the pipe itself. If n_2 is greater than n_1, the pipe will not transmit light; otherwise, the critical angle can be determined from the following equation:

$$\sin \theta_c = \frac{n_2}{n_1}$$

Write a MATLAB program that determines whether or not light will be transmitted by two materials that form a pipe. Assume that the ASCII data file named indicies.dat contains a number of possible light pipes, with each line containing the index of refraction for the rod followed by the index of refraction for the surrounding medium. The program should determine if the materials will indeed form a light pipe and, if so, for what angles of light entering the pipe.

 1. PROBLEM STATEMENT

Determine whether or not specified materials will form a light pipe. If they do form a light pipe, compute the angles at which light can enter the pipe and be transmitted.

 2. INPUT/OUTPUT DESCRIPTION

As shown in Figure 4.2, the input to the program is a data file containing the indices of refraction for the potential light pipes. The output is a message indicating whether or not light is transmitted and the angles at which it can enter the pipe.

 3. HAND EXAMPLE

The index of refraction of air is 1.0003 and the index of glass is 1.5. If we form a light pipe of glass surrounded by air, the critical angle θ_c can be com-

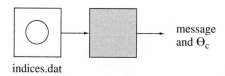

indices.dat

Fig. 4.2 I/O diagram.

puted as follows:

$$\theta_c = \sin^{-1}\left(\frac{n_2}{n_1}\right) = \sin^{-1}\left(\frac{1.0003}{1.5}\right)$$
$$= \sin^{-1}(0.66687) = 41.82°$$

This light pipe will transmit light for all angles of incidence greater than 41.82°.

4. MATLAB SOLUTION

```
%
%            This program reads the indices of refraction
%            for materials forming a light pipe. For each
%            pair of materials, we determine if the pipe
%            will transmit light, and at what angles of
%            incidence in degrees.
%
factor = 180/pi;
load indices.dat
n1 = indices(:,1);
n2 = indices(:,2);
for k = 1:length(n1)
   if n2(k) > n1(k)
      fprintf('Light is not transmitted for \n')
      fprintf ('rod index %g and medium index %g \n\n',...
              n1(k), n2(k))
   else
      critical_angle = asin(n2(k)/n1(k))*factor;
      fprintf('Light is transmitted for \n')
      fprintf('rod index %g and medium index %g \n',...
              n1(k),n2(k))
      fprintf ('for angles greater than %g degrees \n\n',...
              critical_angle)
   end
end
```

5. TESTING

We first test this program using an indices.dat file with the data shown:

```
1.31 1.473
1.5  1.0003
1.49 1.33
```

The corresponding output is

```
Light is not transmitted for
rod index 1.31 and medium index 1.473

Light is transmitted for
rod index 1.5 and medium index 1.0003
for angles greater than 41.8257 degrees

Light is transmitted for
rod index 1.49 and medium index 1.33
for angles greater than 63.204 degrees
```

See if you can find materials for three light pipes that transmit light with angles of incidence greater than 45°.

Additional problems related to this application are included in the end-of-chapter problems.

4.3   WHILE LOOPS

The while loop is an important structure for repeating a set of statements as long as a specified condition is true. The general format for this control structure is as follows:

```
while expression
    statement group A
end
```

If the expression is true, then statement group A is executed. After these statements are executed, the condition is retested. If the condition is still true, the group of statements is executed again. When the condition is false, control skips to the statement following the end statement. The variables modified in statement group A should include the variables in the expression, or the value of the expression will never change. If the expression is always true (or is a value that is nonzero), the loop becomes an infinite loop. (Remember that you can use the ^c keys to abort an infinite loop.)

infinite loop

We illustrate the use of a while loop in a set of statements that add the values in a vector to a sum until a negative value is reached. Since all the values in the vector could be positive, the expression on the `while` statement must accommodate that situation also:

```
sum = 0;
k = 1;
while x(k) >= 0 & k <= length(x)
    sum = sum + x(k);
    k = k+1;
end
```

In the following section, we develop a problem solution that uses nested `for` loops inside a `while` loop.

PROBLEM SOLVING APPLIED:
TEMPERATURE EQUILIBRIUM

The design of new materials for improving the air flow characteristics around vehicles involves analyzing the materials for not only air flow but also for properties such as temperature distribution. In this problem we consider the temperature distribution in a thin metal plate as it reaches a point of thermal equilibrium [2]. The plate is designed to be used in an application in which the temperatures of all four sides of the metal plate will be maintained at constant or isothermal temperatures. The temperature at other points on the plate is a function of the temperature of the surrounding points. If we consider the plate to be similar to a grid, then a matrix could be used to store the temperatures of the corresponding points on the plate. Figure 4.3 contains a grid for a plate that is being analyzed with six temperature measurements along the sides and eight temperature measurements along the top and bottom. The isothermal points on all four sides are shaded. A total of 48 temperature values are represented.

thermal equilibrium

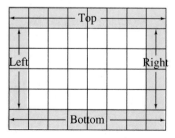

Fig. 4.3   Grid for a metal plate.

Fig. 4.4   Points surrounding T_0.

interior point

The isothermal temperatures on the four sides would be specified; we assume that the top and the two sides are maintained at the same temperature while the bottom of the plate is maintained at a different temperature. The rest of the points are initially set to some arbitrary temperature, usually zero. The new temperature of each interior point is calculated as the average of its four surrounding points (see Figure 4.4) with the following equation:

$$T_0 = \frac{T_1 + T_2 + T_3 + T_4}{4}$$

After computing the new temperature for an interior point, the difference between the old temperature and the new temperature is computed. If the temperature change is greater than some specified tolerance value, the plate is not yet in thermal equilibrium, and the entire process is repeated.

We use two matrices for the temperatures, one for the old temperatures and one for the new temperatures. We need two matrices because we assume that the temperature changes for all the points are occurring simultaneously, even though we compute them one at a time. If we used only one matrix, we would be updating information before we were through with the old information. For example, suppose that we are computing the new temperature at position (3,3). The new value is the average of temperatures in positions (2,3), (3,2), (3,4), and (4,3). When we compute the new temperature at position (3,4), we again compute an average, but we want to use the old value in position (3,3), not its updated value.

tolerance

Thus, we use a matrix of old temperatures to compute a matrix of new temperatures and to determine if any of the temperature changes exceed the tolerance. We then move the new temperatures to the old array. When none of the temperature changes exceed the tolerance, we assume that equilibrium has been reached, and we print the final temperatures.

 1. PROBLEM STATEMENT
Determine the equilibrium values for a metal plate with isothermal sides.

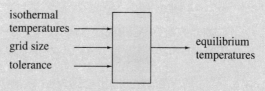

Fig. 4.5   I/O diagram.

 2. INPUT/OUTPUT DESCRIPTION

As shown in Figure 4.5, the input is the size of the plate grid, the isothermal temperatures and the tolerance value. The output is the set of equilibrium temperatures for the metal plate.

 3. HAND EXAMPLE

To be sure that we understand the process, we examine a simple case, studying each iteration. Assume that the matrix contains four rows and four columns. The isothermal temperatures are 100° and 50°, and we initialize all other points to zero. We use a tolerance value of 40°. The initial set of temperatures and the successive iterations to thermal equilibrium are shown next:

Initial Temperatures

100	100	100	100
100	0	0	100
100	0	0	100
50	50	50	50

First Iteration

100	100	100	100
100	50	50	100
100	37.5	37.5	100
50	50	50	50

Second Iteration

100	100	100	100
100	71.875	71.875	100
100	59.375	59.375	100
50	50	50	50

Since none of the temperature changes between the first and second iteration exceeded the tolerance value of 40°, the temperatures in the second iteration are also the equilibrium temperatures.

4. MATLAB SOLUTION

```
%
%          This program initializes the temperatures in
%          metal plate and determines the equilibrium
%          temperatures based on a tolerance value.
%
nrows = input('Enter number of rows ');
ncols = input('Enter number of columns ');
iso1 = input('Enter temperature for top and sides ');
iso2 = input('Enter temperature for bottom ');
tolerance = input('Enter equilibrium tolerance ');
%
%     Initialize and print temperature matrix.
%
old = zeros(nrows,ncols);
old(1,:) = iso1 + zeros(1,ncols);
old(:,1) = iso1 + zeros(nrows,1);
old(:,ncols) = iso1 + zeros(nrows,1);
old(nrows,:) = iso2 + zeros(1,ncols);
disp('Initial Temperatures');
disp(old)
new = old;
equilibrium = 0;
%
%     Update temperatures and test for equilibrium
%
while ~equilibrium
   for m=2:nrows-1
      for n=2:ncols-1
         new(m,n) =  (old(m-1,n) + old(m,n-1) +...
                         old(m,n+1)  + old(m+1,n))/4;
      end
   end
   if all(new-old <= tolerance)
      equilibrium = 1;
      disp('Equilibrium Temperatures');
      disp(new)
   end
   old = new;
end
```

5. TESTING

If we use the data from our hand-worked example in this program, the output is the following:

```
Enter number of rows 4
Enter number of columns 4
Enter temperature for top and sides 100
Enter temperature for bottom 50
Enter equilibrium tolerance 40
Initial Temperatures
    100    100    100    100
    100      0      0    100
    100      0      0    100
     50     50     50     50
Equilibrium Temperatures
   100.0000   100.0000   100.0000   100.0000
   100.0000    71.8750    71.8750   100.0000
   100.0000    59.3750    59.3750   100.0000
    50.0000    50.0000    50.0000    50.0000
```

The tolerance is an important variable in this problem solution. Try running this example using a much smaller tolerance, such as $1°$ and observe the difference in the equilibrium temperatures. It would also be interesting to have the solution print the number of iterations necessary to achieve equilibrium.

Additional problems related to this application are included in the end-of-chapter problems.

SUMMARY

We presented three different types of control structures in this chapter. An if statement, along with optional else and elseif clauses, allows us to execute different sets of commands depending on the value of a logical expression. The for statement and the while statement allow us to generate loops that repeat a specified number of times or until a logical expression becomes false. These control structures are necessary for solving many types of engineering problems. Example engineering applications that used optical fibers and that discussed temperature distributions in thin metal plates were presented, and solutions were developed for related problems.

MATLAB SUMMARY

This MATLAB summary lists all the special symbols, commands, and functions that were defined in this chapter. A brief description is also included for each one.

Special Characters

<	less than
<=	less than or equal
>	greater than
>=	greater than or equal
==	equal
~=	not equal
&	and
|	or
~	not

Commands and Functions

all	determines if all values are true
any	determines if any values are true
break	exits a loop
else	optional clause in if structure
elseif	optional clause in if structure
end	defines end of a control structure
exist	determines if a variable exists
find	locates nonzero values
finite	determines if values are finite
for	generates loop structure
if	tests logical expression
isempty	determines if matrix is empty
isnan	determines if values are NaNs
isstr	determines if value is a string
strcmp	compares two strings
while	generates a loop structure

PROBLEMS

Problems 1 to 8 relate to the engineering applications presented in this chapter. Problems 9 to 30 relate to new engineering applications.

Optical Fibers These problems relate to the table of refraction indices for materials given in "Problem Solving Applied" on optical fibers.

1. Print a table containing the information in the indices.dat file in the following format:

```
     Refraction Indices
Rod Index      Medium Index
```

2. Add a column to the table printed in problem 1 that contains the critical angle in degrees.

3. Add a column to the table printed in problem 1 that contains the critical angle in radians.

4. Print a table of the information from the `indices.dat` file that relates to light pipes that will transmit light. Use the following format:

    ```
              Light Transmitting Pipes
    Rod Index        Medium Index     Angles of Incidence
                                            (degrees)
    ```

5. Modify the table printed in problem 4 so that it uses radians for the angles of incidence.

Temperature Distribution   These problems relate to the temperature distribution problem from "Problem Solving Applied" on temperature equilibrium.

6. Modify the program developed in "Problem Solving Applied" on temperature equilibrium so that the four sides of the metal plate can have different temperatures.

7. Modify the program developed in "Problem Solving Applied" on temperature equilibrium so that it prints the number of iterations necessary to achieve equilibrium.

8. Modify the program developed in "Problem Solving Applied" on temperature equilibrium so that only the top of the plate is held at a constant temperature. The rest of the temperature values are dependent on the their neighboring temperatures; a value may have two, three, or four neighboring temperatures.

Rocket Trajectory   A small rocket is being designed to make wind shear measurements in the vicinity of thunderstorms [2]. Before testing begins, the designers are developing a simulation of the rocket's trajectory. They have derived the following equation that they believe will predict the performance of the test rocket, where t is the elapsed time in seconds:

$$\text{height} = 60 + 2.13t^2 - 0.0013t^4 + 0.000034t^{4.751}$$

The equation gives the height above ground level at time t. The first term (60) is the height in feet above ground level of the nose of the rocket.

9. Write a program to compute and print the time and height of the rocket from $t = 0$ to the time that it hits the ground, in increments of 2 seconds. If the rocket has not hit the ground within 100 seconds, stop the program.

10. Modify the program in problem 9 such that instead of a table, the program prints the time at which the rocket begins falling back to the ground and the time at which the rocket impacts.

Suture Packaging   Sutures are strands or fibers used to sew living tissue together after an injury or an operation [2]. Packages of sutures must be sealed carefully before they are shipped to hospitals so that contaminants cannot enter the packages. The object that seals the package is referred to as a sealing die. Generally, sealing dies are heated with an electric heater. For the sealing process to be a success, the sealing die is maintained at an established temperature and must contact the package with a predetermined pressure for an established time period. The time period in which the sealing die contacts the package is called the dwell time. Assume that the acceptable range of parameters for an acceptable seal are the following:

Temperature:	150–170°C
Pressure:	60–70 psi
Dwell Time:	2–2.5 s

11. A data file named `suture.dat` contains information on batches of sutures that have been rejected during a one-week period. Each line in the data file contains the batch number, the temperature, the pressure, and the dwell time for a rejected batch. The quality control engineer would like to analyze this information and needs a report that computes the percentage of the batches rejected due to temperature, the percentage rejected due to pressure, and the percentage rejected due to dwell time. It is possible that a specific batch may have been rejected for more than one reason and should be counted in all applicable totals. Write a program to compute and print these three percentages.

12. Modify the program developed in problem 11 such that it also prints the number of batches in each rejection category and the total number of batches rejected. (Remember that a rejected batch should appear only once in the total but could appear in more that one rejection category.)

13. Write a program to read the data file `suture.dat`, and make sure that the information relates only to batches that should have been rejected. If any batch should not be in the data file, print an appropriate message with the batch information.

Timber Regrowth   A problem in timber management is to determine how much of an area to leave uncut so that the harvested area is reforested in a certain period of time [2]. It is assumed that reforestation takes place at a known rate per year depending on climate and soil conditions. A reforestation equation expresses this growth as a function of the amount of timber standing and the reforestation rate. For example, if 100 acres are left standing after harvesting and the reforestation rate is 0.05, then $100 + 0.05 \times 100$, or 105 acres, are forested at the end of the first year. At the end of the second year, the number of acres forested is $105 + 0.05 \times 105$, or 110.25 acres.

14. Assume that there are 14,000 acres total with 2,500 acres uncut, and that the reforestation rate is 0.02. Print a table showing the number of acres reforested at the end of each year, for a total of 20 years.

15. Modify the program developed in problem 14 so that the user can enter the number of years to be used for the table.

16. Modify the program developed in problem 14 so that the user can enter a number of acres, and the program will determine how many years are required for the number of acres to be forested.

Weather Patterns   In Chapter 1 we discussed the types of information that is collected by the National Weather Bureau [2]. Figure 1.3 contained a sample of the reports that are available with weather information. A group of data files included in the diskette that accompanies this text contain weather information for Stapleton International Airport for the period January 1991 to December 1991. Each file contains one month of data; each line in the file contains 32 pieces of information, in the order shown in Figure 1.3. The data have been edited so that they are totally numeric. If a field of information contained T, for trace amount, the corresponding value in the data file contains 0.001. There are nine possible weather types, and since several weather types can occur during a single day, nine fields are used to store this information. For example, if weather type 1 occurred, the first of the nine fields will contain a 1; otherwise, it will contain a 0. If weather type 2 occurred, the second of the nine fields will contain a 1; otherwise, it will contain a 0. The peak wind gust direction has been converted to an integer using the following table:

N	1
NE	2
E	3
SE	4
S	5
SW	6
W	7
NW	8

The values on each line in the data file are separated by blanks, and the data files are named `Jan91.dat`, `Feb91.dat`, and so on.

17. Write a program to determine the number of days that had temperatures in the following categories for January 1991:

    Below 0

    0–32

    33–50

    51–60

    61–70

    Over 70

18. Modify the program developed in problem 17 so that it prints percentages instead of the number of days.

19. Modify the program developed in problem 17 so that it uses the time period May to August 1991.

20. Modify the program developed in problem 17 so that it computes the average temperature for days with fog in November 1991.

21. Modify the program developed in problem 17 so that it determines the date in December 1991 with the largest difference between the maximum temperature and the minimum temperature. Print the date, both temperatures, and the difference.

Critical Path Analysis   A critical path analysis is a technique used to determine the time schedule for a project [2]. This information is important in the planning stages before a project is begun, and it is also useful to evaluate the progress of a project that is partially completed. One method for this analysis starts by breaking a project into sequential events, then breaking each event into various tasks. Although one event must be completed before the next one is started, various tasks within an event can occur simultaneously. The time it takes to complete an event, therefore, depends on the number of days required to finish its longest task. Similarly, the total time it takes to finish a project is the sum of time it takes to finish each event. Assume that the critical path information for a major construction project has been stored in a data file named path.dat. Each line of the data file contains an event number, a task number, and the number of days required to complete the task. The data have been stored such that all the task data for event 1 are followed by all the task data for event 2, and so on.

22. Write a program to read the critical path information and print a project completion timetable that lists each event number, the maximum number of days for a task within the event, and the total number of days for the project completion.

23. Modify the program developed in problem 22 such that the program prints the event number and task number for all tasks requiring more than five days.

24. Modify the program developed in problem 22 such that the program prints the number of each event and a count of the number of tasks within the event.

Sensor Data   Suppose that a data file named sensor.dat contains information collected from a set of sensors. Each row contains a set of sensor readings, with the first row containing values collected at 0.0 seconds, the second row containing values collected at 1.0 seconds, and so on.

25. Write a program to read the data file and print the number of sensors and the number of seconds of data contained in the file.

26. Write a program to preprocess the sensor data such that all values that are greater than 10.0 are set to 10.0, and all values less than −10.00 are set to −10.0

27. Write a program to plot the data from the first sensor for the time period represented by the data file.

28. Write a program to print the locations of sensor data values that are not finite.

29. Write a program to print the locations of sensor data values with absolute values greater than 20.0.

30. Write a program to print the count of the number of sensor data values that are zero.

5

Courtesy of United Airlines

GRAND CHALLENGE: Speech Recognition

The modern jet cockpit has literally hundreds of switches and gauges. Several research programs have been looking at the feasibility of using a speech recognition system in the cockpit to serve as a pilot's assistant. The system would respond to verbal requests from the pilot for information such as fuel status or altitude. The pilot would use words from a small vocabulary that the computer had been trained to understand. In addition to understanding only a limited number of words, the system would also have to be trained using the speech for the pilot who would be using the system. This training information could be stored on a diskette, and inserted into the computer at the beginning of a flight so that the system could recognize the current pilot. The computer system would also use speech synthesis to respond to the pilot's request for information.

Statistical Measurements

Introduction

Analyzing data collected from engineering experiments is an important part of evaluating the experiment. This analysis ranges from simple computations on the data, such as calculating the average value, to more complicated analysis that computes measures or metrics, such as the standard deviation and variance of the data. These measurements are statistical measurements because they have statistical properties that are not exact. For example, the sine of 60° is an exact value that is the same every time we compute it, but the number of miles to the gallon that we get with our car is a statistical measurement because it varies somewhat depending on parameters such as the temperature, the speed that we travel, the type of road, and whether we are in the mountains or the desert. Just as we can measure properties and characteristics of statistical data, we can also use the computer to generate sequences of values (random numbers) with specified characteristics. In this chapter we learn how to use the data analysis functions in MATLAB, and we learn how to generate sequences of random numbers with specified characteristics.

5.1   DATA ANALYSIS FUNCTIONS

Analyzing data is an important part of evaluating test results. MATLAB contains a number of functions to make it easier to evaluate and analyze data. We first present a number of simple analysis functions, and then we present functions that compute more complicated measures or metrics related to a data set.

metric

Simple Analysis

The following groups of functions are frequently used in evaluating a set of test data.

Maximum and Minimum   This set of functions can be used to determine maximums and minimums, and their locations:

max(x)	If x is a vector, this function returns the largest value in x. If x is a matrix, this function returns a row vector containing the maximum element from each column.
max(x, y)	This function returns a matrix the same size as x and y. Each element in the matrix contains the maximum value from the corresponding positions in x and y.
[y, i] = max(x)	In this form, the function stores the maximum values from x in the vector y, and the indices of the maximum values in the vector i. If there are several identical maximum values, the index of the first one found is returned.
min(x)	If x is a vector, this function returns the smallest value in x. If x is a matrix, this function returns a row vector containing the minimum element from each column.
min(x, y)	This function returns a matrix the same size as x and y. Each element in the matrix contains the minimum value from the corresponding positions in x and y.
[y, i] = min(x)	In this form, the function stores the minimum values from x in the vector y, and the indices of the minimum values in the vector i. If there are several identical minimum values, the index of the first one found is returned.

Mean and Median   The mean of a group of values is the average. The Greek symbol μ (mu) is used to represent the mean value, as shown in the following equation,

mean

which uses summation notation to define the mean:

$$\mu = \frac{\sum_{k=1}^{N} x_k}{N} \tag{5.1}$$

where $\sum_{k=1}^{N} x_k = x_1 + x_2 + \cdots + x_N$

median

The median is the value in the middle of the group, assuming that the values are sorted. If there is an odd number of values, then assuming that the values in a vector x of length N are sorted, the median is the value in position ceil(N/2). (Thus, if N = 5, then the median is in position 3.) If there is an even number of values, then again assuming that the values in a vector x of length N are sorted, the median is the average of the values in position (N/2)-1 and position N/2.

The functions for computing the mean and median are the following:

mean(x)   If x is a vector, this function computes the mean value (or average value) of the elements of the vector x. If x is a matrix, this function computes a row vector that contains the mean value of each column.

median(x)   If x is a vector, this function computes the median value of the elements in the vector x. If x is a matrix, this function computes a row vector that contains the median value of each column. The values of x can be in any order.

Sums and Products   MATLAB contains functions for computing the sums and products of columns in a matrix, and functions for computing the cumulative sums and products of the elements in a matrix.

sum(x)   If x is a vector, this function returns the sum of the elements in x. If x is a matrix, this function returns a row vector that contains the sum of each column.

prod(x)   If x is a vector, this function returns the product of the elements in x. If x is a matrix, this function returns a row vector that contains the product of each column.

cumsum(x)   If x is a vector, this function returns the sum of the elements in x; sum(x) and cumsum(x) return the same value for a vector. If x is a matrix, this function returns a row vector that contains the cumulative sum of values in each column; thus, the first value in the row vector will contain the sum of the elements in the first column, the second value in the row vector will contain the sum of the elements in the first two columns, and so on.

cumprod(x)   If x is a vector, this function returns the product of the elements in x; prod(x) and cumprod(x) return the same value for a vector. If x is a matrix, this function returns a row vector that contains the cumulative product of each column; thus, the first value in the row vector will contain the product of the elements in the first column, the second value in the row vector will contain the product of the elements in the first two columns, and so on.

sort   MATLAB contains a function for sorting values into ascending order.

sort(x)   If x is a vector, this functions sorts the values into ascending order. If x is a matrix, this function sorts each column into ascending order.

[y,i] = sort(x)   In this form, the sorted values are returned to the matrix y and the reordered indices are stored in the matrix i. If x contains complex values, the elements are sorted by absolute value.

Practice!

Determine the matrices represented by the following function references. Then use MATLAB to check your answers. Assume that w, x, and y are the following matrices:

$$w = [0 \quad 3 \quad -2 \quad 7]$$
$$x = [3 \quad -1 \quad 5 \quad 7]$$
$$y = \begin{bmatrix} 1 & 3 & 7 \\ 2 & 8 & 4 \\ 6 & -1 & -2 \end{bmatrix}$$

1. max(w)

2. min(y)

3. min(w,x)

4. [z,i] = max(y)

5. mean(y)

6. median(w)

```
7. cumprod(y)

8. sum(x)

9. sort(2*w+x)

10. sort(y)
```

Variance and Standard Deviation

One of the most important statistical measurements for a set of data is the variance. Before we give the mathematical definition for variance, it is useful to develop an intuitive understanding. Consider the values of vectors data1 and data2 that are plotted in Figure 5.1. (These plots were generated using the plot command along with the subplot and axis commands from Chapter 7.) If we look at both plots, we can see that if we attempted to draw a line through the middle of the values, this line would be at approximately 3.0 in both plots. Thus, we would assume that both

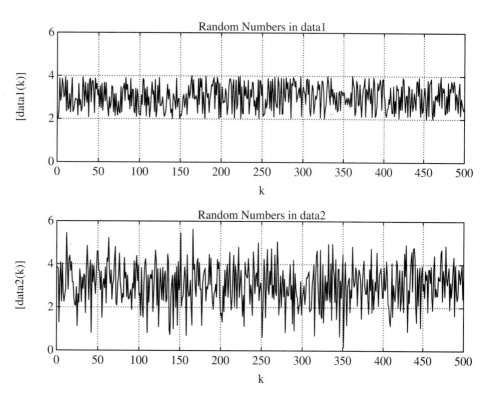

Fig. 5.1   Random sequences.

variance

vectors have approximately the same mean value of 3.0. However, the data in the two vectors clearly have some distinguishing characteristics. The values in data2 *vary* more from the mean, or *deviate* more from the mean. Thus, the measures of *variance* and *deviation* for the values in data2 should be greater than the variance and deviation for the values in data1. Hence, an intuitive understanding for variance or deviation relates to the variance or deviation of the values from the mean. The larger the variance (or the deviation), the more the values fluctuate around the mean value.

Mathematically, the variance is represented by σ^2, where σ is the Greek symbol sigma. The variance for a set of data values (which we will assume are stored in a vector x) can be computed using the following equation:

$$\sigma^2 = \frac{\sum_{k=1}^{N} (x_k - \mu)^2}{N - 1} \tag{5.2}$$

This equation is a bit intimidating at first, but if you look at it closely, it becomes much simpler. The term $x_k - \mu$ is the difference or deviation of x_k from the mean. This value is squared so that we will always have a positive value. We then add the squared deviations for all the data points. This sum is then divided by $N - 1$, which approximates an average. (The equation for the variance sometimes uses a denominator of N, but the form of the equation used here has statistical properties that make it generally more desirable.) Thus, Equation (5.2) computes the average squared deviation of the data from the mean.

The standard deviation is defined to be the square root of the variance, as follows:

$$\sigma = \sqrt{\sigma^2} \tag{5.3}$$

MATLAB includes a function to compute the standard deviation. To compute the variance, simply square the standard deviation.

std(x)   If x is a vector, this function computes the standard deviation of the values in x. If x is a matrix, this function computes a row vector containing the standard deviation of each column.

Practice!

Determine the matrices represented by the following function references. Then use MATLAB to check your answers. Assume that w, x, and y are the following matrices.

$$w = [0 \quad 3 \quad -2 \quad 7]$$
$$x = [3 \quad -1 \quad 5 \quad 7]$$

$$y = \begin{bmatrix} 1 & 3 & 7 \\ 2 & 8 & 4 \\ 6 & -1 & -2 \end{bmatrix}$$

1. `std(w)`

2. `std(x)^2`

3. `std(y(:,2))`

4. `std(y)`

5. `std(y).^2`

Histograms

The plotting capabilities of MATLAB are discussed in detail in Chapter 7. However, the histogram is a special type of graph that is particularly relevant to the statistical measurements discussed in this section. A histogram is a plot showing the distribution of a set of values. In MATLAB, the histogram computes the number of values falling in 10 bins that are equally spaced between the minimum and maximum numbers from the set of values. For example, if we plot the histograms of the values in vectors `data1` and `data2` (see Figure 5.1), we obtain the histograms in Figure 5.2. Note that the information from a histogram is different from the information obtained from the mean and variance. The histogram shows us not only the range of values, but also how they are distributed. For example, the values in `data1` tend to be equally distributed across the range of values from 2 to 4. (In Section 5.2 we will see that these types of values are called uniformly distributed values.) The values in `data2` are not equally distributed across the range of values. In fact, most of the values are centered around the mean. (In Section 5.2, we will see that this type of distribution is a Gaussian or normal distribution.)

The MATLAB command to generate and plot a histogram is the following:

```
hist(x)
```

where x is a vector containing the data to be used in the histogram. If we want to plot the second column in a matrix, we can use the colon operator as shown in this statement:

```
hist(data(:,2))
```

The `hist` command also allows us to select the number of bins. Therefore, if we want to increase the resolution of the histogram so that 25 bins are used, instead of 10, we use the following equation:

```
hist(x,25)
```

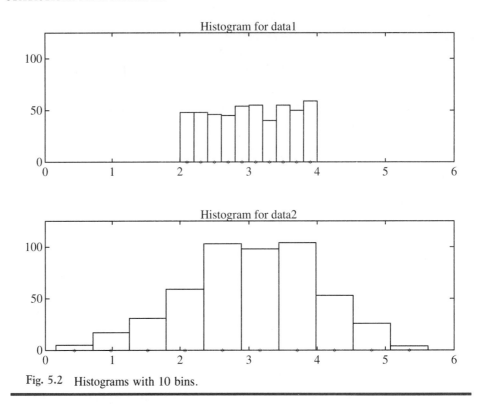

Fig. 5.2   Histograms with 10 bins.

The corresponding plots with 25 bins using the data1 and data2 vectors are shown in Figure 5.3 (The plots in Figures 5.2 and 5.3 were generated using the hist command along with the subplot and axis commands from Chapter 7.)

The information used to plot the histogram can also be stored in vectors. Consider the following commands:

```
[n, x] = hist(data1);
[n, x] = hist(data1,25);
```

Neither of these commands plots a histogram. The first command computes values for two vectors, n and x. The n vector contains the counts for the 10 bins, and the x vector contains the midpoint for each of the bin ranges. The second command is similar, but it stores 25 counts in n, and 25 bin midpoints in x. These vectors are useful in generating bar graphs, which will be discussed in Chapter 7.

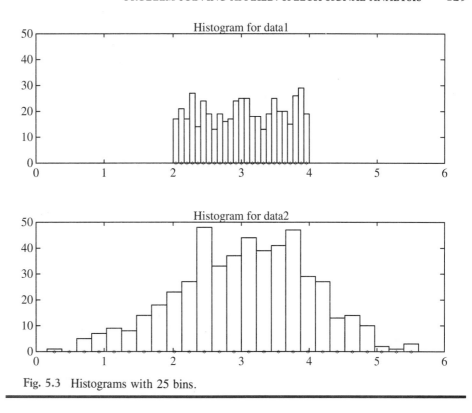

Fig. 5.3   Histograms with 25 bins.

 PROBLEM SOLVING APPLIED: SPEECH SIGNAL ANALYSIS

Suppose that we want to design a system to recognize the words for the 10 digits: "zero," "one," "two," . . . , "nine." One of the first things that we might do is analyze the data values for the ten corresponding sequences (or signals) to see if there are some statistical measurements that would allow us to distinguish these digits. The MATLAB data analysis functions allow us to easily compute these measurements. We can then print a table of the measurements and look for those measurements that allow us to distinguish values. For example, one measurement might allow us to narrow the possible digit to three digits, and then another measurement might allow us to then identify the correct digit from the three digits.

We use actual speech signals to compute a number of statistical measurements that could then be used as part of a digit recognition algorithm. The data file for each digit contains 1,000 values. Write a MATLAB program to read an ASCII data file zero.dat and compute the following information: mean, standard deviation, variance, average power, average magnitude, and number of zero crossings.

We have already discussed the mean, standard deviation, and variance. The
average power is the average squared value and will be discussed in more detail in
Section 5.3. The average magnitude is the average absolute value. The number of
zero crossings is the number of times that the values transition from a negative to
a positive value or from a positive value to a negative value.

average power
average magnitude
zero crossings

1. PROBLEM DESCRIPTION

Compute the following statistical measurements for a speech utterance: mean,
standard deviation, variance, average power, average magnitude, and number
of zero crossings.

2. INPUT/OUTPUT DESCRIPTION

Figure 5.4 contains a diagram showing the file containing the utterance as the
input and the various statistical measurements as output.

3. HAND EXAMPLE

For a hand example, assume that an utterance contains the following sequence
of values:

$$[2.5 \quad 8.2 \quad -1.1 \quad -0.2 \quad 1.5]$$

Using a calculator, we can compute the following values:

$$\text{mean} = (2.5 + 8.2 - 1.1 - 0.2 + 1.5)/5$$
$$= 2.18$$
$$\text{variance} = [(2.5 - \mu)^2 + (8.2 - \mu)^2 + (-1.1 - \mu)^2$$
$$+ (-0.2 - \mu)^2 + (1.5 - \mu)^2]/4$$
$$= 13.307$$

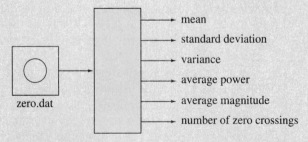

Fig. 5.4   I/O diagram.

$$\text{standard deviation} = \sqrt{13.307}$$
$$= 3.648$$
$$\text{average power} = (2.5^2 + 8.2^2 + (-1.1)^2 + (-0.2)^2 + 1.5^2)/5$$
$$= 15.398$$
$$\text{average magnitude} = (|2.5| + |8.2| + |-1.1| + |0.2| + |1.5|)/5$$
$$= 2.7$$
$$\text{number of zero crossings} = 2$$

4. MATLAB SOLUTION

In this solution, we use the MATLAB data analysis functions as much as possible. A for loop is used to compute the number of zero crossings.

```
%
%          This program computes a number of statistics
%          for an utterance stored in a data file.
%
load zero.dat;
x = zero;
fprintf('Digit Statistics \n\n')
fprintf('mean: %f \n',mean(x))
fprintf('standard deviation: %f \n',std(x))
fprintf('variance: %f \n',std(x)^2)
fprintf('average power: %f \n',mean(x.^2))
fprintf('average magnitude: %f \n',mean(abs(x)))
crossings = 0;
for j=1:length(x)-1
    if x(j)*x(j+1) < 0
        crossings = crossings + 1;
    end
end
fprintf('zero crossings: %f \n',crossings)
```

5. TESTING

The following set of values on the next page were computed for the utterance "zero" using the file zero.dat, which is contained on the diskette that accompanies this text.

```
        Digit Statistics

        mean: 0.002931
        standard deviation: 0.121763
        variance: 0.014826
        average power: 0.014820
        average magnitude: 0.089753
        zero crossings: 106.000000
```

5.2   RANDOM NUMBERS

Random numbers are not defined by an equation. Instead, they have certain characteristics that define them. There are many engineering problems that require the use of random numbers in the development of a solution. In some cases, the random numbers are used to develop a simulation of a complex problem. The simulation can be tested over and over to analyze the results, and each test represents a repetition of the experiment. We also use random numbers to approximate noise sequences. For example, the static that we hear on a radio is a noise sequence. If we are testing a program that uses an input data file that represents a radio signal, we may want to generate noise and add it to a speech signal or a music signal in order to provide a more realistic signal.

Random Number Function

rand

seed value

The rand function in MATLAB generates random numbers from the interval [0,1]. The random numbers can be uniformly distributed in the interval [0,1] or they can be Gaussian random numbers (also called normal random numbers), which have a mean of 0 and a variance of 1. A seed value is used to initiate a random sequence of values. We will discuss both uniform and Gaussian numbers in detail after we present the various forms of the rand function:

rand(n)	This function returns a matrix with n rows and n columns. Each value in the matrix is a random number between 0 and 1.
rand(m,n)	This function returns a matrix with m rows and n columns. Each value in the matrix is a random number between 0 and 1.
rand('uniform') rand('normal')	This function specifies whether uniform or Gaussian numbers are desired. The default is uniform random numbers.

rand('dist')	This function returns the string that describes the current distribution, which is either 'uniform' or 'normal'.
rand('seed',n)	This function sets the value of the seed to the value n. The value of n is initially set to 0.
rand('seed')	This function returns the current value of the seed of the random number generator.

Uniform Random Numbers

probability density function (pdf)

Random numbers can be characterized by a density function (also called a probability density function, or pdf). This function is very similar to a histogram, and shows the range of values of the random numbers and indicates the likelihood of the occurrence of particular values. The MATLAB function for generating uniform random numbers generates values that are equally likely to occur at any value between 0.0 and 1.0; the density function for these random numbers is shown in Figure 5.5. The density function shows the upper and lower bounds (0 and 1) of the random number on the abscissa (x-axis); the rectangular nature of the function (y-axis) indicates that all values between 0 and 1 are equally likely to occur. The area of a density function is always unity, and since the width of this uniform density function is 1.0, its height must also be 1.

The following commands generate and print 10 random numbers uniformly distributed between 0 and 1:

```
rand('seed',0)      % set seed
rand('uniform')     % specify uniform numbers
rand(10,1)          % generate 10 values
```

The rand function generates the same sequence of random values in each work session if the same seed value is used. Therefore, these commands will print the fol-

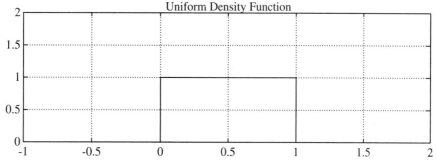

Fig. 5.5   PDF for uniform numbers between 0 and 1.

lowing values in the column vector:

```
0.2190
0.0470
0.6789
0.6793
0.9347
0.3835
0.5194
0.8310
0.0346
0.0535
```

Random sequences with values which range between values other than 0 and 1 are often needed. For example, Figure 5.6 contains a density function for random numbers that are uniformly distributed between -5.0 and 5.0. Again the area of the density function is unity.

Assume that we have a random number named r that has been generated from a uniform distribution between 0 and 1. We can then generate a random number that is uniformly distributed between a lower bound and an upper bound by modifying r. We first multiply (or scale) r by the width of the density function of the desired random number. This width is computed by subtracting the lower bound from the upper bound. Then the scaled value is added to the lower limit to adjust the new value to the proper range. For example, suppose that we want to generate values between -5 and 5. We first generate a random number r (which is between 0 and 1) and then multiply it by 10, which is the difference between the upper and lower bounds $[5 - (-5)]$. We then add the lower bound (-5), giving a resulting value that is equally likely to be any value between -5 and 5. Thus, if we want to convert a value r that is uniformly distributed between 0 and 1 to a value uniformly distributed between a lower bound a and an upper bound b, we use the following equation:

$$x = (b - a) \cdot r + a \qquad (5.4)$$

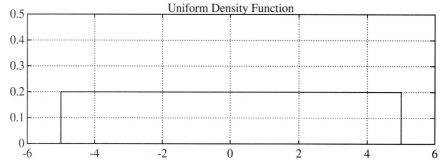

Fig. 5.6   PDF for uniform numbers between -5 and 5.

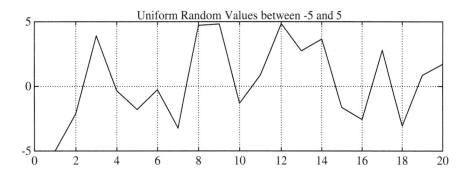

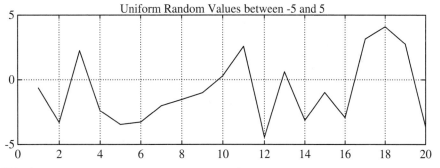

Fig. 5.7   Random sequences between −5 and 5.

Most random number generators, including the one used by MATLAB, use a seed parameter that is used to initiate the sequence. Then if two random sequences are generated with the same number of samples and the same lower and upper bounds, but with different seed values, the two sequences will be different. This is illustrated in the two plots in Figure 5.7, which each connect 20 random values that were generated with a lower bound of −5.0 and an upper bound of 5.0, but with different seed values. When we run an engineering simulation, we may want to generate the same sequence of random numbers for several test runs; using the same seed allows us to do that.

Practice!

Give the MATLAB statements to generate 10 random numbers with the specified characteristics. Check your answers by executing the statements and printing the values generated in the vectors.

1. Uniform random numbers between 0 and 10.0.

2. Uniform random numbers between −1 and +1.

> 3. Uniform random numbers between -20 and -10.
>
> 4. Uniform random numbers between 4.5 and 5.0.
>
> 5. Uniform random numbers between π and $-\pi$.

random signal

A sequence of random numbers is also called a random signal. We are often interested in the average value or mean of the signal, which can easily be computed using the mean function. Theoretically, the average value for a uniform random sequence is also equal to the midpoint of the density function, or as follows:

$$\mu = \frac{\text{upper bound} + \text{lower bound}}{2} \tag{5.5}$$

The variance of a random signal is another value that we are often interested in computing. This value can be computed by squaring the result of the std function in MATLAB. Using results from probability theory, it can be shown that the variance can also be expressed as the average squared value minus the square of the average:

$$\sigma^2 = \frac{\sum_{k=1}^{N} x_k^2}{N} - \mu^2 \tag{5.6}$$

The variance of a uniform distribution can also be expressed (again using results from probability theory) in terms of the upper and lower bounds:

$$\sigma^2 = \frac{(\text{upper bound} - \text{lower bound})^2}{12} \tag{5.7}$$

Since the difference between the upper bound and the lower bound is also the width of the density function, Equation 5.7 can also be written as:

$$\sigma^2 = \frac{\text{width}^2}{12} \tag{5.8}$$

Note that Equation 5.6 applies to all sequences of random numbers, while Equations 5.7 and 5.8 apply only to sequences of uniformly distributed random numbers. It is also important to recognize that when using actual data, the average and variance do not match exactly the theoretical values; however, as we use more and more points from a random distribution, the computed mean and variance should approach the theoretical values. To illustrate this point, we used MATLAB to generate uniform random numbers in the interval $[-5,5]$. The theoretical mean is 0, and the theoretical variance is $100/12$, or 8.3333 (from Equation (5.8)). The following table shows the number of values generated, and the mean and variance computed using the MATLAB functions mean and std:

number of values	mean value	variance value
10	0.1286	9.1868
100	0.3421	8.0959
1000	0.0036	7.8525
5000	−0.0022	8.2091

As we generate more random values, the computed statistics approach the theoretical statistics.

If a different seed value is used, or if the statements are executed again without resetting the seed, a new sequence of random values is generated. Therefore, slight changes occur in the values computed for the means and variances, but the computed values should still approach the theoretical values as the number of samples in the sequence increases.

Practice!

Compute the theoretical mean and variance for random numbers with the specified characteristics. Use MATLAB to generate 1,000 values with the specified characteristics. Calculate the mean and variance of the 1,000 values, and compare to the theoretical values using Equations (5.5) and (5.8). The calculated and theoretical values should be similar.

1. Uniform random numbers between 0 and 10.0.

2. Uniform random numbers between −1 and +1.

3. Uniform random numbers between −20 and −10.

4. Uniform random numbers between 4.5 and 5.0.

5. Uniform random numbers between π and $-\pi$.

Gaussian Random Numbers

When we generate a random sequence with a uniform distribution, all values are equally likely to occur. We sometimes need to generate random numbers using distributions in which some values are more likely to be generated than others. For example, suppose that a random sequence represents outdoor temperature measurements taken over a period of time. We would find that temperature measurements have some variation but typically are not equally likely. For example, we might find that the values vary over only a few degrees, although larger changes could occasionally occur due to storms, cloud shadows, and day-to-night changes.

Random sequences that have some values that are more likely to occur than others can often be modeled with a Gaussian random variable (also called a normal random variable). The density function for a Gaussian random variable is given by the following equation:

$$f(x) = \frac{1}{\sqrt{2\pi\sigma^2}} e^{-(x-\mu)^2/(2\sigma^2)}$$

(5.9)

where μ is the mean and σ^2 is the variance. For example, the density function for a Gaussian random variable with $\mu = 0$ and $\sigma^2 = 1$ is shown in Figure 5.8.

The mean value of this random variable corresponds to the x-coordinate of the peak of the density function. From the plot of the density function, you can see that values close to the mean are more likely to be generated than values far from the mean because the density function has higher values near the mean. Note that while a uniform random variable has specific upper and lower bounds, a Gaussian random variable does not have bounds. A Gaussian random variable typically occurs in a small region around the mean, but it can occasionally be far from the center value (or mean value). The two plots in Figure 5.9 show two different sequences generated from a random number generator with the density function shown in Figure 5.8.

In Figure 5.10, we present three different Gaussian density functions. Note that all three density functions have a mean value of 5; however, random numbers from distribution g1 are more likely to be close to 5 than are values from g2 or g3. Also, random numbers from distribution g3 are more likely to be further from the mean than are values from g2 or g1. More specifically, the three distributions have the same mean value but different standard deviations and variances. Distribution g1 has the smallest variance, and distribution g3 has the largest variance.

For Gaussian random numbers, it can be shown that approximately 68% of the values will fall within one standard deviation of the mean, 95% will fall within two standard deviations of the mean, and 99% will fall within three standard deviations of the mean. These statistics can be useful in working with Gaussian random numbers.

MATLAB will generate Gaussian values with a mean of zero and a variance of 1.0 if we specify a normal distribution. To modify these values to another Gaussian

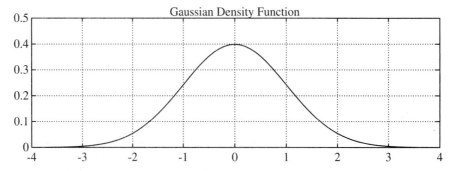

Fig. 5.8   PDF for Gaussian numbers with $\mu=0$, $\sigma^2=1$.

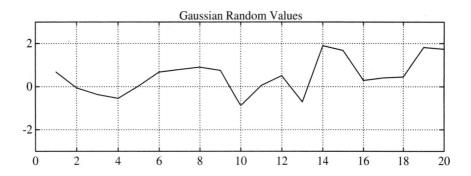

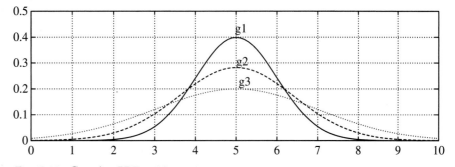

Fig. 5.9   Random sequences with $\mu=0$, $\sigma^2=1$.

distribution, multiply the values by the standard deviation of the desired distribution, and add the mean of the desired distribution. Thus, if r is a random number with a mean of zero and a variance of 1.0, the following equation will generate a new random number x with a standard deviation of a and a mean of b:

$$x = a \cdot r + b \qquad (5.10)$$

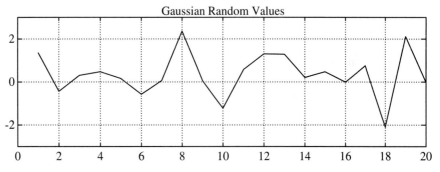

Fig. 5.10   Gaussian PDFs with same mean.

The following MATLAB statements generate Gaussian (or normal) random numbers with a mean of 5 and a variance of 2:

```
rand('seed',0)
rand('normal')
s = sqrt(2)*rand(10,1) + 5
```

The values generated for the column vector and printed by this program are the following:

```
6.6475
5.8865
5.1062
5.4972
4.0150
7.3987
5.0835
7.5414
5.3734
6.2327
```

Density Functions

histogram

In Section 5.1 we presented the `hist` command, which generated a histogram, or plot of the distribution of a set of values. This command can be used to estimate the density function of a random signal. If we have 1,000 random numbers, and we plot the histogram of the values, we are plotting a scaled version of the density function for the random numbers. For example, suppose that we generate 1,000 uniform random numbers between zero and one, and store them in the column vector `u_values`. We can then use the `hist` command to plot the density function using 25 bins:

```
rand('uniform')
u_values = rand(1000,1);
hist(u_values,25)
```

The corresponding plot is shown in Figure 5.11. As we would expect, the numbers are distributed between zero and one, and the distribution is relatively flat. If we generate 1,000 Gaussian random numbers with a mean of zero and a variance of one, and store them in the column vector `g_values`, we can then plot the density function with these statements:

```
rand('normal')
g_values = rand(1000,1);
hist(g_values,25)
```

The corresponding plot is shown in Figure 5.12. Again, as we would expect, the distribution peaks around the mean (zero), and most of the values are within two standard deviations (-2 to 2).

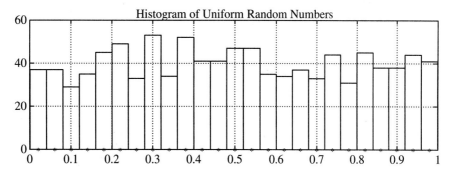

Fig. 5.11   Histogram of uniform numbers.

Practice!

Use MATLAB to generate 1,000 values with the specified characteristics. Calculate the mean and variance of the 1,000 values, and compare to the specified values. The calculated and the specified values should be similar. Also plot the histogram of the values using 25 bins.

1. Gaussian random numbers with a mean of 1.0 and a variance of 0.5.

2. Gaussian random numbers with a mean of -5.5 and a standard deviation of 0.25.

3. Gaussian random numbers with a mean of -5.5 and a standard deviation of 1.25.

4. Gaussian random numbers with a mean of π and a standard deviation of $\pi/8$.

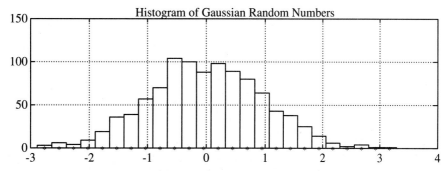

Fig. 5.12   Histogram of Gaussian numbers.

PROBLEM SOLVING APPLIED: FLIGHT SIMULATOR

Computer simulations are used to generate situations that model or emulate a real-world situation. Some computer simulations are written to play games such as checkers, poker, or chess. To play the game, you indicate your move, and the computer will then select an appropriate response. Other animated games use computer graphics to develop an interaction as you use the keys or a mouse to play the game. In more sophisticated computer simulations, such as those in a flight simulator, the computer not only responds to the input from the user but also generates values such as temperatures, wind speeds, and the locations of other aircraft. The simulators also simulate emergencies that occur during the flight of an aircraft. If all this information generated by the computer was always the same set of information, then the value of the simulator would be greatly reduced. It is important that there be "randomness" to the generation of the data.

Monte Carlo simulations

Monte Carlo simulations use random numbers to generate values that model events.

Write a program to generate a random sequence to simulate one hour of wind speed data that is updated every 10 seconds. (One hour of data is then represented by 361 values.) From analysis of actual wind patterns, it has been determined that the wind speed can be modeled as a Gaussian random number. The mean and variance are unique to a specific region and time of year and are entered as input parameters. In addition, it is assumed that the plane has a 1% chance of flying into a small storm. The length of time that the plane is in a small storm is three minutes. When the plane is in a small storm, the wind speed increases by 10 mi/hr. Also, there is a 0.01% chance that the plane will fly into a microburst, which lasts for one minute and increases the wind speed by 50 mi/hr. Plot the time and speed data and save them in an ASCII file named windspd.dat.

 1. PROBLEM STATEMENT

Generate one hour of simulated wind speed data using the statistics developed for the area of the flight path.

 2. INPUT/OUTPUT

As shown in Figure 5.13, the input to the program is the statistics of the weather in the flight path, which are represented by the mean and variance of the wind speed in normal weather. The output is the plot and the data file containing the simulated wind speeds.

 3. HAND EXAMPLE

This simulation uses several different random number sequences. The wind speed is a Gaussian random number sequence with a specified mean and vari-

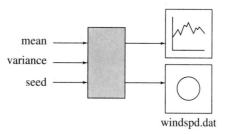

Fig. 5.13 I/O diagram.

ance. The probabilities of encountering a storm or a microburst are given as percentage values and can be modeled as uniform random numbers. We will assume that a storm occurs if a uniform random number between 0 and 1 has a value between 0.0 and 0.01, and that a microburst occurs if the uniform random number is between 0.01 and 0.0101.

4. MATLAB SOLUTION

```
%
%           This program generates one hour of simulated
%           wind speeds based on inputs for the mean and
%           variance of the wind speed and for the random
%           number seed.
%
mn_speed = input('Enter mean of wind speed ');
var_speed = input('Enter variance of wind speed ');
std_speed = sqrt(var_speed);
seed = input('Enter seed for random numbers ');
%
%     Generate simulated wind speed without storms.
%
rand('seed',seed)
rand('normal')
speed = std_speed*rand(1,361) + mn_speed;
%
%     Add simulated storms and microbursts
%
rand('uniform')
t = [0:1:360]*(1/360);
k=1;
```

```
while k <= 361
    random_x = rand(1);
    if random_x <= 0.01
        end_storm = min(361,k+17);
        speed(k:end_storm) = speed(k:end_storm) + 10;
        k = k + 18;
    elseif 0.01 < random_x & random_x <= 0.0101
        end_micro = min(361,k+5);
        speed(k:end_micro) = speed(k:end_micro) + 50;
        k = k + 6;
    else
        k = k + 1;
    end
end
%
%      Plot the data and save it to a file.
%
```

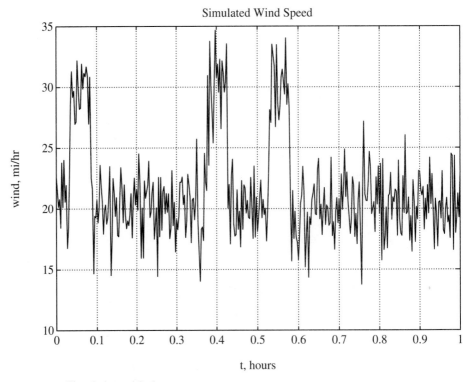

Fig. 5.14   Simulation with three storms.

```
plot(t,speed),...
title('Simulated Wind Speed'),...
xlabel('t, hours'),...
ylabel('wind, mi/hr'),...
grid
data(:,1) = t';
data(:,2) = speed';
save windspd.dat data /ascii
```

5. TESTING

Figure 5.14 contains a typical plot of the wind speed for one hour that was simulated using this program with the following interaction:

```
Enter mean of wind speed 20
Enter variance of wind speed 5
Enter seed for random numbers 0
```

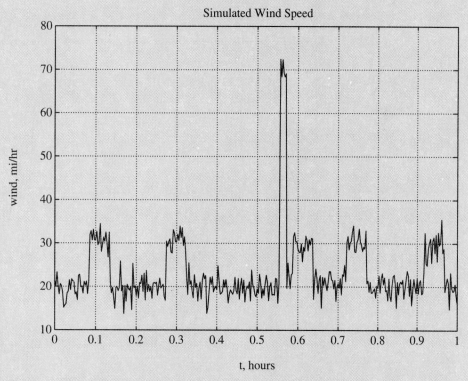

Fig. 5.15   Simulation with microburst.

For a more realistic wind pattern, we could make several changes. During a storm, the increase in speed could be a uniform random number with a mean of 10 and a large variance to simulate turbulence. We might want to add statements that provides a less abrupt wind speed increase and decrease for the beginning and ending of a storm. We could also make the length of the storm a uniform random variable, that might vary from 1 minute to 10 minutes.

After running the program a number of times, the wind speed data contained a microburst, as shown in Figure 5.15. Another improvement in the wind speed simulation would be to make the microburst last from one minute to three minutes, and to let it occur at any time uniformly distributed within a storm.

5.3   SIGNAL-TO-NOISE RATIOS

In generating signals to use in testing engineering techniques, we often want to generate sequences such as the one in Figure 5.16, which contains a sinusoid with noise

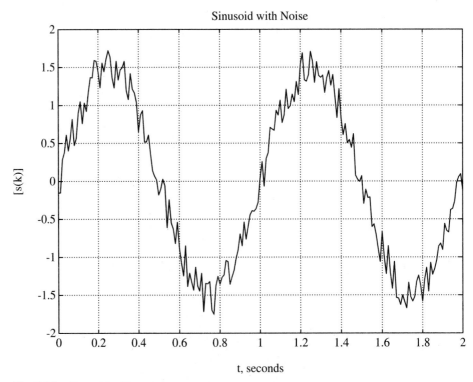

Fig. 5.16   Sinusoid with noise.

added to it. The amount of noise added to a signal can be specified with a signal-to-noise ratio, or SNR. The SNR is defined in terms of the power of a signal, so we discuss signal power, and then return to a mathematical definition of SNR.

Signal Power

Intuitively, the power is a measure of the amplitude of a signal. The larger the amplitude of the signal values, the larger the power. Since an amplitude can be positive or negative, power is defined in terms of squared amplitudes, so that the values are always positive. The power in a signal x (represented by a vector x) can be estimated by the average squared value of the signal:

$$\text{power} \approx \frac{\sum_{k=1}^{N} x_k^2}{N} \tag{5.11}$$

Note that this can easily be computed in MATLAB using the sum function:

```
power = sum(x.^2)/N;
```

It can also be shown (but the proof requires results from calculus) that the power in a signal is equal to the variance plus the mean squared:

$$\text{power} = \sigma^2 + \mu^2 \tag{5.12}$$

In MATLAB, this can be computed with the following statement:

```
power = std(x)^2 + mean(x)^2;
```

If a signal is a sinusoid, it can also be shown (again using calculus) that the signal power is equal to one-half the squared amplitude of the sinusoid. Thus, the power of the sinusoid $4 \sin 2\pi t$ is equal to $16/2$, or 8.

Practice!

Give the MATLAB statements to generate 100 values from the sequence indicated. Compute the power using the average squared value and then compute it using the mean and variance; the values should be similar.

1. Uniform values between 0 and 10.

2. Uniform values between -2 and 4.

3. Uniform values with mean of 0 and variance of 1.0.

4. Uniform values with mean of -0.5 and variance of 4.0.

5. Gaussian values with a mean of 0 and a variance of 2.0

6. Gaussian values with a mean of 0 and a variance of 0.5

7. Gaussian values with a mean of -2.5 and a variance of 0.5

Computation of SNR

SNR

A signal-to-noise ratio is the ratio of the power in a signal to the power in the noise. For example, a SNR of 1 specifies that the ratio of signal power to noise power is $1:1$. If we have a sinusoid with an amplitude of 3 added to uniform noise between -1 and 1, we can compute the SNR using the power measurements of the two signals. The power of the sinusoid is $9/2$, and the power of the noise is equal to $2^2/12$ or $1/3$ (from Equation (5.8)). Therefore, the SNR can be computed as shown:

$$\text{SNR} = \frac{9/2}{1/3} = 13.5$$

For a general sinusoid with amplitude A, and uniform noise between a and b, the SNR can be computed using the following MATLAB statement:

```
SNR = ((A^2)/2)/((b-a)^2/12);
```

sampling time

To illustrate with another example, suppose that we want to generate 201 points of a signal that contains a 1 Hz sinusoid with zero-mean noise in an SNR of 46. The sinusoid should have an amplitude of 1.5 and a phase angle of 0.0, and be sampled at 100 Hz (which means a sampling time of $1/100$ or 0.01 seconds). The SNR is equal to the power of the sinusoid divided by the power of the noise:

$$\text{SNR} = \frac{\text{signal power}}{\text{noise power}}$$
$$= \frac{(1.5^2)/2}{\text{noise power}}$$
$$= 46$$

Solving for the noise power, we have:

$$\text{noise power} = \frac{(1.5^2)/2}{46}$$
$$= 0.024$$

Since the noise is specified to be zero-mean, the noise power is equal to the variance. Thus, the noise variance is 0.024. Since the noise is uniform and zero-mean, it ranges between a and $-a$. Therefore, the variance is equal to $(2a)^2/12$, and

$$0.024 = (2a)^2/12$$

or

$$a = 0.27$$

We can now generate the desired noise signal and add it to the sinusoid to obtain the desired SNR ratio. The commands to generate this signal are shown below:

```
%
%      Generate and plot sine plus noise.
%
rand('seed',0);
rand('uniform');
t = 0:0.01:2.0;
s = 1.5*sin(2*pi*t) + (0.54*rand(1,201) - 0.27);
plot(t,s),...
title('Sinusoid with Noise'),...
xlabel('t, seconds'),...
ylabel('[s(k)]'),...
grid
```

This sinusoid plus noise is shown in Figure 5.16. Note that there are two periods of the signal in 2.0 seconds. This corresponds to the fact that the frequency is 1 Hz, so the period is 1.0 second.

Adding Noise to an Existing Signal

Suppose that we want to add noise to a signal that has already been collected and stored in a data file. If we want to add noise that maintains a specified SNR, we need to estimate the power of the signal so that we can determine the appropriate power for the noise signal. A good estimate of the power of a signal is the average squared signal value, which can easily be computed in MATLAB. Then we can determine the necessary power for the noise. We know that the power is a function of mean and variance, so we need one of these values specified in order to determine the other value. It is often desirable for noise to be zero-mean, so this is usually assumed if no other information is available. We can then compute the variance needed and generate the noise and add it to the existing signal.

Practice!

Generate and plot a signal composed of 100 points of a 5-Hz sinusoid sampled at 50 Hz plus zero-mean noise as specified:

1. Uniform noise with SNR of 5.

2. Uniform noise with SNR of 1.

3.   Uniform noise with SNR of 0.2.

4.   Gaussian noise with SNR of 5.

5.   Gaussian noise with SNR of 1.

6.   Gaussian noise with SNR of 0.2.

SUMMARY

In this chapter we presented a number of functions for performing a statistical analysis of engineering data. We also presented examples showing how to generate simulated data for engineering applications. The evaluation and interpretation of statistical measurements and metrics are very important in engineering, and thus a number of examples were presented. Specific applications included analyzing speech signals, performing Monte Carlo simulations, and generating noise in specified signal-to-noise ratios.

MATLAB SUMMARY

This MATLAB summary lists all the special symbols, commands, and functions that were defined in this chapter. A brief description is also included for each one.

Commands and Functions

cumprod	determines cumulative products
cumsum	determines cumulative sums
hist	plots histogram
max	determines maximum value
mean	determines mean value
median	determines median value
min	determines minimum value
prod	determines product of values
rand	generates random numbers
sort	sorts values
std	computes standard deviation
sum	determines sum of values

PROBLEMS

Problems 1 to 8 relate to the engineering applications presented in this chapter. Problems 9 to 18 relate to new engineering applications.

Speech Signals   These problems relate to the speech analysis problem given in "Problem Solving Applied" on speech signal analysis.

1.   The diskette contained at the end of this text contains files `zero.dat`, `one.dat`, `two.dat`, ... , `nine.dat` that contain the speech signals for the utterances of the words "zero," "one," "two," ... , "nine." Run the program developed in "Problem Solving Applied" on speech signal analysis using each of these files, and generate a table with the statistics for each digit.

2.   Run the program developed in "Problem Solving Applied" on speech signal analysis using the file `digit.dat`. This is one of the digits, spoken by the same person who created the data files with the digits used in problem 1. Using the table created in problem 1, what digit do you think is contained in the file `digit.dat`?

3.   Run the program developed in "Problem Solving Applied" on speech signal analysis using the files `two_a.dat` and `two_b.dat`. These files also contain utterances of the digit "two", but the utterances are spoken by two additional people. Make a table comparing the statistics for these same words spoken by three different people. The differences illustrate some of the difficulty in designing speech recognition systems that are speaker independent.

Flight Simulator Wind Speed   These problems relate to the wind speed simulation in "Problem Solving Applied" on the flight simulator that generated one hour of simulated wind speeds.

4.   Modify the program developed in "Problem Solving Applied" on the flight simulator such that it uses uniform random numbers to generate the initial wind. Compare plots of the simulated wind using uniform numbers and Gaussian numbers. Which do you think is the better simulation? Why?

5.   Modify the program developed in "Problem Solving Applied" on the flight simulator such that the increase in speed during a storm is a uniform number with a mean of 10 miles per hour and a standard deviation equal to 1.5 miles per hour.

6.   Modify the program developed in "Problem Solving Applied" on the flight simulator such that the increase in speed during a storm is a uniform number with a mean of 10 miles per hour and a variance that is twice the value input for the variance of the wind speed when there is not a storm.

7.   Modify the program in "Problem Solving Applied" on the flight simulator such that the user enters the probability of encountering a storm and of encountering a microburst.

8.   Modify the program in "Problem Solving Applied" on the flight simulator such that the length of a storm is a uniform random variable between 1 minute and 10 minutes.

Component Reliability   Equations for analyzing reliability of components can be developed from the study of statistics and probability. Reliability can also be determined using computer simulations if an individual component reliability is known. Consider the diagrams in Figure 5.17. In the series design, three components are connected serially. In order for information to flow from point a to point b, all three components must work properly. In the parallel design, only one of the three components must work properly for the information to flow from point a to point b. If we know the reliability of a component (which is the proportion of the time that the component works properly), we could determine equations to compute the reliability of the serial design or the parallel design [2]. We can also estimate the reliability of the designs using computer simulations. For example, if the reliability of each of the components in the series design is 0.8 (which means that the component works 80% of the time), we could generate three uniform numbers between 0 and 1. If all three numbers are less than or equal to 0.8, then the design works for one simulation; if any of the numbers are greater than 0.8, then the design does not work for one simulation. If we run hundreds or thousands of simulations, we can compute the proportion of the time that the overall design works. To compute the reliability of the parallel design with a component reliability of 0.8, we again generate three uniform numbers between 0 and 1. If any of the three numbers are less than or equal to 0.8,

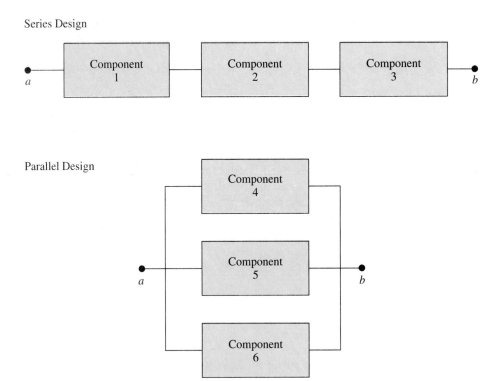

Fig. 5.17

then the design works for one simulation; if all of the numbers are greater than 0.8, then the design does not work for one simulation.

9. Write a program that simulates the series design in Figure 5.17 using a component reliability of 0.8. Compare the reliability computed using 500 simulations and 1,000 simulations.

10. Write a program that simulates the parallel design in Figure 5.17 using a component reliability of 0.8. Compare the reliability computed using 500 simulations and 1,000 simulations. Should the reliability for the parallel design be higher or lower than the reliability for the series design? Explain.

11. Write a program that simulates the design shown in Figure 5.18 using a component reliability of 0.8 for component 1 and a component reliability of 0.92 for component 2. Print the reliability computed using 5,000 simulations.

12. Write a program that simulates the design shown in Figure 5.19 using a component reliability of 0.8 for component 1 and a component reliability of 0.92 for component 2. Print the reliability computed using 5,000 simulations. Again, compare the results for this parallel design to the series design. Which design should have a larger reliability?

13. Write a program that simulates the design shown in Figure 5.20 using a component reliability of 0.8 for component 1, 0.85 for component 2, and 0.95 for component 3. Print the reliability computed using 5,000 simulations.

14. Write a program that simulates the design shown in Figure 5.21 using a component reliability of 0.8 for components 1 and 2, and 0.95 for components 3 and 4. Print the reliability computed using 5,000 simulations.

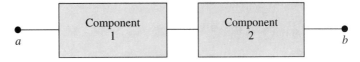

Fig. 5.18

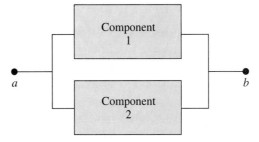

Fig. 5.19

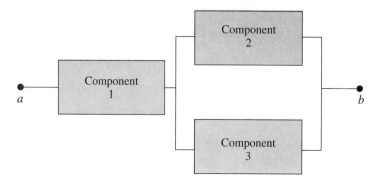

Fig. 5.20

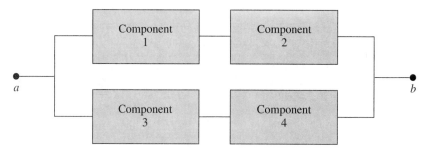

Fig. 5.21

15. Write a program that simulates the design shown in Figure 5.22 using a component reliability of 0.95 for all components. Print the reliability computed using 5,000 simulations.

Power Plant Output   The power output in megawatts from a power plant over a period of eight weeks has been stored in a data file named plant.dat. Each line in the data file represents data for one week, and contains the output for day 1, day 2, . . . , day 7.

16. Write a program that uses the power plant output data and prints a report that lists the number of days with greater-than-average power output. The report should give the week number and the day number for each of these days, in addition to printing the average power output for the plant during the eight-week period.

17. Write a program that uses the power plant output data and prints the day and week during which the maximum and minimum power output occurred. If the maximum or minimum occurred on more than one day, print all the days involved.

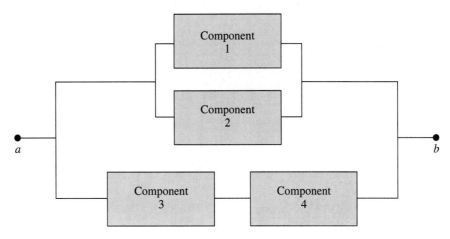

Fig. 5.22

18. Write a program that uses the power plant output data and prints the average power output for each week. Also print the average power output for day 1, for day 2, and so on.

6

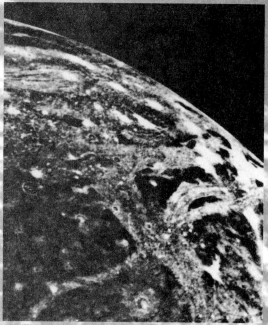

Courtesy of National Aeronautics and Space Administration

GRAND CHALLENGE: Machine Vision

Communications equipment and microprocessor-controlled instrumentation can collect high-resolution images, such as the one in the photo above of Ganymede, a moon of Jupiter, that was taken from the Voyager Spacecraft. However, it is a very difficult task to program the computer to interpret the images. Successful algorithms usually require some knowledge about the image. For example, a computer algorithm to distinguish between the image of a building and the image of a tree is much easier to develop than an algorithm to distinguish between a car, a bus, a truck, and a tank. When working with images, we would also like to be able to track an object from one image to another image to determine the direction and speed of a moving object. Tracking algorithms are also very complicated algorithms.

Matrix Computations

Introduction

A matrix is a convenient way to represent engineering data. In previous chapters, we discussed mathematical computations and functions that could be applied element-wise to values in matrices. In this chapter we present a set of matrix operations and functions that apply to the matrix as a unit, as opposed to individual elements in the matrix. We first consider a set of mathematical computations that compute new values from a matrix (or matrices). We then consider a group of functions that allow us to manipulate a matrix and change it from one form to another.

6.1   MATRIX OPERATIONS

Many engineering computations use a matrix as a convenient way to represent a set of data. In this section, we present several computations that are frequently performed with data stored in a matrix. In this chapter, we are generally concerned with matrices that have more than one row and more than one column. Recall that scalar multiplication and matrix addition and subtraction are performed element-wise and were covered in Chapter 3 in the discussion of array operations. Matrix multiplication is covered in this chapter. The topic of matrix division is covered in Chapter 8, since it applies specifically to the solution of simultaneous linear equations.

Transpose of a Matrix

The transpose of a matrix is a new matrix in which the rows of the original matrix are the columns of the new matrix. We use a superscript T after a matrix name to refer to the transpose. For example, consider the following matrix and its transpose:

$$A = \begin{bmatrix} 2 & 5 & 1 \\ 7 & 3 & 8 \\ 4 & 5 & 21 \\ 16 & 13 & 0 \end{bmatrix} \quad A^T = \begin{bmatrix} 2 & 7 & 4 & 16 \\ 5 & 3 & 5 & 13 \\ 1 & 8 & 21 & 0 \end{bmatrix}$$

If we consider a couple of the elements, we see that the value in position (3,1) has now moved to position (1,3), and the value in position (4,2) has now moved to position (2,4). In fact, we have interchanged the row and column subscript so that we are moving the value in position (i,j) to position (j,i).

In MATLAB the transpose of the matrix A is denoted by A'. Remember that if the matrix is not a square matrix (one with the same number of rows and columns), then the transpose will have a different size than the original matrix. We frequently use the transpose operation to convert a row vector to a column vector, or a column vector to a row vector.

If the matrix A contains complex values, then A' is the complex conjugate transpose. To obtain an unconjugated transpose, use A.' or conj(A').

Dot Product

The dot product is a scalar computed from two vectors of the same size. This scalar is the sum of the products of the values in corresponding positions in the vectors, as shown in the summation equation, which assumes that there are N elements in the vectors A and B:

$$\text{dot product} = A \cdot B = \sum_{i=1}^{N} a_i b_i$$

To illustrate, assume that A and B are the following vectors:

$$A = \begin{bmatrix} 4 & -1 & 3 \end{bmatrix} \quad B = \begin{bmatrix} -2 & 5 & 2 \end{bmatrix}$$

The dot product is then:

$$A \cdot B = 4 \cdot (-2) + (-1) \cdot 5 + 3 \cdot 2$$
$$= (-8) + (-5) + 6$$
$$= -7$$

inner product

The dot product is also called an inner product.

In MATLAB, we can compute the dot product with the following statement:

```
dot_product = sum(A.*B);
```

Recall that A.*B contains the results of an element-wise multiplication of A and B. When A and B are both row vectors or are both column vectors, A.*B is also a vector. We then sum the elements in this vector, thus yielding the dot product. If A is a row vector and B is a column vector, then the computation could be done with either of the following statements:

```
dot_product = sum(A'.*B);
dot_product = sum(A.*B');
```

After we discuss matrix multiplication, we will show you another way to compute the dot product of two vectors.

Matrix Multiplication

Matrix multiplication is not computed by multiplying corresponding elements of the matrices. The value in position $c(i,j)$ of the product C of two matrices A and B is the dot product of row i of the first matrix and column j of the second matrix, as shown in the summation equation:

matrix product

$$c_{i,j} = \sum_{k=1}^{N} a_{ik} b_{kj}$$

Since the dot product requires that the vectors have the same number of elements, then the first matrix (A) must have the same number of elements in each row as there are in the columns of the second matrix (B). Thus, if A and B both have five rows and five columns, their product has five rows and five columns. Furthermore, for these matrices, we can compute both AB and BA, but in general, they will not be equal.

If A has two rows and three columns, and B has three rows and three columns, the product AB will have two rows and three columns. To illustrate, consider the following matrices:

$$A = \begin{bmatrix} 2 & 5 & 1 \\ 0 & 3 & -1 \end{bmatrix} \quad B = \begin{bmatrix} 1 & 0 & 2 \\ -1 & 4 & -2 \\ 5 & 2 & 1 \end{bmatrix}$$

The first element in the product C = AB is

$$c_{1,1} = \sum_{k=1}^{3} a_{1k}b_{k1}$$
$$= a_{1,1}b_{1,1} + a_{1,2}b_{2,1} + a_{1,3}b_{3,1}$$
$$= 2 \cdot 1 + 5 \cdot (-1) + 1 \cdot 5$$
$$= 2$$

Similarly, we can compute the rest of the elements in the product of A and B:

$$AB = C = \begin{bmatrix} 2 & 22 & -5 \\ -8 & 10 & -7 \end{bmatrix}$$

In this example, we cannot compute BA, because B does not have the same number of elements in each row as A has in each column.

size of product An easy way to decide if a matrix product exists is to write the sizes of the two matrices side by side. Then if the two inside numbers are the same, the product exists, and the size of the product is determined by the two outside numbers. To illustrate, in the previous example, the size of A is 2×3 and the size of B is 3×3. Therefore, if we want to compute AB we write the sizes side by side:

$$2 \times 3, 3 \times 3$$

The two inner numbers are both the value 3, so AB exists, and its size is determined by the two outer numbers, 2×3. If we want to compute BA we again write the sizes side by side:

$$3 \times 3, 2 \times 3$$

The two inner numbers are not the same, so BA does not exist.

In MATLAB, matrix multiplication is denoted by an asterisk. Thus, the commands to generate the matrices in our previous example, and then to compute the matrix product are:

```
A = [2,5,1;  0,3,-1];
B = [1,0,2;  -1,4,-2;  5,2,1];
C = A*B;
```

If we execute the MATLAB command C = B*A, we get a warning message that C does not exist.

dot product A simple way to compute a dot product between two row vectors F and G is to use the following statement:

```
dot_product = F*G'
```

If we assume that the length of the vectors is N, the sizes of the components in the matrix product FG' are:

$$1 \times N, N \times 1$$

Thus, the product matrix exists, and its size is 1×1, which is a scalar; this result is reassuring, since we know that the dot product is a scalar. If F and G are column vectors, then the dot product can be computed as shown:

```
dot_product = F'*G
```

Similar expressions can be derived for the dot product of a column vector and a row vector.

outer product
An outer product can be defined for two vectors. For example, assume that F and G are the vectors defined below:

$$F = [2 \quad 5 \quad -1] \qquad G = [0 \quad 1 \quad -3]$$

Consider the following statement:

```
outer_product = F'*G
```

First we look at the sizes of the two matrices F' and G:

$$3 \times \underline{1, 1} \times 3$$

The product exists, since the inner numbers are the same, and the size of the product is 3×3. The outer product can be defined using the definition of matrix multiplication as shown:

$$F'G = \begin{bmatrix} 2 \\ 5 \\ -1 \end{bmatrix} \quad [0 \quad 1 \quad -3] = \begin{bmatrix} 0 & 2 & -6 \\ 0 & 5 & -15 \\ 0 & -1 & 3 \end{bmatrix}$$

To correctly multiply two matrices, we must carefully consider the sizes of the matrices and then select the proper order and the proper use of the transpose function. As illustrated in the previous examples, when F and G are row vectors, FG' is a scalar, F'G is a 3×3 matrix, and FG is not defined.

Assume that I is a square identity matrix. (Recall from Chapter 3 that an identity matrix has ones on the main diagonal and zeros elsewhere.) If A is a square matrix of the same size, then AI and IA are both equal to A. Use a small matrix A, and verify by hand that these matrix products are both equal to A. These computations demonstrate that the product of a matrix with the identity matrix is always the matrix itself.

Matrix Powers

Recall that if A is a matrix, then A.^2 is the operation that squares each element in the matrix. If we want to square the matrix—that is, if we want to compute A*A—we use the operation A^2. A^4 is equivalent to A*A*A*A. If we compute the power of a matrix, and the power is not an integer, as in A^2.5, then more complicated calculations involving eigenvalues and eigenvectors are used. To perform a matrix multiplication between two matrices, the number of rows in the first matrix must be the same value as the number of columns in the second matrix; therefore, to raise a matrix to a power, the number of rows must equal the number of columns, and the matrix must be a square matrix.

Matrix Inverse

By definition, the inverse of a square matrix A is the matrix A^{-1} such that the matrix products AA^{-1} and $A^{-1}A$ are both equal to the identity matrix. For example, consider the following two matrices A and B:

$$A = \begin{bmatrix} 2 & 1 \\ 4 & 3 \end{bmatrix} \quad B = \begin{bmatrix} 1.5 & -.5 \\ -2 & 1 \end{bmatrix}$$

If we compute the products AB and BA, we obtain the following matrices (do the matrix multiplications by hand to be sure you follow the steps):

$$AB = \begin{bmatrix} 1 & 0 \\ 0 & 1 \end{bmatrix} \quad BA = \begin{bmatrix} 1 & 0 \\ 0 & 1 \end{bmatrix}$$

Therefore, A and B are inverses of each other, or $A = B^{-1}$ and $B = A^{-1}$.

inv

Computing the inverse is a tedious process; fortunately MATLAB contains an inv function that performs the computations for us. (We do not present the algorithm for computing an inverse in this text; refer to a linear algebra text if you are interested in the techniques for computing an inverse.) Thus, if we execute inv (A), using the A matrix defined above, the result will be the matrix B. Similarly, if we execute inv (B), the result should be the matrix A. Try this yourself.

The inverse of a matrix is used in solving a number of types of engineering problems. We will discuss some of these applications in chapters in Part 3.

Practice!

Use MATLAB to define the following matrices, and to then compute the specified matrices.

$$A = \begin{bmatrix} 2 & 1 \\ 0 & -1 \\ 3 & 0 \end{bmatrix} \quad B = \begin{bmatrix} 1 & 3 \\ -1 & 5 \end{bmatrix}$$

$$C = \begin{bmatrix} 3 & 2 \\ -1 & -2 \\ 0 & 2 \end{bmatrix} \quad D = [1 \quad 2]$$

$$I = \begin{bmatrix} 1 & 0 \\ 0 & 1 \end{bmatrix}$$

1. AB

2. DB

3. BC'

4. (CB)D'

5. B⁻¹

6. BB⁻¹

7. B⁻¹B

8. AC'

9. (AC')⁻¹

10. (AC')⁻¹(AC')

11. IB

12. BI

Determinants

A determinant of a matrix is a scalar computed from the entries in the matrix. Determinants have various engineering applications in engineering, including computing inverses and solving systems of simultaneous equations. For a 2×2 matrix A, the determinant is the following:

$$\text{determinant of A} = |A| = a_{1,1}a_{2,2} - a_{2,1}a_{1,2}$$

Therefore, the determinant of A, or $|A|$ is equal to 8 for the following matrix:

$$A = \begin{bmatrix} 1 & 3 \\ -1 & 5 \end{bmatrix}$$

For a 3×3 matrix A, the determinant is the following:

$$|A| = a_{1,1}a_{2,2}a_{3,3} + a_{1,2}a_{2,3}a_{3,1} + a_{1,3}a_{2,1}a_{3,2} - a_{3,1}a_{2,2}a_{1,3} - a_{3,2}a_{2,3}a_{1,1}$$
$$- a_{3,3}a_{2,1}a_{1,2}$$

If A is the following matrix:

$$A = \begin{bmatrix} 1 & 3 & 0 \\ -1 & 5 & 2 \\ 1 & 2 & 1 \end{bmatrix}$$

then $|A|$ is equal to $5 + 6 + 0 - 0 - 4 - (-3)$, or 10.

A more involved process is necessary for computing determinants of matrices with more than three rows and columns. We do not include the discussion of the

det

process for computing a general determinant here, because MATLAB will automatically compute a determinant using the det function, with a square matrix as its argument, as in det(A).

◆ PROBLEM SOLVING APPLIED:
✦ MOLECULAR WEIGHTS OF PROTEINS

A protein sequencer is a sophisticated piece of equipment that plays a key role in genetic engineering [1,2]. The sequencer can determine the order of amino acids that make up a chainlike protein molecule. This order of amino acids then aids genetic engineers in identifying the gene that made the protein. Enzymes are used to dissolve bonds to the neighboring genes, thus separating the valuable gene out of the DNA. This gene is then inserted into another organism, such as a bacterium, that will multiply itself along with the foreign gene.

Although there are only 20 different amino acids, protein molecules have hundreds of amino acids linked in a specific order. In this problem, we assume that the sequence of amino acids in a protein molecule has been identified, and that we want to compute the molecular weight of the protein molecule. Table 6.1 contains an alphabetical listing of the amino acids, their three-letter references, and their molecular weights.

TABLE 6.1   Amino Acids

AMINO ACID	REFERENCE	MOLECULAR WEIGHT
1. Alanine	Ala	89
2. Arginine	Arg	175
3. Asparagine	Asn	132
4. Aspartic	Asp	132
5. Cysteine	Cys	121
6. Glutamic	Glu	146
7. Glutamine	Gln	146
8. Glycine	Gly	75
9. Histidine	His	156
10. Isoleucine	Ile	131
11. Leucine	Leu	131
12. Lysine	Lys	147
13. Methionine	Met	149
14. Phenylalanine	Phe	165
15. Proline	Pro	116
16. Serine	Ser	105
17. Threonine	Thr	119
18. Tryptophan	Trp	203
19. Tyrosine	Tyr	181
20. Valine	Val	117

The input to this problem is a data file that contains the number and type of amino acid molecules in each protein molecule. Assume that the data file is generated by the protein sequencer instrumentation. Each line in the data file corresponds to one protein, and each line contains 20 integers that correspond to the 20 amino acids in the alphabetical order shown in Table 6.1. Therefore, a line containing the following values is generated from the protein LysGluMetAspSer-Glu:

0 0 0 1 0 2 0 0 0 0 0 0 1 1 0 0 1 0 0 0 0

The data file is named `protein.dat`.

 1. PROBLEM STATEMENT
Compute the molecular weights for a group of protein molecules.

2. INPUT/OUTPUT DESCRIPTION
Figure 6.1 contains a diagram showing that the input is a file containing the amino acids identified in a group of protein molecules. The output of the program is the corresponding set of protein molecular weights.

3. HAND EXAMPLE
Suppose that the protein molecule is the following:

LysGluMetAspSerGlu

The corresponding molecular weights for the amino acids are as follows:

147,146,149,132,105,146

Therefore, the protein molecular weight is 825. The line from the data file that indicates these six amino acids contains the following values:

0 0 0 1 0 2 0 0 0 0 0 0 1 1 0 0 1 0 0 0 0

The molecular weight for the protein is the sum of the product of the number

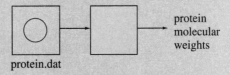

protein
molecular
weights

protein.dat

Fig. 6.1   I/O diagram.

of amino acids and the corresponding weights. This sum of products can be considered to be a dot product between the protein vector and the weight vector. If we compute the protein molecular weight for a group of proteins, the computations can be computed as a matrix product, as shown in the example below for two proteins:

$$
\begin{bmatrix}
0 & 0 & 0 & 1 & 0 & 2 & 0 & 0 & 0 & 0 & 0 & 1 & 1 & 0 & 0 & 1 & 0 & 0 & 0 & 0 \\
0 & 1 & 0 & 0 & 0 & 1 & 1 & 0 & 0 & 3 & 0 & 0 & 0 & 0 & 0 & 0 & 0 & 1 & 0 & 0
\end{bmatrix}
\begin{bmatrix}
89 \\
175 \\
132 \\
132 \\
121 \\
146 \\
146 \\
75 \\
156 \\
131 \\
131 \\
147 \\
149 \\
165 \\
116 \\
105 \\
119 \\
203 \\
181 \\
117
\end{bmatrix}
=
\begin{bmatrix}
825 \\
1063
\end{bmatrix}
$$

4. MATLAB SOLUTION

By recognizing that this problem can be posed as a matrix multiplication, we have simplified the MATLAB solution. The amino acid information is read from the data file into a matrix protein, a column vector mw with the amino acid molecular weights is defined, and the protein molecular weights are then contained in the vector that is the result of the matrix product of protein and mw.

```
%
%      This program computes the molecular weights for
%      a group of protein molecules. A data file contains
%      the occurrence and number of amino acids in each
%      protein molecule.
```

```
%
load protein.dat
mw = [ 89 175 132 132 121 146 146  75 156 131 ...
       131 147 149 165 116 105 119 203 181 117 ];
%
%     Compute protein weights.
%
weights = protein*mw';
%
%     Print protein weights.
%
[rows cols] = size(protein);
for k=1:rows
   fprintf('protein %3.0f: molecular weight = %5.0f \n',...
         k, weights(k))
end
```

5. TESTING

Assume that the proteins that have been identified are the following:

GlyIleSerThrTrp

AspHisProGln

ThrTyrSerTrpLysMetHisMet

AlaValLeuValMet

LysGluMetAspSerGluLysGluGlyGlu

Then the corresponding data file is the following:

```
0 0 0 0 0 0 0 1 0 1 0 0 0 0 0 0 1 1 1 0 0
0 0 0 1 0 0 1 0 1 0 1 0 0 0 0 0 1 0 0 0 0 0
0 0 0 0 0 0 0 0 1 0 0 1 2 0 0 1 1 1 1 0
1 0 0 0 0 0 0 0 0 0 1 0 1 0 0 0 0 0 0 2
0 0 0 1 0 4 0 1 0 0 0 2 1 0 0 1 0 0 0 0
```

The output for this test file is the following:

```
protein   1: molecular weight =  633
protein   2: molecular weight =  550
protein   3: molecular weight = 1209
protein   4: molecular weight =  603
protein   5: molecular weight = 1339
```

Additional problems related to this application are included in the end-of-chapter problems.

6.2   MATRIX MANIPULATIONS

MATLAB contains a number of functions that manipulate the contents of a matrix. We will discuss each of these functions and give an example to illustrate.

Rotation

rot90

A matrix A can be rotated 90° in a counterclockwise direction using the rot90 function. Let the matrix A be the following:

$$A = \begin{bmatrix} 2 & 1 & 0 \\ -2 & 5 & -1 \\ 3 & 4 & 6 \end{bmatrix}$$

If we execute the command:

B = rot90(A);

then the value of B is:

$$B = \begin{bmatrix} 0 & -1 & 6 \\ 1 & 5 & 4 \\ 2 & -2 & 3 \end{bmatrix}$$

The rot90 function will also accept a second argument that specifies how many 90° rotations are desired. Thus, the first two commands are equivalent to the third command:

B = rot90(A);
C = rot90(B);

C = rot90(A,2);

Flip

fliplr
flipud

Two functions are used to flip a matrix. The fliplr function flips a matrix left to right, and the flipud function flips a matrix up to down. For example, consider the following MATLAB commands:

A = [1, 2; 4, 8;   -2, 0];
B = fliplr(A);
C = flipud(B);

After executing these commands, the matrices A, B, and C contain the following values:

$$A = \begin{bmatrix} 1 & 2 \\ 4 & 8 \\ -2 & 0 \end{bmatrix} \qquad B = \begin{bmatrix} 2 & 1 \\ 8 & 4 \\ 0 & -2 \end{bmatrix} \qquad C = \begin{bmatrix} -2 & 0 \\ 4 & 8 \\ 1 & 2 \end{bmatrix}$$

Reshape

reshape

The reshape function allows you to reshape a matrix into one with a different number of rows and columns. The number of elements in the original matrix and in the reshaped matrix must be the same, or an error message will be printed. The reshape function has three arguments: The first argument contains the original matrix, and the next two arguments contain the number of rows and the number of columns for the new matrix. The numbers are selected in column order from the old matrix and are then used to fill the new matrix in column order. Consider the following MATLAB statements:

```
A  =  [ 2 5 6 -1;  3 -2 10 0];
B  =  reshape (A, 4, 2) ;
C  =  reshape (A, 8, 1) ;
```

After executing these statements, the values in matrices A, B, and C are the following:

$$A = \begin{bmatrix} 2 & 5 & 6 & -1 \\ 3 & -2 & 10 & 0 \end{bmatrix} \qquad B = \begin{bmatrix} 2 & 6 \\ 3 & 10 \\ 5 & -1 \\ -2 & 0 \end{bmatrix} \qquad C = \begin{bmatrix} 2 \\ 3 \\ 5 \\ -2 \\ 6 \\ 10 \\ -1 \\ 0 \end{bmatrix}$$

Extraction

The functions diag, triu, and tril allow you to extract elements from a matrix. The definitions of all three functions use the definition of the main diagonal, which is the diagonal that starts in the upper left corner of a matrix and contains the values with equal row and column subscripts, such as $a_{1,1}$, $a_{2,2}$, and $a_{3,3}$. Even nonsquare matrices have main diagonals. For example, in the previous example, the elements on the main diagonal of A are 2,-2; the elements on the main diagonal of B are

diag

2,10; and the element on the main diagonal of C is 2. The function diag(A) will extract the elements on the main diagonal of A and store them in a column vector.

A second argument k can be used with the diag function to specify other diagonals. If k is zero, the main diagonal is selected. If k is greater than zero, then the kth diagonal above the main diagonal is selected. If k is less than zero, then the kth diagonal below the main diagonal is selected.

If the argument of the diag function is a vector instead of a matrix with rows and columns, then the function will generate a square matrix with the given vector as the diagonal. For example, the following statements:

```
V = [1 2 3];
A = diag(V);
```

will generate a matrix with the following elements:

$$A = \begin{bmatrix} 1 & 0 & 0 \\ 0 & 2 & 0 \\ 0 & 0 & 3 \end{bmatrix}$$

triu

The function reference triu(A) generates a matrix containing the values from A that are on the main diagonal, or are above the main diagonal, with zeros elsewhere. The function triu can also have a second integer argument. The function reference triu(A, k) is a matrix the same size as A containing elements from A on or above the kth diagonal, and zeros elsewhere. Consider these MATLAB statements:

```
A = [1:2:7; 3:3:12; 4:-1:1; 1:4];
B = triu(A);
C = triu(A,-1);
D = triu(A,3);
```

The resulting matrices are the following:

$$A = \begin{bmatrix} 1 & 3 & 5 & 7 \\ 3 & 6 & 9 & 12 \\ 4 & 3 & 2 & 1 \\ 1 & 2 & 3 & 4 \end{bmatrix} \quad B = \begin{bmatrix} 1 & 3 & 5 & 7 \\ 0 & 6 & 9 & 12 \\ 0 & 0 & 2 & 1 \\ 0 & 0 & 0 & 4 \end{bmatrix}$$

$$C = \begin{bmatrix} 1 & 3 & 5 & 7 \\ 3 & 6 & 9 & 12 \\ 0 & 3 & 2 & 1 \\ 0 & 0 & 3 & 4 \end{bmatrix} \quad D = \begin{bmatrix} 0 & 0 & 0 & 7 \\ 0 & 0 & 0 & 0 \\ 0 & 0 & 0 & 0 \\ 0 & 0 & 0 & 0 \end{bmatrix}$$

tril

The tril function is similar to the triu function, but the tril function generates lower triangular matrices instead of upper triangular matrices. If we replace references to triu in the previous example with references to tril, we have the following:

```
A = [1:2:7; 3:3:12; 4:-1:1; 1:4];
B = tril(A);
```

```
C = tril(A,-1);
D = tril(A,3);
```

The resulting matrices are the following:

$$A = \begin{bmatrix} 1 & 3 & 5 & 7 \\ 3 & 6 & 9 & 12 \\ 4 & 3 & 2 & 1 \\ 1 & 2 & 3 & 4 \end{bmatrix} \quad B = \begin{bmatrix} 1 & 0 & 0 & 0 \\ 3 & 6 & 0 & 0 \\ 4 & 3 & 2 & 0 \\ 1 & 2 & 3 & 4 \end{bmatrix}$$

$$C = \begin{bmatrix} 0 & 0 & 0 & 0 \\ 3 & 0 & 0 & 0 \\ 4 & 3 & 0 & 0 \\ 1 & 2 & 3 & 0 \end{bmatrix} \quad D = \begin{bmatrix} 1 & 3 & 5 & 7 \\ 3 & 6 & 9 & 12 \\ 4 & 3 & 2 & 1 \\ 1 & 2 & 3 & 4 \end{bmatrix}$$

Practice!

Determine the matrices generated by the following function references. Then check your answers using MATLAB. Assume that A and B are the following matrices:

$$A = \begin{bmatrix} 0 & -1 & 0 & 3 \\ 4 & 3 & 5 & 0 \\ 1 & 2 & 3 & 0 \end{bmatrix} \quad B = \begin{bmatrix} 1 & 3 & 5 & 0 \\ 3 & 6 & 9 & 12 \\ 4 & 3 & 2 & 1 \\ 1 & 2 & 3 & 4 \end{bmatrix}$$

1. rot90(B)

2. rot90(A,3)

3. fliplr(A)

4. flipud(fliplr(B))

5. reshape(A,4,3)

6. reshape(A,6,2)

7. reshape(A,2,6)

8. reshape(flipud(B),8,2)

9. triu(B)

10. triu(B,-1)

11. tril(A,2)

12. diag(rot90(B))

⬟ PROBLEM SOLVING APPLIED: IMAGE ALIGNMENT

An image such as the one in the photograph at the beginning of this chapter is initially represented by a matrix of values that correspond to light intensities. Each point in the image is called a pixel, or picture element. A high-resolution image will be represented by a matrix with many values (or pixels); a low-resolution image will be represented by a matrix with fewer values. For example, a high-resolution image might be represented by a matrix with 1,024 rows and 1,024 columns, or a total of over a million numbers. Each value in the image is a code that represents a light intensity. The light intensity can be coded to represent color, or it can be coded to represent shades of gray.

In the following example, assume that an image is represented by a matrix with six rows and six columns. Also assume that each value in the matrix contains a value from 0 to 7 that represents a shade of gray, with 0 representing white, 7 representing black, and the values between 0 and 7 representing grays from light gray to dark gray. A sample image is the following:

$$\begin{bmatrix} 0 & 0 & 2 & 6 & 2 & 0 \\ 0 & 1 & 0 & 6 & 6 & 0 \\ 1 & 0 & 0 & 2 & 6 & 0 \\ 0 & 0 & 0 & 0 & 0 & 0 \\ 0 & 0 & 1 & 0 & 0 & 0 \end{bmatrix}$$

Assume that we have two images of the same object and that the images have the same resolution and gray scale coding. Also, assume that we do not know if the two images are aligned the same way. To determine the correct alignment, we can hold one image constant, perform operations to rotate or flip the other image, and then compare the two images. The images will be aligned when the values in corresponding locations are exactly the same. Note that there might be more that one alignment in which the corresponding locations are exactly the same. For example, let the matrices A and B represent the same image:

$$A = \begin{bmatrix} 0 & 0 & 0 & 0 & 2 & 0 \\ 0 & 0 & 0 & 0 & 0 & 2 \\ 0 & 0 & 0 & 0 & 0 & 0 \\ 0 & 0 & 0 & 0 & 0 & 0 \\ 0 & 0 & 0 & 0 & 0 & 2 \\ 0 & 0 & 0 & 0 & 2 & 0 \end{bmatrix} \quad B = \begin{bmatrix} 0 & 2 & 0 & 0 & 2 & 0 \\ 2 & 0 & 0 & 0 & 0 & 2 \\ 0 & 0 & 0 & 0 & 0 & 0 \\ 0 & 0 & 0 & 0 & 0 & 0 \\ 0 & 0 & 0 & 0 & 0 & 0 \\ 0 & 0 & 0 & 0 & 0 & 0 \end{bmatrix}$$

To align B with A, we could rotate B 270° in a counterclockwise direction (or 90° in a clockwise direction); we could also flip B up to down and rotate it 90° in a counterclockwise direction. Go through these steps to be sure that you agree that these steps will align the images.

To determine if two images contain the same values (or are aligned), we might consider computing the difference between corresponding elements in the two matrices (which we call image 1 and image 2), and then adding the differences. This value can be computed with the following MATLAB statement:

```
dif = sum(sum(image1-image2));
```

We need to use the sum function twice because the first reference generates a vector containing the column sums of differences, and we want to add the columns sums together. Unfortunately, this sum can equal zero even if the images are not the same. Consider the following pair of matrices:

$$\begin{bmatrix} 1 & 7 \\ 5 & 1 \end{bmatrix} \quad \begin{bmatrix} 1 & 2 \\ 5 & 6 \end{bmatrix}$$

If we compute the sum of all the differences between elements in these two matrices, we find that the sum is equal to $5 + (-5)$, or zero. These two matrices are clearly not the same, and yet the difference sum is zero. This problem occurred because positive and negative differences can cancel each other. If we square the differences or compute the absolute values before we add them, then the values cannot cancel each other. The new difference sum, which we will call a distance measurement, can then be expressed in the following MATLAB statement:

```
distance = sum(sum((image1-image2).^2));
```

We can compute this distance for each possible alignment. The two images are aligned when the distance is equal to zero. If we assume that the numbers in the two aligned images might have slightly different values (due to instrumentation variations or noise in the communication channels), we can compute the distance for all possible alignments, and then choose the one with the minimum distance.

 1. PROBLEM STATEMENT
Determine the best 90° rotation alignment between two images.

 2. INPUT/OUTPUT DESCRIPTION
Figure 6.2 contains a diagram showing that the two images are read from two files. The output is the best 90° alignment between two images.

 3. HAND EXAMPLE
Suppose that the two images are the following:

$$C = \begin{bmatrix} 4 & 3 \\ 2 & 1 \end{bmatrix} \quad D = \begin{bmatrix} 1 & 3 \\ 3 & 4 \end{bmatrix}$$

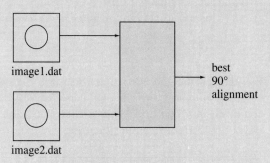

Fig. 6.2   I/O diagram.

If we rotate D 0°, 90°, 180°, and 270° respectively, we have the following four matrices:

$$\begin{bmatrix} 1 & 3 \\ 3 & 4 \end{bmatrix} \quad \begin{bmatrix} 3 & 4 \\ 1 & 3 \end{bmatrix} \quad \begin{bmatrix} 4 & 3 \\ 3 & 1 \end{bmatrix} \quad \begin{bmatrix} 3 & 1 \\ 4 & 3 \end{bmatrix}$$

If we then compute the distance (or sum of the squares of the differences between elements) between C and these four rotated versions of D, we get the values 19, 7, 1, and 13, respectively. Therefore, the minimum distance is 1 and an alignment of 180° is the best alignment using rotations of 90° in a counterclockwise direction.

4. MATLAB SOLUTION

We assume that the two images are stored in ASCII data files, with the values in row order. A for loop is used to compute the four different rotations and distance measurements. The min function is used to select both the minimum distance and the position of the distance measurement in the vector computed by the min function. The position of the minimum is then used to compute the rotation that was associated with the minimum value.

```
%
%       This program determines the best alignment
%       between two images using rotations of 90 degrees.
%
load image1.dat
load image2.dat
%
```

```
%       Compute rotational distances.
%
for k=0:3
    a = rot90(image2,k);
    distance(k+1) = sum(sum((image1-a).^2));
end
%
%       Print best alignment.
%
[minval, minloc] = min(distance);
fprintf('Image alignment best at %3.0f degrees \n',...
        (minloc-1)*90)
fprintf(' (counterclockwise) \n')
```

5. TESTING

Note that this solution will work for images of any resolution. The only requirement is that the images be the same size. If we test this solution using images A and B defined earlier in this section, the output is the following:

```
Image alignment best at 270 degrees
(counterclockwise)
```

If we test this solution using images C and D, the output is the following:

```
Image alignment best at 180 degrees
(counterclockwise)
```

Additional problems related to this application are included in the end-of-chapter problems.

SUMMARY

In this chapter we summarized a number of matrix computations and manipulations. We defined the transpose of a matrix and the inverse of a matrix. We also defined the computation of a dot product (between two vectors) and a matrix product (between two matrices). A number of MATLAB functions can be used to change the form or structure of a matrix. The rot90 function can be used to rotate the values in a matrix in a counterclockwise direction. The reshape function allows you to define a new matrix with the same number of elements. Matrix extraction functions allow you to extract elements from a matrix to generate new vectors or matrices.

MATLAB SUMMARY

This MATLAB summary lists all the special symbols, commands, and functions that were defined in this chapter. A brief description is also included for each one.

Special Characters

'	indicates a matrix transpose
*	indicates matrix multiplication

Commands and Functions

diag	extracts the elements on the main diagonal
det	computes the determinant of a matrix
fliplr	flips a matrix left to right
flipud	flips a matrix up to down
inv	computes the inverse of a matrix
reshape	reshapes a matrix
rot90	rotates a matrix 90° counterclockwise
tril	generates a lower triangular matrix
triu	generates an upper triangular matrix

PROBLEMS

Problems 1 to 10 relate to the engineering applications presented in this chapter. Problems 11 to 21 relate to new engineering applications.

Protein Molecular Weights   These problems relate to the "Problem Solving Applied" section on protein molecular weight.

1. Modify the program so that it also determines and prints the protein number and molecular weight for the protein with the maximum molecular weight and for the protein with the minimum molecular weight.

2. Modify the program so that it also determines and prints the average molecular weight of the proteins analyzed.

3. Modify the program so that it prints the total number of each amino acid for the proteins in the file. Use a format similar to the following:

```
        Summary of Amino Acid Occurrences
Amino Acid Number           Number of Occurrences
      1.                            xxx
      2.                            xxx
      ...
      20.                           xxx
```

4. Modify the program so that it prints the protein number and protein molecular weight for the protein with the largest number of amino acids.

5. Modify the program so that it prints the average number of amino acids for the proteins analyzed.

Image Alignment   These problems relate to section "Problem Solving Applied" on image alignment.

6. Modify the program so that it also prints the distances computed for the various rotations in addition to selecting the best alignment.

7. Modify the program so that it prints the alignment in terms of degrees clockwise.

8. Modify the program so that it also compares the alignment with the second image flipped left to right and up to down.

9. Modify the program so that it also compares the alignment with the second image flipped left to right and with the flipped image rotated 90°, 180°, and 270° in a clockwise direction. (Why do we not also include the up-to-down flip as an additional case?)

10. Modify the program so that it computes the distance as the sum of the absolute values of the differences between corresponding points. Run the program using the sample data and determine if the same alignment is selected.

Amino Acids   The amino acids in proteins contain molecules of oxygen(O), carbon(C), nitrogen(N), sulfur(S), and hydrogen(H), as shown in Table 6.2. Assume that the numeric information in this table is contained in a data file named `elements.dat`. The molecular weights for oxygen, carbon, nitrogen, sulfur, and hydrogen are as follows:

Oxygen   15.9994
Carbon   12.011
Nitrogen   14.00674
Sulfur   32.066
Hydrogen   1.00794

11. Write a program that computes the molecular weight of each amino acid, and generates a data file that contains the information from the `elements.dat` file plus the molecular weights of the amino acids. Name the new file `aaweights.dat`.

12. Modify the program developed in problem 11 so that it computes and prints the average amino acid molecular weight.

13. Modify the program developed in problem 11 so that it computes and prints the numbers of the amino acids with the minimum and maximum molecular weights.

TABLE 6.2 Amino Acid Molecules

AMINO ACID	O	C	N	S	H
Alanine	2	3	1	0	7
Arginine	2	6	4	0	15
Asparagine	3	4	2	0	8
Aspartic	4	4	1	0	6
Cysteine	2	3	1	1	7
Glutamic	4	5	1	0	8
Glutamine	3	5	2	0	10
Glycine	2	2	1	0	5
Histidine	2	6	3	0	10
Isoleucine	2	6	1	0	13
Leucine	2	6	1	0	13
Lysine	2	6	2	0	15
Methionine	2	5	1	1	11
Phenylalanine	2	9	1	0	11
Proline	2	5	1	0	10
Serine	3	3	1	0	7
Threonine	3	4	1	0	9
Tryptophan	2	11	2	0	11
Tyrosine	3	9	1	0	11
Valine	2	5	1	0	11

Matrix Analysis   Each of the following problems reads a matrix from a data file `array.dat` and then analyzes the matrix to determine its properties.

14. Write a program to read a matrix from a data file `array.dat`. Determine if the matrix is an upper triangular matrix, and then print "Upper Triangular" or "Not Upper Triangular."

15. Write a program to read a matrix from a data file `array.dat`. Determine if the matrix is a lower triangular matrix, and then print "Lower Triangular" or "Not Lower Triangular."

16. Write a program to read a matrix from a data file `array.dat`. Determine if the matrix is a diagonal matrix, and then print "Diagonal" or "Not Diagonal." If the diagonal matrix is also an identity matrix, print "Identity" instead of "Diagonal."

17. A symmetric matrix is a square matrix that is symmetric around the main diagonal. Note that the transpose of a symmetric matrix is also equal to the original matrix. Write a program to read a matrix from a data file `array.dat`. Determine if the matrix is a symmetric matrix, and then print "Symmetric" or "Not Symmetric."

18. In a Toeplitz matrix, each diagonal contains the same value, but different diagonals can contain different values. Write a program to read a matrix from a data file `array.dat`. Determine if the matrix is a Toeplitz matrix, and then print "Toeplitz" or "Not Toeplitz."

19. A tridiagonal matrix is a matrix that can contain nonzero elements only on the main diagonal, the diagonal above the main diagonal, and the diagonal below the main diagonal. Write a program to read a matrix from a data file `array.dat`. Determine if the matrix is tridiagonal, and then print "Tridiagonal" or "Not Tridiagonal."

20. Some numerical techniques require that the rows of a matrix be reordered such that the row with the largest absolute value in column 1 be moved to the first row. Then, considering the rest of the rows after row 1, the row with the largest absolute value in column 2 is moved to the second row. This process continues until the rows have been reordered. This process is called row pivoting. Write a function to perform row pivoting.

21. Column pivoting is defined in a similar manner as row pivoting (see problem 20), but the columns are reordered such that column 1 contains the value with the largest absolute value in row 1, and so on. Write a function to perform column pivoting.

*Courtesy of National Center for Atmospheric Research/National
Science Foundation*

GRAND CHALLENGE: Prediction of Global Change

To understand the interactions of the biosphere system created by the atmosphere
and the ocean, we must understand its behavior with a great amount of detail. This
detail includes understanding CO_2 dynamics in the atmosphere and ocean, ozone depletion, and climatological changes due to the releases of chemicals or energy. One
way to collect some of this data is with sounding rockets, which are often used to
study the upper regions of the earth's atmosphere. A telemetry system, mounted in
the nose of the rocket, transmits scientific data to a receiver at the launch site as the
rocket passes through different levels of the atmosphere. These data are then collected for later analysis using the computer.

Plotting Capabilities

Introduction

Engineers use plots to analyze and evaluate data. Therefore, it is very important that you learn to use a powerful graphics tool so that you can easily generate plots in a variety of formats. MATLAB allows you to generate x–y plots, polar plots, bar graphs, contour plots, and 3-D plots. In this chapter we summarize the plot commands and many of the options that can be used with them. Being able to easily generate professional-looking plots with MATLAB will be useful not only in your engineering courses, but also in your math and science courses.

7.1   X-Y PLOTS

The most common plot used by engineers and scientists is the x–y plot. The data that we plot are usually read from a data file or computed in our programs, and stored in vectors that we will call x and y. We generally assume that the x values represent the independent variable, and that the y values represent the dependent variable. The y values can be computed using a function of x, or the x and y values might be measured in an experiment.

Rectangular Coordinates

Cartesian coordinate system

The data points that we use in engineering applications are usually either rectangular data points or polar data points. Rectangular data points identify points in a Cartesian coordinate system with distance values along a horizontal axis and a vertical axis as shown in Figure 7.1. Polar data points identify points in a polar coordinate system with an angle and a distance from the center of the space as shown in Figure 7.2. In this section we consider plots of points using rectangular coordinates, and then in the next section, we consider plots of points using polar coordinates. Before we present the commands for generating plots, we review the commands presented in Chapter 2 that add labels and grids to our plots. A plot should not be considered complete unless it has an informative title and x and y labels that describe the values being measured along with the units of measurement if they are known. We use a grid in example plots because the grid makes it easier to estimate values on the curves that are plotted.

Labels

The commands for adding labels and grid lines to plots were presented in Chapter 2, but we review them here. We also include commands for inserting text within the plot.

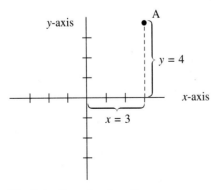

Fig. 7.1   Cartesian Coordinates.

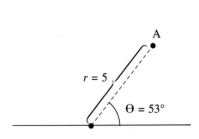

Fig. 7.2   Polar Coordinates.

title('text')	This command writes the text as a title at the top of the current plot.
xlabel('text')	This command writes the text beneath the x-axis on the current plot.
ylabel('text')	This command writes the text beside the y-axis on the current plot.
text(x,y,'text')	This command writes the text on the graphics screen at the point specified by the coordinates (x,y) using the axes from the current plot. If x and y are vectors, the text is written at each point.
text(x,y,'text','sc')	This command writes the text on the graphics screen at the point specified by the coordinates (x,y) assuming that the lower-left corner is (0,0) and the upper-right corner is (1,1).
gtext('text')	This command writes the text at the position on the graphics screen indicated by the mouse or arrow keys.
grid	This command adds grid lines to the current plot.

Plot Commands

linear scale
logarithmic scale
Most plots that we generate assume that the x and y axes are divided into equally spaced intervals; these plots are called linear plots. Occasionally, we may like to use a logarithmic scale on one or both of the axes. A logarithmic scale (base 10) is convenient when a variable ranges over many orders of magnitude because the wide range of values can be graphed without compressing the smaller values. The problems at the end of this chapter include problems that illustrate some of the advantages of logarithmic scales.

The MATLAB commands for generating linear and logarithmic plots of the vectors x and y are the following:

plot(x,y)	This command generates a linear plot of the values of x and y with the values of x representing the independent variable and the values of y representing the dependent variable.
semilogx(x,y)	This command generates a plot of the values of x and y using a logarithmic scale for x and a linear scale for y.
semilogy(x,y)	This command generates a plot of the values of x and y using a linear scale for x and a logarithmic scale for y.

loglog(x,y)   This command generates a plot of the values of x and y using logarithmic scales for both x and y.

It is important to recognize that the logarithm of a negative value or of zero does not exist. Therefore, if the data to be plotted on semilog axes or log-log axes contain negative values or zeros, a warning message will be printed by MATLAB informing you that these data points have been omitted from the data plotted.

If either x or y is a matrix in one of the plotting commands, the vector is plotted with the rows or columns of the matrix, whichever lines up in terms of the number of values. This is one way to generate a plot with multiple functions plotted on it.

If both x and y are matrices of the same size, the columns of x are plotted with the columns of y. This again will generate a plot containing multiple functions.

Each of these commands can also be executed with one argument, as in plot(y). In these cases, the plots are generated with the values of the indices of the vector y used as the x values.

In Chapter 2 we generated a plot of the flight path angle versus the coefficient of lift. Figure 7.3 contains plots of this data using a linear scale, a log x axis, a log y axis, and log-log axes. Each of these plots was generated with a group of com-

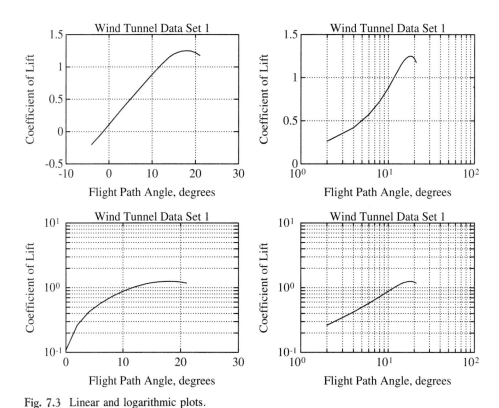

Fig. 7.3 Linear and logarithmic plots.

mands similar to the following, with the plot command changed to identify the type of axis scale desired:

```
plot(x,y),...
title('Wind Tunnel Data Set 1'),...
xlabel('Flight Path Angle, degrees'),...
ylabel('Coefficient of Lift'),...
grid
```

Recall that the commas and ellipses (. . .) are used so that the plot will not be drawn until all the relative commands have been entered. Otherwise, the data will be plotted, then the title will be added and the data replotted, and so on. Also, since the first few points contain negative angles or negative lift values, we will get a warning with the logarithmic plots and the corresponding points will be omitted from the graph. Since only the first few points are involved, the logarithmic plots are still useful; however, if many points throughout the data were omitted from the graph, it would not be representative of the data.

When we are deciding the best way to display a set of data, it is useful to generate plots using all four types of plots. We can then compare them to see which type illustrates the information that we want to highlight.

Practice!

Generate 100 points of the function below for values of x beginning at $x = 0$ and incrementing by 0.5:

$$y = 5x^2$$

1. Generate a linear plot of these data.

2. Generate a plot of these data with a logarithmic x scale.

3. Generate a plot of these data with a logarithmic y scale.

4. Generate a log-log plot of these data.

5. Compare the various plots and point out advantages and disadvantages of the different forms.

7.2   POLAR PLOTS

Polar plots are useful when the data values are represented by an angle and a magnitude. For example, if we are measuring light intensity around a light source, we might represent the information with an angle from a fixed axis and a magnitude that represents the intensity. Polar plots are also used when plotting complex values and in some digital signal processing applications.

Polar Coordinates

A point is represented in polar coordinates by an angle θ (theta) and a magnitude r. The value of θ is generally given as a value between 0 and 2π radians (which is equivalent to a value between 0° and 360°). The magnitude is a positive value that represents the distance from the axis at the given angle to the point.

Polar Command

The MATLAB command for generating a polar plot of the vectors `theta` and `r` is the following:

> `polar(theta,r)`   This command generates a polar plot of the angles `theta` (in radians) with the corresponding magnitudes `r`.

If either `theta` or `r` is a matrix, the vector is plotted with the rows or columns of the matrix, whichever lines up in terms of the number of values. This is one way to generate a polar plot with multiple functions plotted on it.

If both `theta` and `r` are matrices of the same size, the columns of `theta` are plotted with the columns of `r`. This again will generate a plot containing multiple functions.

The command `polar(r)` will generate a plot with the indices of the vector `r` used as the θ values.

Suppose that we want to generate points on a curve with increasing radius. We could generate angle values from 0 to 2π, with a corresponding radius that increases from 0 to 1. Figure 7.4 contains a polar plot generated with the following statements:

```
theta = 0:2*pi/100:2*pi;
r = theta/(2*pi);
polar(theta,r),...
title('Polar Plot'),...
grid
```

Rectangular/Polar Transformations

It is sometimes useful to transform coordinates from one coordinate system to another. The equations for performing these transformations can be easily derived with a little trigonometry. Draw a sketch and derive them yourself.

$$
\begin{aligned}
\text{Polar to Rectangular} \quad & x = r \sin \theta \\
& y = r \cos \theta \\
\text{Rectangular to Polar} \quad & r = \sqrt{x^2 + y^2} \\
& \theta = \tan^{-1}\left(\frac{y}{x}\right)
\end{aligned}
$$

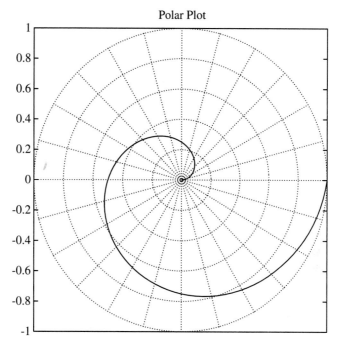

Fig. 7.4   Polar plot with increasing radius.

Be careful when computing the inverse tangent that you compute the angle in the correct quadrant. If you are using MATLAB to make the conversions, you can be assured of getting the correct quadrant by using the `atan2` function.

Practice!

Convert the following points from rectangular coordinates to polar coordinates.

1. $(3, -2)$

2. $(0, -1)$

3. $(-2, 0)$

4. $(0.5, 1)$

Convert the following points from polar coordinates to rectangular coordinates. (Assume that the first value in each pair is the angle in radians.)

5. $(\pi, 1)$

6. $(\pi/2,0)$

7. $(2.3,0.5)$

8. $(0.5,0.5)$

7.3   BAR GRAPHS AND STAIR GRAPHS

Figure 7.5 contains a bar graph and a stair graph for the wind tunnel data initially presented in Chapter 2 and then used again in Section 7.1. The two graphs are similar, but the vertical lines dropping to the x-axis are omitted in the stair graph. The options within the bar and stairs commands are summarized here:

bar (y)   This command draws a bar graph of the elements of the vector y.

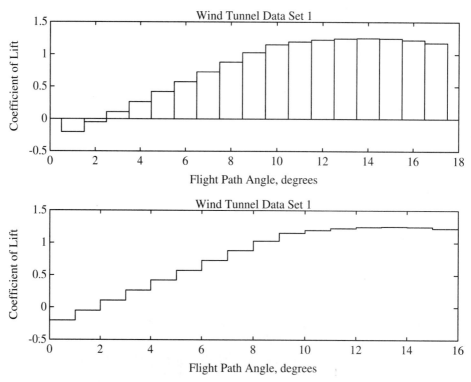

Fig. 7.5   Bar graph and stair graph.

bar(x, y)	This command draws a bar graph of the elements of the vector y at the locations specified in the vector x, which must contain equally spaced ascending values.
stairs(y)	This command draws a stair graph of the elements of the vector y.
stairs(x, y)	This command draws a stair graph of the elements of the vector y at the locations specified in the vector x, which must contain equally spaced ascending values.

Histograms are similar to bar graphs, but the values plotted by a histogram are computed from the distribution of values in a vector, while the bar charts and stair graphs are generated directly from the values themselves. You may want to review the hist command that was presented in Chapter 5.

The figures in Figure 7.5 were generated using the following MATLAB commands:

```
bar(y),...
title('Wind Tunnel Data Set 1'),...
xlabel('Flight Path Angle, degrees'),...
ylabel('Coefficient of Lift');
pause;
stairs(y),...
title('Wind Tunnel Data Set 1'),...
xlabel('Flight Path Angle, degrees'),...
ylabel('Coefficient of Lift');
```

While bar graphs and stair graphs are not used as commonly in engineering as x-y plots, there are times when they are useful.

7.4   PLOTTING OPTIONS

In Section 7.1 we reviewed the options for labeling a plot and for adding the grid lines. MATLAB also includes a number of additional options for enhancing plots. It is useful to read through these options so that you are familiar with them, and can then refer back to this section when you want to add one of them to a plot.

Multiple Plots

There are three ways to plot multiple curves on the same graph. One way is to use arguments in the plot command that contain matrices, as discussed in the section on x-y plots.

The second way to plot multiple curves on the same graph is to use multiple arguments in a plot command, as in the following:

```
plot(x, y, w, z)
```

where the variables x, y, w, and z are vectors. When this command is executed, the curve corresponding to x and y will be plotted, and then the curve corresponding to w and z will be plotted on the same graph. The advantage of this technique is that the number of points in the two plots do not have to be the same. MATLAB will automatically select different line types so that you can distinguish between the two plots.

hold

A third way to plot multiple curves on the same graph is with the hold command. The hold command will hold the current graph on the screen. Subsequent plot commands will add to the graph, using the already established axes limits. The execution of another hold command will turn off the hold on the current graph. The hold on and hold off commands can also be used to turn on and turn off the hold on the current graph.

Line and Mark Style

The command plot(x,y) generates a line plot that connects the points represented by the vectors x and y with line segments. You can also select other line types—dashed, dotted, and dashdot. You can also select a point plot instead of a line plot. With a point plot, the points represented by the vectors will be marked with a point instead of connected by line segments. You can also select characters other than a point to indicate the points; the other choices are plus signs, stars, circles, and x-marks. The following list contains these different options for lines and marks:

line type	indicator	point type	indicator
solid	-	point	.
dashed	--	plus	+
dotted	:	star	*
dashdot	-.	circle	O
		x-mark	x

The following command illustrates the use of line and mark styles; it generates a solid line plot of the points represented by the vectors x and y, and then plots the points themselves with x-marks:

```
plot(x,y,x,y,'x')
```

In addition to selecting the line type or point type, you can also select color if you are using a color terminal. The options are the following:

color	indicator
red	r
green	g
blue	b
white	w
invisible	i

The following command generates a solid blue line plot of the points represented by the vectors x and y, and then plots the points themselves with red x-marks:

```
plot(x,y,'b',x,y,'xr')
```

Scaling

MATLAB automatically scales the axes to fit the data values. However, you can override this scaling with the `axis` command. There are several forms of the `axis` command:

axis	This command freezes the current axis scaling for subsequent plots. A second execution of the command returns the system to automatic scaling.
axis(v)	v is a four-element vector that contains the scaling values, [xmin, xmax, ymin, ymax].
axis('square') axis('normal')	This command specifies the aspect ratio, which can be 'square' or 'normal'.

These commands are especially useful when you want to compare curves from different plots because it can be difficult to visually compare curves plotted with different axes.

Subplot

The `subplot` command allows you to split the graph window into subwindows. The possible splits can be to two subwindows or four subwindows. Two subwindows can be arranged as either top and bottom or left and right: Figure 7.5 was generated using a top-and-bottom window split. A four-window split has two subwindows on the top and two subwindows on the bottom; Figure 7.3 was generated using a four-window split. The argument to the `subplot` command is a three-digit number mnp. The digits m and n specify that the graph window is to be split into an m-by-n grid of smaller windows, and the digit p specifies the pth window for the current plot. The windows are numbered from left to right, top to bottom. Therefore, the following commands specify that the graph window is to be split into a top plot and a bottom plot, and the current plot is to be placed in the top subwindow:

```
subplot(211), plot(x,y)
```

Screen Control

MATLAB has two display windows—a command window and a graph window. The hardware configuration determines whether or not both windows are seen simultaneously. The following commands allow you to select the windows and to clear the windows:

shg	show graph window

any keystroke	bring back command window
clc	clear command window
clg	clear graph window
home	home command cursor

Graph Screen Input

The following commands allow you to obtain the coordinates of points from the graph window using a mouse or arrow keys:

[x,y] = ginput	This command allows you to select an unlimited number of points from the graph window using a mouse or arrow keys, and store the values in the vectors x and y. Pressing the return key terminates the input.
[x,y] = ginput(n)	This command allows you to select n points from the graph window using a mouse or arrow keys, and store the values in the vectors x and y. Pressing the return key terminates the input.

Graphics Hard Copy

The following three commands provide hard copy capabilities in MATLAB. However, the student edition of MATLAB does not support the meta command. Also, the performance of these commands may vary from computer to computer.

prtsc	This command initiates a print screen command that is performed on a pixel-by-pixel basis, resulting in a hard copy with the same resolution as the computer screen. On most computers, holding the shift key down and pressing the prtSc key also dumps the screen to a printer.
print	This command sends a high-resolution copy of the current plot to the printer.
meta file	This command opens a high-resolution graphics metafile using the filename specified, and writes the current graph to it. Subsequent meta commands without a filename append the graph to the previously specified filename.

PROBLEM SOLVING APPLIED: SOUNDING ROCKET TRAJECTORY

Sounding rockets are used to probe different levels of the atmosphere to collect information such as that used to monitor the levels of ozone in the atmosphere. In addition to carrying the scientific package for collecting data on the upper atmosphere, the rocket also carries a telemetry system in its nose to transmit data to a receiver at the launch site. In addition to the scientific data, performance measurements on the rocket itself are also transmitted to be monitored by range safety personnel and to be later analyzed by engineers. These performance data include altitude, velocity, and acceleration measurements.

In this section, we assume that we have a data file [2] containing the altitude, velocity, and acceleration data for the data collection portion of the trajectory of a two-stage sounding rocket that was launched to perform high-altitude atmospheric research on the ionosphere. The first stage of the rocket was supposed to burn for approximately 35 seconds in order to accelerate the rocket to a velocity of 1,250 meters per second. The rocket should then coast for almost 2 minutes before reaching the lower region of the ionosphere at about 100 kilometers. By then gravity should slow the rocket's ascent to about 100 meters per second. The second stage should then ignite and accelerate the rocket through the ionosphere and into space. We want to generate plots of these data in order to determine if the actual performance was similar to the predicted performance.

 1. PROBLEM STATEMENT

Generate plots of the altitude, velocity, and acceleration from the data generated by a sounding rocket.

 2. INPUT/OUTPUT DESCRIPTION

Figure 7.6 contains a diagram showing that the input is a file containing the data values, and the output is a set of plots of the data.

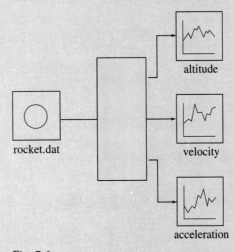

Fig. 7.6

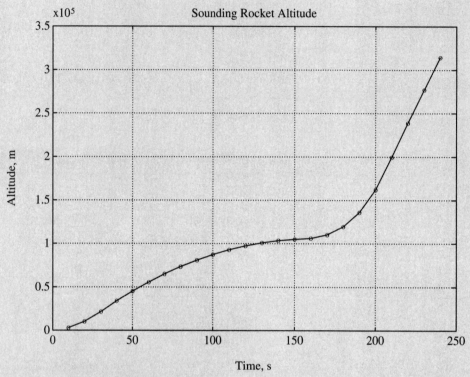

Fig. 7.7   Plot of sounding rocket altitude.

3. HAND EXAMPLE

Since we are not really computing anything in this problem, there is nothing to compute by hand. However, we can skim down the data in the data files so that we have a general idea of how the plots should appear. Then, if the appearance is not as expected, we know to check further.

4. MATLAB SOLUTION

Since the plotting commands in MATLAB are so powerful, the MATLAB solution is very simple. We will plot both a line plot of the data and then a point plot of the original data points.

```
%
%        This program generates plots of the altitude,
%        velocity, and acceleration of a sounding rocket.
%
load rocket.dat
```

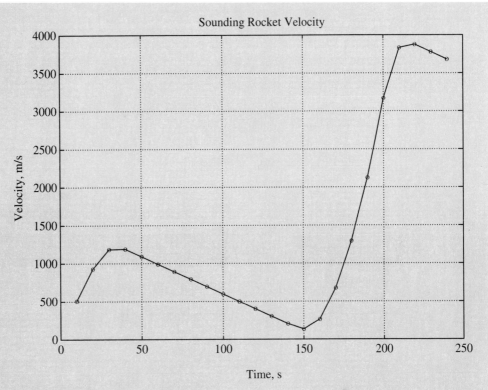

Fig. 7.8   Plot of sounding rocket velocity.

```
time = rocket(:,1);
alt = rocket(:,2);
vel = rocket(:,3);
acc = rocket(:,4);
%
%       These commands generate and label
%       a plot of the altitude data.
%
plot(time,alt,time,alt,'o'),...
title('Sounding Rocket Altitude'),...
xlabel('Time, s'),...
ylabel('Altitude, m'),...
grid
pause
%
%       These commands generate and label
%       a plot of the velocity data.
```

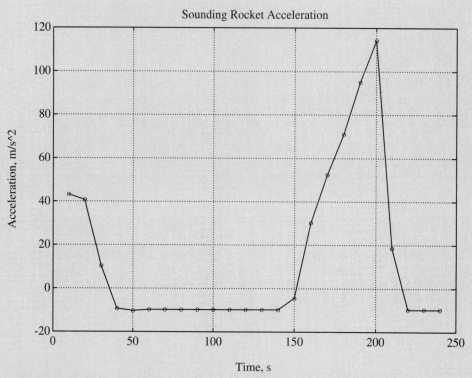

Fig. 7.9 Plot of sounding rocket acceleration.

```
%
plot(time,vel,time,vel,'o'),...
title('Sounding Rocket Velocity'),...
xlabel('Time, s'),...
ylabel('Velocity, m/s'),...
grid
pause
%
%      These commands generate and label
%      a plot of the acceleration data.
%
plot(time,acc,time,acc,'o'),...
title('Sounding Rocket Acceleration'),...
xlabel('Time, s'),...
ylabel('Acceleration, m/s^2'),...
grid
```

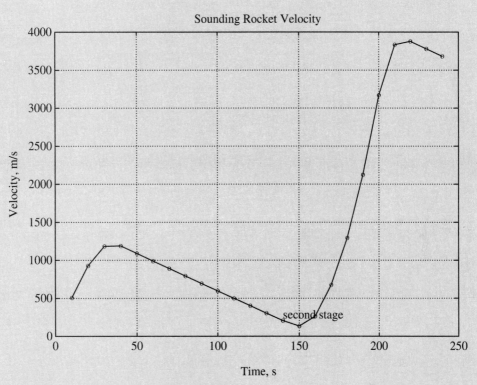

Fig. 7.10   Velocity with second stage indicated.

5. TESTING

The data used for this test case are included in the diskette that accompanies this text. The three plots are shown in Figures 7.7 through 7.9. We can observe from the graph of the velocity data that the rocket's velocity is initially increasing, and then slowly decreases after the first stage of the rocket burns out. When the second stage is fired, the velocity begins to increase again. From the graph of the acceleration data, you can easily see the accelerations caused by the firing of the first stage and of the second stage. In the intervals after the firings, the acceleration is -9.8 m/sec^2, which is the downward acceleration due to gravity. In Figures 7.10 and 7.11, we added the text commands to the other plotting commands to identify the second stage firing in both the velocity and the acceleration plots:

```
%
%        These commands generate and label
%        a plot of the velocity data.
```

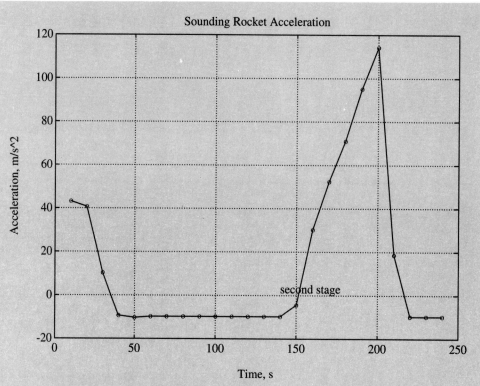

Fig. 7.11   Acceleration with second stage indicated.

```
%
plot(time,vel,time,vel,'o'),...
title('Sounding Rocket Velocity'),...
xlabel('Time, s'),...
ylabel('Velocity, m/s'),...
text(140,200,'second stage'),...
grid
pause
%
%       These commands generate and label
%       a plot of the acceleration data.
%
plot(time,acc,time,acc,'0'),...
title('Sounding Rocket Acceleration'),...
xlabel('Time, s'),...
```

```
ylabel('Acceleration, m/s^2'),...
text(140,0,'second stage'),...
grid
```

Additional problems related to this application are included in the end-of-chapter problems.

7.5   3-D PLOTS

A 3-D surface can be plotted in two ways with MATLAB. A mesh surface can be plotted that gives a 3-D image of the surface. The mesh surface can be viewed from a variable viewpoint, and the scale on the axes can be specified. A 3-D surface can also be plotted as a contour plot that contains slices of the surface at different heights. Both of these plots provide important information in analyzing 3-D data.

Mesh Surfaces

mesh   A mesh surface is generated by a set of values in a matrix; each point in the matrix represents the value of a surface that corresponds to that point in the grid.

To generate data that represent a three-dimensional surface, we first compute a set of x and y values that represent the independent variables, and then compute the z values that represent the surface values. In order to have a uniform looking plot, we typically choose the x and y values to be from a grid that is uniformly spread across the x-y plane. For example, suppose that we want to evaluate the following function for $-0.5 \leq x \leq 0.5$ and for $-0.5 \leq y \leq 0.5$:

$$f(x,y) = z = \left| \sqrt{1 - x^2 - y^2} \right|$$

This function is a form of the following function:

$$x^2 + y^2 + z^2 = 1$$

which is the equation for a sphere centered at (0,0) with radius 1. Since the function $f(x,y)$ uses only the positive square root, $f(x,y)$ is the top half of the sphere. To generate and plot the values of this 3-D surface, we could use the following:

```
for m=1:11
    x = (m-6)*0.1;
    for n=1:11
        y = (n-6)*0.1;
        z(m,n) = sqrt(abs(1 - x.^2 - y.^2));
    end
end
```

```
mesh(z),...
title('3-D Plot')
```

The corresponding plot is shown in Figure 7.12.

meshdom
Another way to generate the values in the z vector is with the meshdom command that has the *x* vector and the *y* vector as arguments. The meshdom command generates special arrays that contain values that correspond to the x and y values of the grid. The portion of the sphere in Figure 7.12 can also be generated using this command with the following statements:

```
[xgrid,ygrid] = meshdom(-0.5:0.1:0.5,-0.5:0.1:0.5);
z = sqrt(abs(1 - xgrid.^2 - ygrid.^2));
mesh(z),...
title('3-D Plot')
```

Viewing Angles   When generating a 3-D mesh surface, we may want to change the viewing location. The viewing position is defined in terms of an azimuth (horizontal rotation) and vertical elevation that is specified in degrees. The viewpoint with a

azimuth, elevation   zero azimuth and a zero elevation is at the lower left corner of the matrix. A positive

3-D Plot

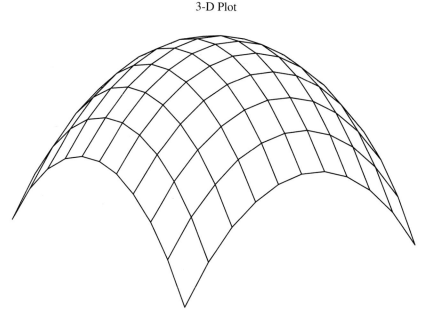

Fig. 7.12   Plot of the top of a sphere.

value for the azimuth indicates counterclockwise rotation of the viewpoint about the vertical or z-axis, which corresponds to a clockwise rotation of the object. Positive values of elevation view the object from above, negative from below.

The default viewpoint has an azimuth of $-37.5°$ and an elevation of $30°$. The viewpoint can be specified by a vector containing the azimuth and elevation that is used as a second argument in the mesh command. To illustrate the viewpoint, we use the matrix z which was used to generate the mesh in Figure 7.12. In Figure 7.13, we view the mesh from a lower elevation, using the following statements:

```
mesh(z,[-37.5,0]),...
title('3-D Plot')
```

In Figure 7.14, we view the mesh from a lower elevation than was used in Figure 7.13, using these statements:

```
mesh(z,[-37.5,-30]),...
title('3-D Plot')
```

Scale Factors   The mesh command also allows you to control the scale factors that set the x, y, and z axes with a three-element vector [sx sy sz]. Therefore, the size of the mesh is based on the relative values of sx, sy, and sz. These scale

3-D Plot

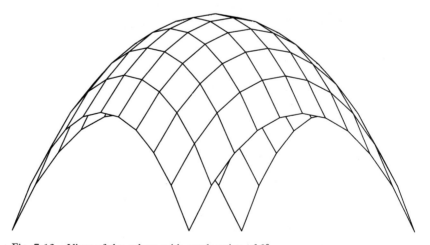

Fig. 7.13   View of the sphere with an elevation of 0°.

3-D Plot

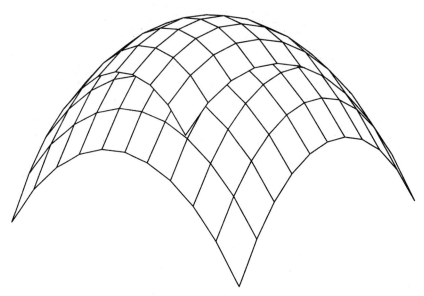

Fig. 7.14   View of the sphere with an elevation of $-30°$.

factors can be used with or without a viewpoint argument. If the mesh command has two arguments, as in mesh(z,a), it will assume that the second argument is a scale factor if it has three elements and that it is a viewpoint designation if it has two elements. Similarly, if the mesh command has three arguments, as in mesh (z,a,b), the second and third arguments represent the scale factor and viewpoint designation, in either order.

Practice!

Generate and plot the 3-D mesh surfaces indicated. Before plotting the mesh, try to figure out how it should look.

1. $f(x,y) = \left| \sqrt{1 - x^2 - y^2} \right|$ for $0 \le x \le 0.5$ and for $0 \le y \le 0.5$

2. $f(x,y) = -\left| \sqrt{1 - x^2 - y^2} \right|$
    for $-0.5 \le x \le 0.5$ and for $-0.5 \le y \le 0.5$

3. $f(x,y) = \left| \sqrt{1 - x^2 - y^2} \right|$ for $-.5 \le x \le 0$ and for $0 \le y \le 0.5$

4. $f(x,y) = \left| \sqrt{2 - x^2 - y^2} \right|$ for $-1 \le x \le 1$ and for $-1 \le y \le 1$

Contour Plots

An elevation map contains a group of lines that connect equal elevations. We can think of a line that connects points of equal elevation as showing a slice of the countryside at-that elevation. If we have a map with many lines showing different elevations, we can determine mountains and valleys from it. This type of map is a contour map generated from the 3-D elevation data. MATLAB can generate a similar map from 3-D data stored in a matrix. The various forms of the contour command are described below:

contour(z)	This command draws a contour plot of the matrix z. The number of contour lines and their values are chosen automatically by MATLAB. The upper left corner of the plot corresponds to the value in position z(1,1).
contour(z,n)	This command produces a contour plot of matrix z with n contour levels.
contour(z,v)	This command produces a contour plot of z with contour levels at the values specified by the vector v.

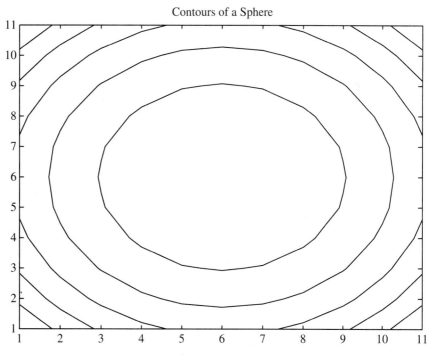

Fig. 7.15 Plot with five contour levels.

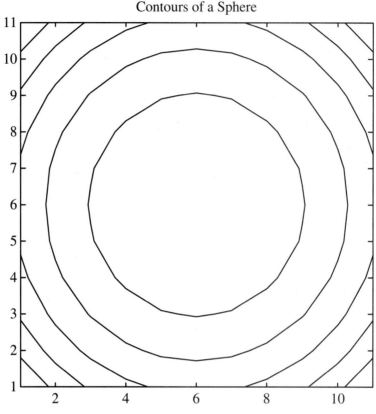

Fig. 7.16   Plot with a square aspect ratio.

The contour command can also be used with arguments that specify the contour levels and the axes scaling.

The following commands generate the contour plot in Figure 7.15 of the spherical surface that we have used in several of our examples in this section:

```
[x,y] = meshdom(-0.5:0.1:0.5,  -0.5:0.1:0.5);
z = sqrt(abs(1 - x.^2 - y.^2));
contour(z,5),...
title('Contours of a Sphere')
```

If we execute this same set of commands after executing the axis ('square') command, we obtain the contour plot shown in Figure 7.16. The square aspect ratio is often used for contour plots of square data in order to more accurately display the data, as illustrated with this data.

It is interesting to look at 3-D surface plots and contour plots for some of the special matrices that can be easily generated in MATLAB, such as identity matrices and magic squares.

 PROBLEM SOLVING APPLIED: TERRAIN NAVIGATION

Terrain navigation is a key component in the design of remotely piloted vehicles (RPVs), such as robots, planes, and missiles, and underwater autonomous vehicles (UAVs). An RPV or UAV system contains an on-board computer that has stored the terrain information for the area in which it is to move. By knowing at any time where it is (perhaps with the aid of a global positioning system, GPS, receiver), it can then select the best path to get to a designated spot. If the destination changes, the vehicle can refer to its internal maps to recompute the new path.

The computer software that guides these vehicles must be tested over a variety of land formations and topologies. Elevation information for large grids of land (or seafloor) is available in computer databases. One way of measuring the "difficulty" of a land grid with respect to terrain navigation is to determine the number of peaks in the grid, where a peak is a point that has lower elevations all around it.

Write a MATLAB program that will read elevation data from a data file and then print the location of peaks in the grid. Also print a contour plot of the elevation data.

 1. PROBLEM STATEMENT

Determine the number and location of the peaks in the elevation data stored in a data file. Also generate a contour plot of the elevation data.

 2. INPUT/OUTPUT DESCRIPTION

Figure 7.17 contains a diagram showing that the input is the file containing the elevation data, and that the output is a listing of the locations of the peaks plus a contour plot.

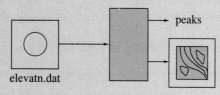

Fig. 7.17   I/O diagram.

3. HAND EXAMPLE

Assume that the following data [2] represents the elevation for a grid that has 6 points along the side and 8 points along the top. We have underlined the peaks within the data.

25	59	63	23	21	34	21	50
32	45	43	30	37	32	30	27
34	38	38	39	36	28	28	35
40	45	42	48	32	30	27	25
39	39	40	42	48	49	25	30
31	31	31	32	32	33	44	35

The location of the peaks are at points (2,5), (4,2), (4,4), and (5,6).

 4. MATLAB SOLUTION

The search for peaks should consider only interior points in the grid of elevation data. A point along the edge cannot be counted as a peak, because we do not know the elevation on one of its sides. If we are considering a point at location elevation(m,n), then the four adjacent points are at the positions indicated in Figure 7.18. Therefore, the point represented by elevation (m,n) is a peak if the following conditions are all true:

```
elevation(m,n) > elevation(m,n-1)
elevation(m,n) > elevation(m-1,n)
elevation(m,n) > elevation(m,n+1)
elevation(m,n) > elevation(m+1,n)
```

We use a nested for loop to evaluate all the necessary points.

```
%
%          This program reads the elevation data for
%          a land grid and determines the peaks.
```

	elevation $(m-1, n)$	
elevation $(m, n-1)$	elevation (m, n)	elevation $(m, n+1)$
	elevation $(m+1, n)$	

Fig. 7.18 Positions adjacent to elevation (m, n).

```
%              It also generates a contour plot.
%
load elevatn.dat
elevation = elevatn;
%
%              Identify the peaks.
%
[rows,cols] = size(elevation);
for m=2:rows-1
     for n=2:cols-1
            if (elevation(m,n)>elevation(m,n-1) & ...
            elevation(m,n)>elevation(m-1,n) & ...
            elevation(m,n)>elevation(m,n+1) & ...
            elevation(m,n)>elevation(m+1,n))
                 fprintf('Peak at (%3.0f, %3.0f) \n',m,n)


        end
     end
end
```

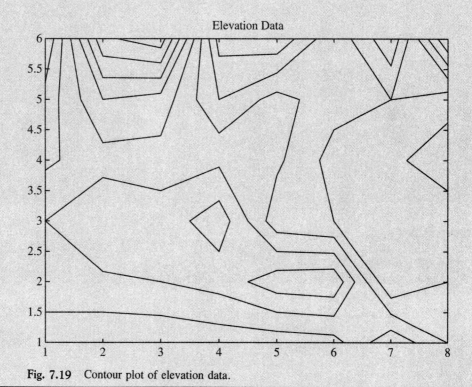

Fig. 7.19 Contour plot of elevation data.

```
%
%          Generate plot.
%
contour(elevation),...
title('Elevation Data')
```

 5. TESTING

The peaks identified from the data in the hand example were:

```
Peak at (    2,    5 )
Peak at (    4,    2 )
Peak at (    4,    4 )
Peak at (    5,    6 )
```

The corresponding plot is shown in Fig 7.19. Since this was not a square grid, we did not use a square aspect ratio.

Additional problems related to this application are included in the end-of-chapter problems.

SUMMARY

In this chapter we summarized the plotting commands and options available in MATLAB. X-Y plots are the type of plot most commonly used in engineering, and MATLAB allows us to easily generate them in linear form or in logarithmic form. Generating several plots on the same graph is also very useful in comparing and contrasting different sets of data. Not only do plots give us a visual interpretation of data, but they also allow us to easily illustrate key points that we want to emphasize in the data. Many engineering applications also use 3-D data, which can be plotted in MATLAB using 3-D mesh surfaces or contour plots.

MATLAB SUMMARY

This MATLAB summary lists all the commands and functions that were defined in this chapter. A brief description is also included for each one.

Commands and Functions

axis	controls axis scaling
bar	generates a bar graph
clc	clears command window
clg	clears graph window

contour	generates a contour plot
ginput	obtains coordinates from graph window
grid	adds grid lines to a plot
gtext	write text on the graphics screen
hold	holds current graph on the screen
home	moves command cursor to home position
loglog	generates a log-log plot
mesh	generates a 3-D surface mesh plot
meshdom	generates vectors for grid computations
meta	stores plot in file (not in student edition)
plot	generates a linear x-y plot
polar	generates a polar plot
print	prints a high-resolution plot
prtsc	initiates a print screen command
semilogx	generates a log-linear plot
semilogy	generates a linear-log plot
shg	shows graph window
stairs	generates a stair graph
subplot	splits graphics window into subwindows
text	writes text on the graphics screen
title	writes a title at the top of a plot
xlabel	writes a label beneath the x-axis
ylabel	writes a label beside the y-axis

PROBLEMS

Problems 1 to 13 relate to the engineering applications presented in this chapter. Problems 14 to 27 relate to new engineering applications.

Sounding Rocket Trajectory   These problems relate to the "Problem Solving Applied" section on sounding rocket trajectory.

1. Plot the velocity data using a linear-log plot. Compare the result with the linear plots given in the section.

2. Plot the acceleration data using a linear-log plot. Compare the results with the linear plot given in the section.

3. Plot the altitude and velocity data on the same graph. Is there any information that was not obvious in the separate graphs?

4. Plot the altitude and acceleration data on the same graph. Is there any information that was not obvious in the separate graphs?

5. Plot the velocity and acceleration data on the same graph. Is there any information that was not obvious in the separate graphs?

6. Add text to show the firing of the first stage of the rocket in the velocity and acceleration plots.

7. Obtain from the plot the intervals in the acceleration plot for which the acceleration is due only to gravity.

Terrain Navigation   These problems relate to the "Problem Solving Applied" terrain section on navigation.

8. Modify the terrain navigation program so that it prints a count of the number of peaks in the elevation grid.

9. Modify the terrain navigation program so that it prints the location of valleys instead of peaks, where a valley is a point with an elevation lower than the elevation of its four neighboring points.

10. Modify the terrain navigation program so that it prints the highest point and the lowest point in the elevation data. Include the grid location.

11. Modify the terrain navigation program so that it prints the average elevation of the elevation data.

12. Assuming that the distance between points in a vertical and horizontal direction is 100 feet, give the location of the peaks in feet from the lower left corner of the grid.

13. Assuming that the distance between points in a vertical and horizontal direction is 100 feet, give the location of the peaks in feet from upper left corner of the grid.

Logarithmic Plots   There are many applications in which we collect data from an experiment, and would then like to determine an equation from the data that could be used as a model for that portion of the experiment. For example, if we plot a set of x–y data and obtain something very close to a straight line, we can easily estimate a slope and y-intercept to obtain an equation that models the data.

14. Consider the following exponential equation:

$$y = 3 \cdot 10^{2x}$$

If we plot x–y data computed from this equation with a linear-logarithmic plot, the data are in a straight line. To see why this is true, we take the logarithm of both sides, giving the following:

$$\log_{10} y = 2x + \log_{10} 3$$

This equation is a linear equation in terms of x and $\log_{10} y$. Use MATLAB to plot data from the equation given at the beginning of this problem using a linear-linear scale and then using a linear-logarithmic scale. Estimate the slope and y-intercept of the graph on the linear-logarithmic plot. The slope should be approximately 2 and the y-intercept should be approximately $\log_{10} 3$.

15. Assume that you have a set of data from an experiment. When you plot the data using a linear-logarithmic scale, the data are nearly linear. What kind of equation is a good model for the data, and how do you compute the constants in the equation?

16. Consider the following power equation:

$$y = 5x^3$$

If we plot x–y data computed from this equation with a log-log plot, the data are in a straight line. To see why this is true, we take the logarithm of both sides, giving the following:

$$\log_{10} y = 3 \log_{10} x + \log_{10} 5$$

This equation is a linear equation in terms of $\log_{10}$ x and $\log_{10}$ y. Use MATLAB to plot data from the equation given at the beginning of this problem using a linear-linear scale and then using a log-log scale. Estimate the slope and y-intercept of the graph on the log-log plot. The slope should be approximately 3 and the y-intercept should be approximately $\log_{10} 5$.

17. Assume that you have a set of data from an experiment. When you plot the data using a log-log scale, the data are nearly linear. What kind of equation is a good model for the data, and how do you compute the constants in the equation?

SINC Function   The sinc function is an interesting function to plot. It is also a very important function in the area of digital signal processing. The equation for this function is the following:

$$f(x,y) = \text{sinc}(r)$$
$$= \sin(r)/r$$

where r is the distance of the point x,y from the origin. At the origin, the function is defined to be 1.0. (This function is also sometimes called a "sombrero" function.)

18. Generate a square matrix with 11 rows and 11 columns containing values of the sinc function for $-10 \leq x \leq 10$ and $-10 \leq y \leq 10$. Create a 3-D plot of the surface.

19. Create a contour plot of the sinc function computed in problem 18.

20. Create a 3-D plot of the sinc function computed in problem 18 from a viewpoint below the surface.

21. Create a 3-D plot of the sinc function computed in problem 18 from a viewpoint at 0° elevation.

22. Create a contour plot of the sinc function computed in problem 18 with contours at 0.1, 0.2, 0.3, 0.4, 0.5.

Paraboloid Function   The following equation defines a paraboloid function of two variables:

$$z = 2x^2 + 2y^2 + 2xy - 14x - 16y + 42.$$

Functions of this type occur frequently in adaptive signal processing problems.

23.   Generate a square matrix with 16 rows and 16 columns containing values of this paraboloid function for $-5 \le x \le 10$ and $-5 \le y \le 10$. Create a 3-D plot of the surface. View it from several different points.

24.   Create a 3-D plot of the surface defined in problem 23 using the meshdom function to generate points of the surface.

25.   Create a contour plot of the surface described in problem 23.

26.   Generate a "slice" of the paraboloid surface by restricting x to be equal to 3. Plot this function of y.

27.   Generate a "slice" of the paraboloid surface by restricting y to be equal to 0. Plot this function of x.

PART III

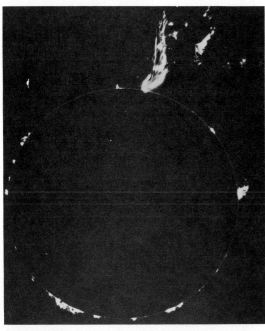

Numerical Techniques

Courtesy of University Corporation for Atmospheric Research/National Center for Atmospheric Research/National Science Foundation

The numerical techniques presented in this part are used for solving special types of problems. In addition to presenting the MATLAB commands, we include a number of diagrams and graphs to help you visualize the concepts behind these numerical methods. When you understand the concepts, you are able to select the correct numerical technique for solving a specific problem. It is also very important to be able to evaluate the computer results, and determine whether or not they provide an appropriate solution for your problem. The powerful graphics capabilities of MATLAB are especially useful in analyzing and interpreting the results of numerical techniques. The material on Solutions to Systems of Linear Equations, Interpolation and Curve Fitting, and Roots of Polynomials will be needed throughout an undergraduate engineering curriculum and should therefore be carefully studied now if time permits. The remaining topics (Polynomial Analysis, Numerical Integration and Differentiation, Ordinary Differential Equations, Matrix Decomposition and Factorization, Signal Processing and Control Systems) are more specialized. We recommend that you skim these sections so that you know what capabilities are available in MATLAB, and you can then return to these sections when you have problems to solve that need these techniques.

213

8

Photo courtesy of General Motors Electric Vehicles

GRAND CHALLENGE: Vehicle Performance

The car in this photograph is General Motors' new electric vehicle prototype, Impact. This car is a zero-emissions vehicle that accelerates from 0 to 60 miles per hour in 8 seconds. It has a practical range of 80 miles per charge; the average daily driving range in the United States is 30 miles a day. The recharge time is approximately 3 hours on a 220 v circuit, and generally would be done overnight to spread the demand for electricity and reduce the burden on power plants. The current battery is a lead-acid battery, but important new battery technology is aiding in adapting nickel metal hydride batteries for this car. Research within the next decade is expected to help give electric vehicles both performance and operating costs that are competitive with today's gasoline-powered cars.

Solutions to Systems of Linear Equations

Introduction

We begin this chapter with a graphical description of the solution to a set of simultaneous equations. Figures are used to illustrate the different possible situations that can occur in solving sets of equations with two and three variables. Solutions to sets of equations with more than three variables are discussed in terms of hyperplanes. We then present two different techniques for solving a system of simultaneous equations using matrix operations. Finally, an example from electrical circuit analysis is presented that uses a set of simultaneous equations to determine the mesh currents in a circuit.

8.1 GRAPHICAL INTERPRETATION

The need to solve a system of simultaneous equations occurs frequently in engineering problems. A number of methods exist for solving a system of equations, but they all involve tedious operations with a number of opportunities to make errors. Therefore, solving a system of equations is an operation that we would like the computer to perform for us. However, we must understand the process so that we can correctly evaluate and interpret the computer results. To understand the process, we start with a graphical interpretation of the solution to a set of equations.

linear equation

     A linear equation with two variables, such as $2x - y = 3$, defines a straight line, and is often written in the form $y = mx + b$, where m represents the slope of the line and b represents the *y-intercept*. Thus, $2x - y = 3$ can also be written as $y = 2x - 3$. If we have two linear equations, they can represent two different lines that intersect in a single point, or they can represent two parallel lines that never intersect, or they can represent the same line; these possibilities are shown in Figure 8.1. Equations that represent two intersecting lines can be easily identified because

slopes

they will have different slopes, as in $y = 2x - 3$ and $y = -x + 3$. Equations that

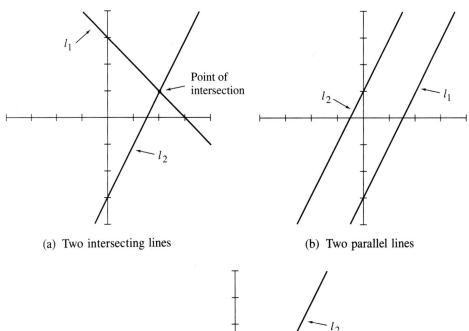

(a) Two intersecting lines   (b) Two parallel lines

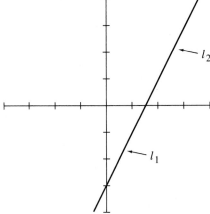

(c) Two identical lines

Fig. 8.1   Two lines.

y-intercept

represent two parallel lines will have the same slope but different y-intercepts, as in $y = 2x - 3$ and $y = 2x + 1$. Equations that represent the same line have the same slope and y-intercept, as in $y = 2x - 3$ and $3y = 6x - 9$.

If a linear equation contains three variables, x, y, and z, then it represents a plane in three-dimensional space. If we have two equations with three variables, they can represent two planes that intersect in a straight line, they can represent two parallel planes, or they can represent the same plane; these possibilities are shown in intersecting planes Figure 8.2. If we have three equations with three variables, the three planes can intersect in a single point, they can intersect in a plane, they can have no common intersection point, or they can represent the same plane. Examples of the possibilities that exist if the three equations define three different planes are shown in Figure 8.3.

These ideas can be extended to more that three variables, although it is harder to visualize the corresponding situations. We call the set of points defined by an hyperplane equation with more than three variables a hyperplane. In general, we can consider a set of M linear equations that contain N unknowns, where each equation defines a unique hyperplane that is not identical to another hyperplane in the system. If $M < N$, then the system is underspecified, and a unique solution does not exist. If

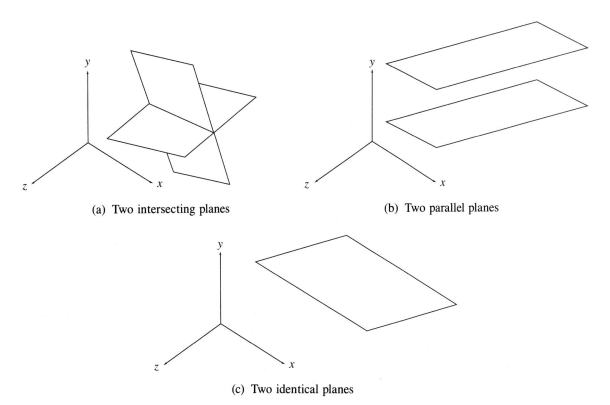

(a) Two intersecting planes   (b) Two parallel planes

(c) Two identical planes

Fig. 8.2   Two planes.

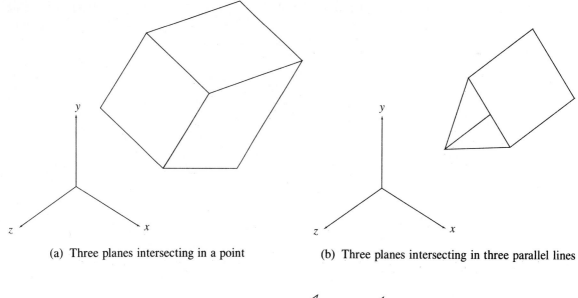

(a) Three planes intersecting in a point   (b) Three planes intersecting in three parallel lines

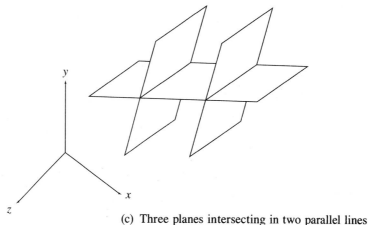

(c) Three planes intersecting in two parallel lines

Fig. 8.3   Three distinct planes.

$M = N$, then a unique solution will exist if none of the equations represent parallel hyperplanes. If $M > N$, then the system is overspecified and a unique solution does not exist. The set of equations is also called a system of equations. A system with a unique solution is called a **nonsingular** system of equations, and a system with no unique solution is called a **singular** set of equations.

    In many engineering problems, we are interested in determining if a common solution exists to a system of equations. If the common solution exists, then we want to determine it. In the next section, we discuss two methods for solving a system of equations using MATLAB.

nonsingular
singular

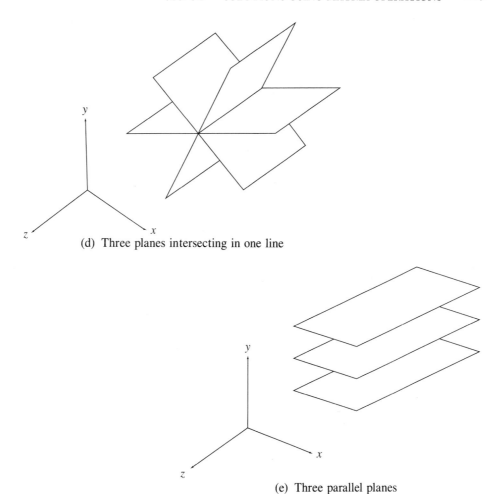

(d) Three planes intersecting in one line

(e) Three parallel planes

Fig. 8.3   (*cont.*)

8.2   SOLUTIONS USING MATRIX OPERATIONS

Consider the following system of three equations with three unknowns:

$$\begin{bmatrix} 3x & +2y & -z = & 10 \\ -x & +3y & +2z = & 5 \\ x & -y & -z = & -1 \end{bmatrix}$$

We can rewrite this system of equations using the following matrices:

$$A = \begin{bmatrix} 3 & 2 & -1 \\ -1 & 3 & 2 \\ 1 & -1 & -1 \end{bmatrix} \qquad X = \begin{bmatrix} x \\ y \\ z \end{bmatrix} \qquad B = \begin{bmatrix} 10 \\ 5 \\ -1 \end{bmatrix}$$

AX = B

Using matrix multiplication (review Section 6.1 if necessary), the system of equations can then be written in this form:

$$AX = B$$

Go through the multiplication to convince yourself that this matrix equation yields the original set of equations.

If we use a different letter for each variable, the notation becomes cumbersome when the number of variables is larger than three. Therefore, we modify our notation so that the variables are designated as x_1, x_2, x_3, and so on. If we rewrite the initial set of equations using this notation, we have:

$$
\begin{bmatrix}
3x_1 & +2x_2 & - x_3 = & 10 \\
- x_1 & +3x_2 & +2x_3 = & 5 \\
x_1 & - x_2 & - x_3 = & -1
\end{bmatrix}
$$

This set of equations is then represented by the matrix equation $AX = B$, where X is the column vector $[x_1, x_2, x_3]^T$.

The matrix equation $AX = B$ is generally used to express a system of equations; however, a system of equations can also be expressed using row vectors for B and X. For example, consider the set of equations used in the previous discussion:

$$
\begin{bmatrix}
3x_1 & +2x_2 & - x_3 = & 10 \\
- x_1 & +3x_2 & +2x_3 = & 5 \\
x_1 & - x_2 & - x_3 = & -1
\end{bmatrix}
$$

XA = B

We can write this set of equations as XA = B if X, A and B are defined as follows:

$$
X = \begin{bmatrix} x_1 & x_2 & x_3 \end{bmatrix} \qquad A = \begin{bmatrix} 3 & -1 & 1 \\ 2 & 3 & -1 \\ -1 & 2 & -1 \end{bmatrix} \qquad B = \begin{bmatrix} 10 & 5 & -1 \end{bmatrix}
$$

Again, go through the multiplication to convince yourself that the matrix equation generates the original set of equations. (Note that the matrix A in this equation is the transpose of the matrix A in the original matrix equation.)

We now present two methods for solving a system of N equations with N unknowns. While MATLAB can be used to solve an underspecified or an overspecified system of equations, the result represents a least squares solution and is not discussed here. We will assume that the system to be solved includes N equations with N unknowns.

Matrix Division

In MATLAB, a system of simultaneous equations can be solved using matrix division. The solution to the matrix equation $AX = B$ can be computed using matrix left division, as in A\B; the solution to the matrix equation $XA = B$ can be computed using matrix right division, as in B/A. (MATLAB uses a Gauss elimination numerical technique to perform both left and right matrix division. For further information on Gauss elimination, refer to a numerical methods text.)

For example, we can define and solve the system of equations in our previous example using the matrix equation $AX = B$ as shown in these statements.:

```
A = [3,2,-1;  -1,  3,  2;  1,  -1,  -1];
B = [10,  5,  -1]';
X = A\B;
```

The vector X then contains the following values: -2, 5, 6. To confirm that the values of X do indeed solve each equation, we can multiply A by X using the expression A*X. The result is a column vector containing the values 10, 5, -1.

We can also define and solve the same system of equations using the matrix equation $XA = B$ as shown in these statements.:

```
A = [3,  -1,1;  2 3 -1;  -1,  2 -1];
B = [10,  5,  -1];
X = B/A;
```

The vector X then contains the following values: -2, 5, 6. To confirm that the values of X do indeed solve each equation, we can multiply X by A using the expression X*A. The result is a row vector containing the values 10, 5, -1.

If a set of equations is singular, an error message is displayed; the solution vector may contain values of NaN or $+\infty$ or $-\infty$, depending on the values of the matrices A and B. It is also possible that a set of equations defines a system containing some equations that describe hyperplanes very close to the same hyperplane, or are very close to being parallel hyperplanes. These systems are called ill-conditioned systems. MATLAB will compute a solution, but a warning message is printed indicating that the results may be inaccurate.

Matrix Inverse

A system of equations can also be solved using the inverse of a matrix. For example, assume that A, X, and B are the matrices defined earlier:

$$A = \begin{bmatrix} 3 & 2 & -1 \\ -1 & 3 & 2 \\ 1 & -1 & -1 \end{bmatrix} \qquad X = \begin{bmatrix} x_1 \\ x_2 \\ x_3 \end{bmatrix} \qquad B = \begin{bmatrix} 10 \\ 5 \\ -1 \end{bmatrix}$$

Then, $AX = B$. Suppose that we multiply both sides of this matrix equation by A^{-1}, as in:

$$A^{-1}AX = A^{-1}B$$

But since $A^{-1}A$ is equal to the identity matrix I, we have:

$$IX = A^{-1}B$$

or

$$X = A^{-1}B$$

In MATLAB, we can compute this solution using the following command:

X = inv (A) *B

This solution is computed using techniques different from the solution using matrix left division, but both solutions will be the same for a system that is not ill-conditioned.

This same system of equations can also be solved using the inverse of a matrix if the system is expressed in the form $XA = B$, where

$$X = [x_1 \quad x_2 \quad x_3] \qquad A = \begin{bmatrix} 3 & -1 & 1 \\ 2 & 3 & -1 \\ -1 & 2 & -1 \end{bmatrix} \qquad B = [10 \quad 5 \quad -1]$$

If we multiply the right sides of the matrix equation by A^{-1}, we have:

$$XAA^{-1} = BA^{-1}$$

But since AA^{-1} is equal to the identity matrix I, we have:

$$XI = BA^{-1}$$

or

$$X = BA^{-1}$$

In MATLAB, we can compute this solution using the following command:

X = B*inv (A)

Be sure to note that B must be defined to be a row vector if this solution form is used.

Practice!

Solve the following systems of equations using matrix division and inverse matrices. Use MATLAB to verify that each solution solves the system of equations using matrix multiplication. For each system that contains equations with two variables, plot the equations on the same graph to show the intersection visually. If a system does not have a unique solution, give a graphical interpretation.

1. $\begin{bmatrix} -2x_1 & + x_2 = & -3 \\ x_1 & + x_2 = & 3 \end{bmatrix}$

2. $\begin{bmatrix} -2x_1 & + x_2 = & -3 \\ -2x_1 & + x_2 = & 1 \end{bmatrix}$

3. $\begin{bmatrix} -2x_1 & + & x_2 & = & -3 \\ -6x_1 & + & 3x_2 & = & -9 \end{bmatrix}$

4. $\begin{bmatrix} -2x_1 & + & x_2 & = & & -3 \\ -2x_1 & + & x_2 & = & -3.00001 \end{bmatrix}$

5. $\begin{bmatrix} 3x_1 & + & 2x_2 & - & x_3 & = & 10 \\ -x_1 & + & 3x_2 & + & 2x_3 & = & 5 \\ x_1 & - & x_2 & - & x_3 & = & -1 \end{bmatrix}$

6. $\begin{bmatrix} 3x_1 & + & 2x_2 & - & x_3 & = & 1 \\ -x_1 & + & 3x_2 & + & 2x_3 & = & 1 \\ x_1 & - & x_2 & - & x_3 & = & 1 \end{bmatrix}$

7. $\begin{bmatrix} 10x_1 & - & 7x_2 & + & 0x_3 & = & 7 \\ -3x_1 & + & 2x_2 & + & 6x_3 & = & 4 \\ 5x_1 & + & x_2 & + & 5x_3 & = & 6 \end{bmatrix}$

8. $\begin{bmatrix} x_1 & + & 4x_2 & - & x_3 & + & x_4 & = & 2 \\ 2x_1 & + & 7x_2 & + & x_3 & - & 2x_4 & = & 16 \\ x_1 & + & 4x_2 & - & x_3 & + & 2x_4 & = & 1 \\ 3x_1 & - & 10x_2 & - & 2x_3 & + & 5x_4 & = & -15 \end{bmatrix}$

✪ PROBLEM SOLVING APPLIED: ELECTRICAL CIRCUIT ANALYSIS

mesh loops

The analysis of an electrical circuit frequently involves finding the solution to a set of simultaneous equations. These equations are often derived using either current equations that describe the currents entering and leaving a node or using voltage equations that describe the voltages around mesh loops in the circuit. For example, consider the circuit shown in Figure 8.4. The three equations that de-

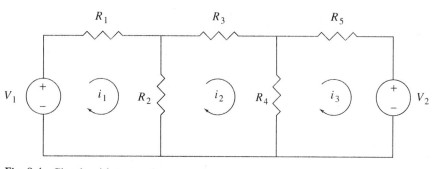

Fig. 8.4 Circuit with two voltage sources.

scribe the voltages around the three loops are the following:

$$\begin{bmatrix} -V_1 & + R_1 i_1 & + R_2(i_1 - i_2) & = 0 \\ R_2(i_2 - i_1) & + R_3 i_2 & + R_4(i_2 - i_3') & = 0 \\ R_4(i_3 - i_2) & + R_5 i_3 & + V_2 & = 0 \end{bmatrix}$$

If we assume that the values of the resistors $(R_1, R_2, R_3, R_4, R_5)$ and the voltage sources (V_1, V_2) are known, then the unknowns in the system of equations are the **mesh currents** (i_1, i_2, i_3). We can then rearrange the system of equations to the following form:

$$\begin{bmatrix} (R_1 + R_2)i_1 & - & R_2 i_2 & + & 0i_3 & = & V_1 \\ -R_2 i_1 & + (R_2 + R_3 + R_4)i_2 & - & R_4 i_3 & = & 0 \\ 0i_1 & - & R_4 i_2 & + (R_4 + R_5)i_3 & = -V_2 \end{bmatrix}$$

Write a MATLAB program that allows the user to enter the values of the five resistors and the values of the two voltage sources. The program should then compute the three mesh currents.

 1. PROBLEM STATEMENT
Using input values for the resistors and voltage sources, compute the three mesh currents in the circuit shown in Figure 8.4.

 2. INPUT/OUTPUT DESCRIPTION
Figure 8.5 contains a diagram illustrating the five resistor value inputs and the two voltage source inputs to the program. The output consists of the three mesh currents.

 3. HAND EXAMPLE
For a hand example, we use the following values:

$$R_1 = R_2 = R_3 = R_4 = R_5 = 1 \text{ ohm}$$
$$V_1 = V_2 = 5 \text{ volts}$$

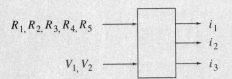

Fig. 8.5 Input/output diagram.

The corresponding set of equations is then:

$$\begin{bmatrix} 2i_1 & - & i_2 & + & 0i_3 & = & 5 \\ - & i_1 & + & 3i_2 & - & i_3 & = & 0 \\ 0i_1 & - & 1i_2 & + & 2i_3 & = & -5 \end{bmatrix}$$

We can use MATLAB to compute the solution using either matrix division or the inverse of a matrix. For this example, we use the following steps:

```
A = [ 2, -1, 0 ; -1, 3, -1 ; 0, -1, 2 ];
B = [ 5, 0, -5 ]';
X = A\B
ERR = SUM(A*X-B)
```

The values in the solution vector are printed, along with the sum of the values of AX-B, which should be zero if the system is nonsingular. For this example, the output is the following:

```
X =
     2.5000
          0
    -2.5000
ERR =
     0
```

 4. MATLAB SOLUTION

In the MATLAB solution we will ask the user to enter the circuit parameters. After solving the system of equations, we print the resulting mesh currents, which can then be used to compute the current or voltage at any point in the circuit. While we used the variable i to match the equation notation, we need to be aware that we could not then use i for complex number computations.

```
%
%          This program reads resistor and voltage values
%          and then computes the corresponding mesh
%          currents for a specified electrical circuit.
%
R = input('Enter resistor values in ohms, [R1...R5] ');
V = input('Enter voltage values in volts, [V1 V2] ');
%
%    Initialize matrix A and vector B using AX = B form.
%
```

```
A = [ R(1)+R(2),            -R(2),            0 ;
          -R(2), R(2)+R(3)+R(4),        -R(4) ;
              0,            -R(4), R(4)+R(5) ];
B = [ V(1) ;
          0 ;
       -V(2) ];
%
fprintf('Mesh Currents \n')
i = A\B
```

5. TESTING

The following program interaction verifies the data used in the hand example:

```
Enter resistor values in ohms, [R1...R5] [1 1 1 1 1]
Enter voltage values in volts, [V1 V2] [5 5]
Mesh Currents
i =
    2.5000
         0
   -2.5000
```

The next interaction computes the mesh currents using a different set of resistor values and voltage values.

```
Enter resistor values in ohms, [R1...R5] [2 8 6 6 4]
Enter voltage values in volts, [V1 V2] [40 20]
Mesh Currents
i =
    5.6000
    2.0000
   -0.8000
```

Verify this solution by multiplying the matrix A by the vector i.

SUMMARY

We began this chapter with a graphical interpretation of the solution of a set of simultaneous equations. We presented graphs illustrating the various cases that could occur with equations of two variables and with equations of three variables. We then extended the discussion to a discussion of N equations with N unknowns, assuming that each equation represented a hyperplane. Two methods for solving a system of N equations with N unknowns using matrix operations were presented. One method uses matrix division, and the other method uses the inverse of a matrix to solve the system of equations. The techniques are illustrated with an example that solves a system of simultaneous equations to determine mesh currents.

MATLAB SUMMARY

This MATLAB summary lists all the special symbols, commands, and functions that were defined in this chapter. A brief description is also included for each one.

Special Characters

\   matrix left division
/   matrix right division

PROBLEMS

Problems 1 to 3 relate to the engineering applications presented in this chapter. Problems 4 to 8 relate to new engineering applications.

Electrical Circuit   These problems relate to the electrical circuit analysis problem given in this chapter.

1. Modify the program so that it accepts the resistor values in kilo-ohms. Be sure to modify the rest of the program accordingly.

2. Modify the program so that the two voltage sources are constrained to always be the same value.

3. Modify the program so that the voltage sources are each 5 volts. Then assume that the resistor values are all equal. Compute the mesh currents for resistor values of 100, 200, 300, . . . , 1000 ohms.

Single Voltage Source Electrical Circuit   These problems present a system of equations generated by an electrical circuit with a single voltage source and five resistors.

4. The following set of equations defines the mesh currents in the circuit shown in Figure 8.6. Write a MATLAB program to compute the mesh currents using resistor values and voltage source value entered by the program user.

$$\begin{bmatrix} -V_1 & + R_2(i_1-i_2) & + R_4(i_1-i_3) & = 0 \\ R_1i_2 & + R_3(i_2-i_3) & + R_2(i_2-i_1) & = 0 \\ R_3(i_3-i_2) & + & R_5i_3 + R_4(i_3-i_1) & = 0 \end{bmatrix}$$

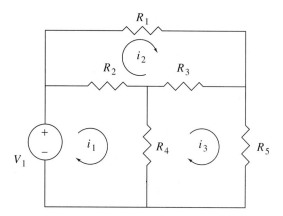

Fig. 8.6 Circuit with one voltage source.

5. Modify the program in problem 4 so that it prints the coefficients and con-
    stants for the system of equations that is being solved.

Simultaneous Equations Assume that a data file named eqns.dat contains the
coeffiecients for a set of simultaneous equations. Each line in the file contains the
coefficients and the constant for one equation; the data file contains data for N equations with N unknowns.

6. Write a program to read the data file eqns.dat. Determine if any of the
    equations represent parallel hyperplanes. (Recall that the equations for two
    parallel hyperplanes have the same coefficients but different constants.) Print
    the data for each group of lines that represent parallel hyperplanes.

7. Write a program to read the data file eqns.dat. Determine if any of the
    equations represent the same hyperplanes. (Recall that the equations for the
    same hyperplanes will be equal or multiples of each other.) Print the data for
    each group of lines that represent the same hyperplane.

8. Combine the steps in the previous two problems to generate a program that
    prints only the data for the equations from the data file that represent distinct
    hyperplanes. Also print the number of the equations that are left.

Inverse Matrix Solutions Use the inverse matrix to determine the solutions
to the following sets of linear equations. Describe each set using AX=B notation and XA=B notation, and compare solutions using both notational forms.
Use the transpose operator to convert the matrices from one form to the other.

9. $x + y + z + t = 4$
    $2x - y + t = 2$
    $3x + y - z - t = 2$
    $x - 2y - 3z + t = -3$

10. $2x + 3y + z + t = 1$
     $x - y - z + t = 1$
     $3x + y + z + 2t = 0$
     $-x + z - t = -2$

11. $x - 2y + z + t = 3$
     $x + z = t$
     $2y - z = t$
     $x + 4y + 2z - t = 1$

12. $x + 2y - w = 0$
     $3x + y + 4z + 2w = 3$
     $2x - 3y - z + 5w = 1$
     $x + 2z + 2w = -1$

Intersecting Hyperplanes   For each of the following points generate two different sets of simultaneous equations that intersect uniquely in the given point. Then solve each set of equations to verify that the solution is the one expected.

13. $[3, -5, 7]$

14. $[0, -2, 1.5, 5]$

15. $[1, 2, 3, -2, -1]$

9

Photo Courtesy of NASA

GRAND CHALLENGE: Nuclear Fusion

The collection of data is an important part of developing new scientific principles and theories. The challenges of nuclear fusion will require greater knowledge of topics such as nuclear stability and nuclear decay. Nuclear energy-level changes are associated with photons called gamma rays. To measure gamma rays from space, a Gamma Ray Observatory was released from the Atlantis space shuttle in 1991. The Gamma Ray Observatory is shown in this picture still in the grasp of the remote manipulator system that is used to move items into and out of the shuttle bay.

Interpolation and Curve Fitting

Introduction

In this chapter we assume that we have a set of data that has been collected from an experiment or from observing a physical phenomena. These data can generally be considered to be coordinates of points of a function $f(x)$. We would like to use these data points to determine estimates of the function $f(x)$ for values of x that were not part of the original set of data. For example, suppose that we have data points $(a, f(a))$ and $(c, f(c))$. If we want to estimate the value of $f(b)$, where $a < b < c$, we could assume that a straight line joined $f(a)$ and $f(c)$, and then use linear interpolation to obtain the value of $f(b)$. If we assume that the points $f(a)$ and $f(c)$ are joined by a cubic (third-degree) polynomial, we would use a cubic spline interpolation method to obtain the value of $f(b)$. Most interpolation problems can be solved using one of these methods. Another type of similar engineering problem requires the computation of an equation for a function that is a "good fit" to the data points. In this type of problem, it is not necessary that the function actually go through all the given points, but it should be a "best fit" in some sense. Least squares methods provide a best fit in terms of minimizing the square of the distances between the given points and the function. We now provide further discussion and examples of interpolation and curve fitting.

9.1   INTERPOLATION

interpolation

In this section we present two types of interpolation: linear interpolation and cubic spline interpolation. In both techniques, we assume that we have a set of data points to represent a set of coordinates $(x, f(x))$. We further assume that we need to estimate a value $f(b)$ which is not one of the original data points, but $a < b < c$, so we know function values $f(a)$ and $f(c)$. In Figure 9.1, we show a set of six data points that have been connected with straight line segments and that have been connected with cubic degree polynomial segments. It should be clear that the values determined for the function between sample points depend on the type of interpolation that we select.

Linear Interpolation

One of the most common techniques for estimating data between two given data points is linear interpolation. Figure 9.2 shows a graph with two arbitrary data points. If we assume that the function between the two points can be estimated by a

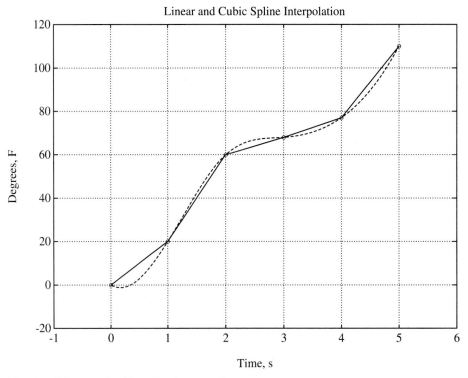

Fig. 9.1   Linear and cubic spline interpolation.

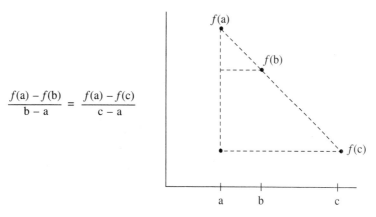

$$\frac{f(a) - f(b)}{b - a} = \frac{f(a) - f(c)}{c - a}$$

Fig. 9.2   Similar triangles.

linear
interpolation

straight line drawn between the two data values, then we can compute the function value at any point between the two data values using an equation derived from similar triangles. This general equation is

$$f(b) = f(a) + \frac{b - a}{c - a}(f(c) - f(a))$$

Given a set of data points, it is relatively straightforward to interpolate for a new point between two of the given points. However, we can use the interpolation equation only after we find the two values in our data between which the desired point falls. It is also possible that the data points do not include points between which the desired point falls. Linear interpolation is performed by the MATLAB interpolation functions table1 and table2.

table1

Table1 Function   The table1 function performs a one dimensional linear interpolation using a table of data. The first argument of the function refers to a table containing the data, which we will refer to as x–y data. The second argument refers to the value of x, for which we want to interpolate a corresponding y value. The function will search the first column of the table of data to find the two consecutive entries between which the desired value falls; the function then performs a linear interpolation for the corresponding function value y. Note that the data in the first column of the table must be in ascending order or in descending order. Also, the value of x for which we want to interpolate a corresponding y value must be between the first and last values in the first column of the table, or an error message will be printed.

   To illustrate the use of this function, we use the following set of temperature measurements taken from the cylinder head in a new engine that is being tested for possible use in a race car:

Time, s	Temperature, °F
0.0	0.0
1.0	20.0
2.0	60.0
3.0	68.0
4.0	77.0
5.0	110.0

We first store this information in a matrix, with the time data in the first column:

```
data1(:,1) = [0.0, 1.0, 2.0, 3.0, 4.0, 5.0]';
data1(:,2) = [0.0, 20.0, 60.0, 68.0, 77.0, 110.0]';
```

We can then use the table1 function to interpolate a temperature to correspond with any time between 0.0 and 5.0 seconds. For example, consider the following commands:

```
y1 = table1(data1,2.6);
y2 = table1(data1,4.9);
```

The corresponding values of y1 and y2 are 64.8 and 106.7.

If the first argument of the table1 function contains more than two columns, the function returns a row vector with a length one less than the width of the table, and each value returned will be interpolated from its corresponding column of data. Therefore, suppose that we measured temperatures at three points around the cylinder head in the engine, instead of at just one point, as in the previous example. The set of data is then the following:

Time, s	Temp1	Temp2	Temp3
0.0	0.0	0.0	0.0
1.0	20.0	25.0	52.0
2.0	60.0	62.0	90.0
3.0	68.0	67.0	91.0
4.0	77.0	82.0	93.0
5.0	110.0	103.0	96.0

We store this information in a matrix, with the time data in the first column:

```
data2(:,1) = [0.0, 1.0, 2.0, 3.0, 4.0, 5.0]';
data2(:,2) = [0.0, 20.0, 60.0, 68.0, 77.0, 110.0]';
```

```
data2(:,3) = [0.0, 25.0, 62.0, 67.0, 82.0, 103.0]';
data2(:,4) = [0.0, 52.0, 90.0, 91.0, 93.0, 96.0]';
```

To determine interpolated values of temperature at the three points in the engine at 2.6 seconds, we use the following command:

```
temps = table1(data2,2.6);
```

Temps is then a vector containing the three values 64.8, 65.0, 90.6.

table2

Table2 Function   The `table2` function performs a two-dimensional interpolation using values from the first column of a table and from the first row of the same table. Therefore, the first column and the first row must be monotonic (either increasing or decreasing) and the value for x and y must be within the limits of the table.

To illustrate the `table2` function, we extend the example from the previous section. Suppose that we start the engine and increase the speed of the engine (specified in revolutions per minute, or rpm) to a constant speed, while measuring the temperature at one point on the cylinder head. Thus, if we start the engine and increase its speed to 2,000 rpm in 5 seconds, we can collect a set of temperature values. If we start the engine and increase its speed to 3,000 rpm in 5 seconds, we can collect another set of temperature values. Similarly, we can continue collecting temperature values for a set of engine speeds. Assume that the data collected are shown in the following table:

	Engine Speed, rpm				
Time, s	2000	3000	4000	5000	6000
0	0	0	0	0	0
1	20	110	176	190	240
2	60	180	220	285	327
3	68	240	349	380	428
4	77	310	450	510	620
5	110	405	503	623	785

With these data, we can now estimate the temperature of the cylinder head at any time between 0 and 5 seconds, for any engine speed between 2,000 and 6,000 rpm. If we wanted to estimate the temperature of the cylinder head at 5 seconds for an engine speed of 3,500 rpm, we can see from looking at the table the value should be somewhere around 450 degrees. If we wanted to estimate the temperature of the cylinder head at 3.1 seconds, for an engine speed of 3,800 rpm, it becomes more difficult to determine an estimate without performing any calculations. Therefore, we now store this data in a matrix data3, and then use the `table2` function to perform this computation for us. Note that we fill the data by rows; in the previous examples we filled the matrices data1 and data2 by columns.

```
data3(1,:) = [0, 2000, 3000, 4000, 5000, 6000];
data3(2,:) = [0, 0, 0, 0, 0, 0];
data3(3,:) = [1, 20, 110, 176, 190, 240];
data3(4,:) = [2, 60, 180, 220, 285, 327];
data3(5,:) = [3, 68, 240, 349, 380, 428];
data3(6,:) = [4, 77, 310, 450, 510, 620];
data3(7,:) = [5, 110, 405, 503, 623, 785];
temp = table2(data3, 3.1, 3800)
```

The value computed is 336.68 °F.

Cubic Spline Interpolation

cubic spline

A cubic spline is a smooth curve constructed to go through a set of points. The curve between each pair of points is a third-degree polynomial that is computed to provide a smooth curve between the two points, and to provide a smooth transition from the third-degree polynomial between the previous two points. Refer to the cubic spline shown in Figure 9.1 that connects six points; a total of five different cubic equations are used to generate this smooth function that joins all six points.

spline

A cubic spline is computed in MATLAB with the spline function. The first two arguments contain the x and y coordinates of the points that define the spline. The third argument contains an x coordinate (or x vector) for the value (or values) that we want to determine from the cubic spline equations. The x values must be in ascending order or an error message is printed. To illustrate the use of the spline function, suppose that we want to use a cubic spline, instead of linear interpolation, to compute the temperature of the cylinder head at 2.6 seconds, we can use the following statements:

```
x = [0, 1, 2, 3, 4, 5];
y = [0.0, 20.0, 60.0, 68.0, 77.0, 110.0];
temp1 = spline(x, y, 2.6)
```

The value of temp1 is 67.3. If we want to use cubic spline interpolation to compute the temperatures at two different times, we can use this statement:

```
temp2 = spline(x, y, [2.6, 4.9])
```

The value of temp2 is [67.3, 105.2].

If we want to plot a cubic spline curve over a range of values, we can generate an x vector with the desired resolution of the curve, and then use the x vector as the third parameter in the spline function. For example, the following statements generated the curve in Figure 9.1:

```
x = [0, 1, 2, 3, 4, 5];
y = [0.0, 20.0, 60.0, 68.0, 77.0, 110.0];
newx = 0:0.1:5;
newy = spline(x,y,newx);
axis([-1,6,-20,120])
plot(x,y,newx,newy,x,y,'o'),...
title('Linear and Cubic Spline Interpolation'),...
xlabel('Time, s'),ylabel('Degrees, F'),grid
```

Note that the plot of x and y joined the coordinates with straight lines, thus representing the linear interpolation; the plot of newx and newy represented the cubic spline interpolation.

Practice!

Assume that we have the following set of data points:

Time, s	Temperature, °F
0.0	72.5
0.5	78.1
1.0	86.4
1.5	92.3
2.0	110.6
2.5	111.5
3.0	109.3
3.5	110.2
4.0	110.5
4.5	109.9
5.0	110.2

1. Generate a plot to compare connecting the temperature points with straight lines with connecting them with a cubic spline.

2. Use MATLAB to compute values for temperatures at the following times using linear interpolation:

    0.3, 1.25, 2.36, 4.48

3. Use MATLAB to compute values for temperatures at the following times using cubic spline interpolation:

    0.3, 1.25, 2.36, 4.48

4. Use MATLAB to compute time values that correspond to the following temperatures using linear interpolation:

81, 96, 100, 106

5. Use MATLAB to compute time values that correspond to the following temperatures using cubic spline interpolation:

81, 96, ,100, 106

 PROBLEM SOLVING APPLIED: ROBOT ARM MANIPULATORS

shuttle remote manipulator

The introductory picture for this chapter shows the shuttle remote manipulator system grasping the Gamma Ray Observatory (GRO) as it begins to move the GRO out of the shuttle bay. Manipulator systems such as this one and many others used on various types of robots use an advanced control system to guide the manipulator arm to desired locations. One of the requirements of such a control system is that it design a path for the arm to move from one location to another that involves a smooth path to avoid sharp jerks in the arm that might cause objects to slip out of its grasp or damage the object or the arm itself. Therefore, the path for the arm is initially designed in terms of a number of points over which the arm is to move. Then interpolation is used to design a smooth curve that will contain all the points. We will consider this problem assuming that the manipulator arm is moving in a plane, but a manipulator arm generally is moving in three-dimensional space instead of two-dimensional space.

An important part of developing an algorithm or solution to a problem is to consider carefully whether or not there are any special cases that need to be considered. In this problem we are assuming that the points over which the arm needs to pass are in a data file, and we would assume that the points are in the necessary order for the arm to move to a location to grasp an object, then move to the location to release the object, and then move back to the original position. We will also assume that intermediate points are included in the path to guide the arm away from obstructions or to guide it over sensors that are collecting information. Therefore, each point will have three coordinates—the x and y coordinates relative to the home position of the manipulator arm and a third coordinate coded as follows:

Code	Interpretation
0	Home position
1	Intermediate positions
2	Location of object to grasp
3	Location to release object

We then want to use a cubic spline to design the path of the manipulator arm to the object, then to the release point, and then back to the original position.

1. PROBLEM STATEMENT
Design a smooth curve using cubic spline interpolation that can be used to guide a manipulator arm to a location to grasp an object, then to another location to release the object, and then back to the original position.

 2. INPUT/OUTPUT DESCRIPTION
Figure 9.3 contains a diagram showing that the input is a file containing the *x-y* coordinates of the points over which the manipulator arm must pass. The output of the program is the smooth curve covering these points.

 3. HAND EXAMPLE
One of the functions of the hand example is to determine if there are any special cases that we must consider in developing the solution to the problem. Therefore, for a small example, we will consider a data file containing the following set of points for guiding the manipulator arm:

x	y	Code	Code Interpretation
0	0	0	Home position
2	4	1	Intermediate position
6	4	1	Intermediate position
7	6	2	Object grasp position
12	7	1	Intermediate position
15	1	3	Object release position
8	−1	1	Intermediate position
4	−2	1	Intermediate position
0	0	0	Home position

These points are shown connected by straight lines in Figure 9.4.

As we consider the steps in designing the cubic spline path using MATLAB, we will break the path into three paths: a path from the home position to the object grasp position, a path from the object grasp position to the object release position, and a path from the object release position to the home position. These three paths are chosen for two reasons. First, the manipulator arm

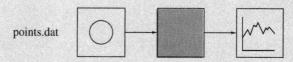

points.dat

Fig. 9.3   I/O diagram.

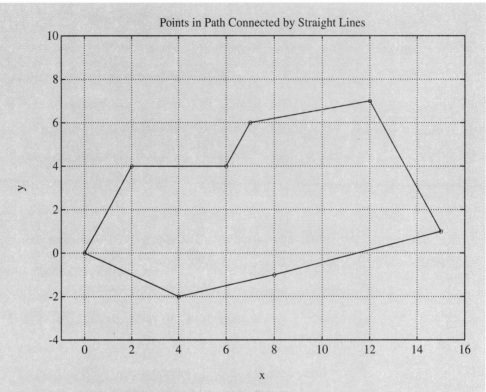

Fig. 9.4 Points in a path connected by straight lines.

must stop at the end of each of these three paths, so they really are three separate paths. Second, since the cubic spline function requires that the x coordinates be increasing, we cannot use a single cubic spline reference to compute a path that essentially goes in a circle. The x coordinates in each path must be in ascending order, but we will assume that that has already been checked by a preprocessing program, such as the ones discussed in the problems at the end of this chapter.

 4. MATLAB SOLUTION

The steps for separating the data file into the three separate paths is straightforward. However, it is not as straightforward to determine the number of points to use in the cubic spline interpolation. Since the coordinates could contain very large values and very small values, we decided to determine the minimum x distance between points in the overall path. We then computed the x increment for the cubic spline to be equal to that minimum distance divided by 10. Hence, there will be at least 10 points interpolated along the cubic spline between every pair of points, but there will be more points between most pairs of points.

```
%
%       This program reads a data file containing the
%       points for a path for a manipulator arm to
%       go to a location to grasp an object, then
%       move to another location to release the
%       object, and then move back to the start position.
%
load points.dat;
x = points(:,1);
y = points(:,2);
code = points(:,3);
%
%       Generate the three separate paths.
%
grasp = find(code==2);
release = find(code==3);
lenx = length(x);
x1 = x(1:grasp);          y1 = y(1:grasp);
x2 = x(grasp:release);    y2 = y(grasp:release);
x3 = x(release:lenx);     y3 = y(release:lenx);
%
%   Compute time sequences.
incr = min(abs(x(2:lenx)-x(1:lenx-1)))/10;
t1 = x(1):incr*sign(x(grasp)-x(1)):x(grasp);
t2 = x(grasp):incr*sign(x(release)-x(grasp)):x(release);
t3 = x(release):incr*sign(x(lenx)-x(release)):x(lenx);
%
%       Compute splines.
%
s1 = spline(x1,y1,t1);
s2 = spline(x2,y2,t2);
s3 = spline(x3,y3,t3);
%
%       Plot spline path.
%
axis([-1, 16, -4, 10])
plot([t1 t2 t3],[s1 s2 s3],[x1' x2' x3'],[y1' y2' y3'],...
  'o'),...
title('Path for Manipulator Arm'),...
xlabel('x'),ylabel('y'),grid
```

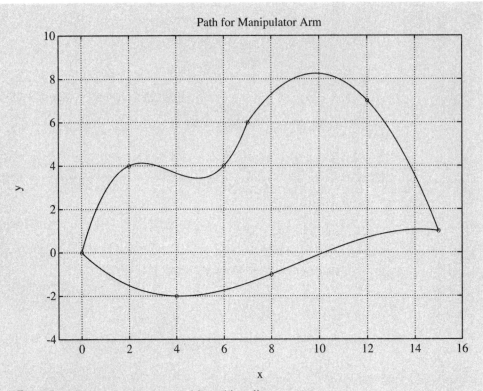

Fig. 9.5 Points in a path connected by cubic splines.

 5. TESTING

If we test this solution using the data file from the hand example, we obtain the cubic spline path shown in Figure 9.5.

9.2   LEAST SQUARES CURVE FITTING

least squares

best fit

Assume that we have a set of data points that were collected from an experiment, and after plotting the data points, it is clear that they generally fall in a straight line. However, if we try to draw a straight line through the points, only a couple of the points would probably fall on the line. A least squares curve fitting method could be used to find the straight line that was the closest to the points by minimizing the squared distance from each point to the straight line. While this line can be considered a "best fit" to the data points, it is possible that none of the points would actually fall on the best fit line. (Note that this is very different from interpolation because the curves used in linear interpolation and cubic spline interpolation actually contained all the data points.) In this section we first present a discussion on fitting a straight line to a set of data points, and then we discuss fitting a polynomial to a set of data points.

Linear Regression

squared distance

Linear regression is the name given to the process that determines the linear equation that is the best fit to a set of data points in terms of minimizing the sum of the squared distances between the line and the data points. To understand this process, we first consider the set of temperature values presented in the previous section that were collected from the cylinder head of a new engine. If we plot these data points, we find that they appear to be close to a straight line. If we plot the points, it appears that a good estimate of a line through the points is $y = 20x$. Figure 9.6 contains a plot of these points along with the linear estimate. The following commands were used to generate this plot:

```
x = [ 0, 1, 2, 3, 4, 5 ];
y = [ 0, 20, 60, 68, 77, 110 ];
y1 = 20*x;
axis([-1,6,-20,120])
```

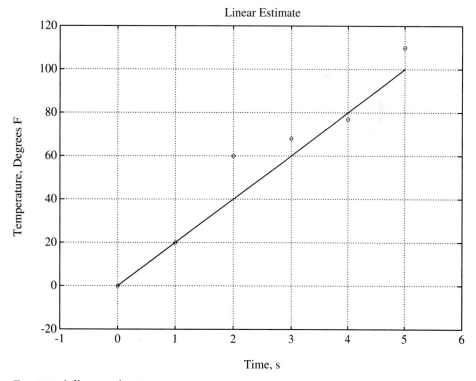

Fig. 9.6   A linear estimate.

```
plot(x,y1,x,y,'o'),title('Linear Estimate'),...
xlabel('Time, s'),ylabel('Temperature, Degrees F'), grid
```

To measure the quality of the fit of this linear estimate to the data, we first determine the distance from each point to the linear estimate; these distances are shown in Figure 9.7. The first two points fall exactly on the line, so d_1 and d_2 are zero. The value of d_3 is equal to $60 - 40$, or 20; the rest of the distances can be computed in a similar way. If we compute the sum of the distances, some of the positive and negative values would cancel each other and give a sum that is smaller than it should be. To avoid this problem we could add absolute values or squared values; linear regression uses squared values. Therefore, the measure of the quality of the fit of this linear estimate is the sum of the squared distances betwen the points and the linear estimates. This sum can be computed using MATLAB with the following commands, assuming that the previous statements have been executed that defined x, y, and y1:

linear regression
sum of squares

```
sum_sq = sum((y-y1).^2)
```

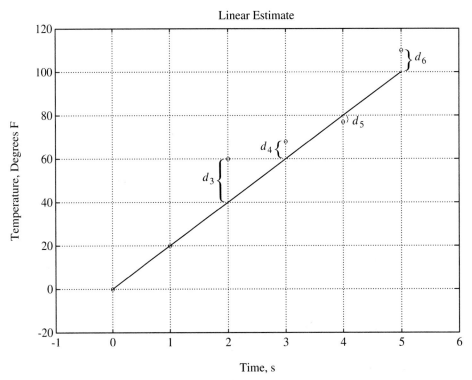

Fig. 9.7   Distances between points and linear estimate.

For this set of data, the value of sum_sq is 573.

If we drew another line through the points, we could compute the sum of squares that corresponds to this new line. Of the two lines, the best fit is provided by the line with the smaller sum of squared distances. To find the line with the smallest sum of squared distances, we can write an equation that computes the distances using a general linear equation $y = mx + b$. We then write an equation that represents the sum of the squared distances; this equation will have m and b as its variables. Using techniques from calculus, we can then compute the derivatives of the equation with respect to m and b, and set the derivatives equal to zero. The values of m and b that are determined in this way represent the straight line with the minimum sum of squared distances. These values can be computed using the polyfit function for a linear polynomial fit that has degree 1. The polyfit function is presented after a discussion on polynomial regression, but you can skip the section on polynomial regression if you are interested only in computing a linear regression. Figure 9.8 contains the set of data points with the linear equation that is a best fit in terms of least squares. The sum of squares of this best fit is 356.82.

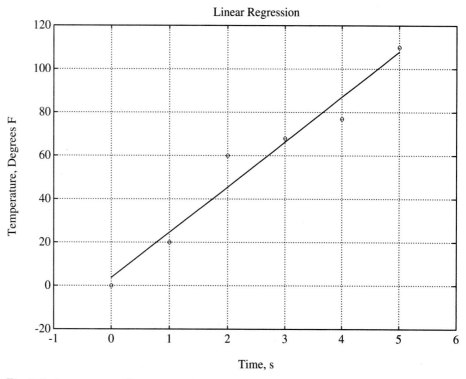

Fig. 9.8   Least squares linear regression.

Polynomial Regression

In the previous discussion we presented a technique for computing the linear equation that best fits a set of data. A similar technique can be developed using a single polynomial (not a set of polynomials as in a cubic spline) to fit the data by minimizing the distance of the polynomial from the data points. First, recall that a polynomial with one variable can be written in the following general formula:

general formula for a polynomial

$$f(x) = a_0 x^N + a_1 x^{N-1} + a_2 x^{N-2} + \cdots + a_{N-1} x + a_N$$

polynomial degree

The degree of a polynomial is equal to the largest value used as an exponent. Therefore, the general form for a cubic polynomial is

$$g(x) = a_0 x^3 + a_1 x^2 + a_2 x + a_3$$

Note that a linear equation is also a polynomial of degree one.

In Figure 9.9 we plot the original set of data points that we used in the linear regression example (and also in the linear interpolation and cubic spline interpolation); we also plot the best fit for polynomials of degree 2 through degree 9. Note that as the degree of the polynomial increases, the number of points that fall on the curve also increase. If a set of N points are used to determine an Nth-degree polyno-

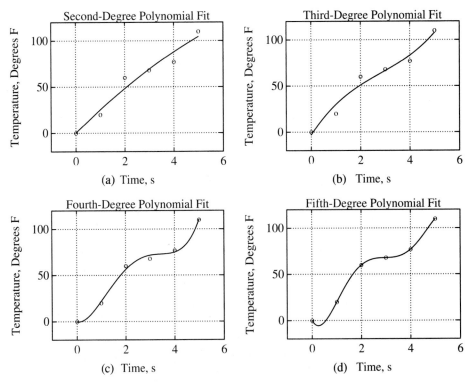

Fig. 9.9 Polynomial fits.

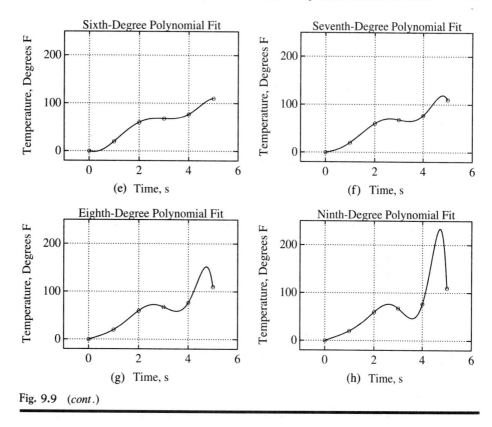

Fig. 9.9   (*cont.*)

mial, all *N* points will fall on the polynomial. Be cautious using high-degree polynomials because they can have large variations between the data points, which can cause erroneous estimates for values in these intervals.

Polyfit and Polyval Functions

polyfit

The MATLAB function for computing the best fit to a set of data with a polynomial with a specified degree is the `polyfit` function. This function has three arguments: the x and y coordinates of the data points and the degree N of the polynomial. The function returns the coefficients, in descending powers of x, of the Nth degree polynomial that fits the vectors x and y. (Note that there are N + 1 coefficients for an Nth degree polynomial.)

The best linear fit for the example data, plotted in Figure 9.8, had a least squares sum of 356.82. The plot and the calculation of this least squares sum were performed by the following statements:

```
x = [ 0, 1, 2, 3, 4, 5];
y = [ 0, 20, 60, 68, 77, 110 ];
coef = polyfit(x,y,1);
```

```
m = coef(1);
b = coef(2);
ybest = m*x + b;
sum_sq = sum((y - ybest).^2);
axis([-1,6,-20,120])
plot(x,ybest,x,y,'o'),title('Linear Regression'),...
xlabel('Time, s'),ylabel('Temperature, Degrees F'),grid
```

polyval

The polyval function is used to evaluate the least squares polynomial at a set of data points. The first argument of the polyval function is the coefficents of the polynomial, and the second argument is the vector of x values for which we want polynomial values. In the previous example, we computed the points of the linear regression using values from the coefficients; we could also have computed them using the polyval function as shown in this statement:

```
ybest = polyval(coef,x);
```

We now use the same data that were used for the linear regression, and fit polynomials of degree 2 through degree 9. From the previous discussion on polynomial fits, we would expect that the lower-degree polynomials would not contain all the data points, but that the sixth-degree polynomial would contain all data points. It is interesting to observe that the polynomials of degree 6 through degree 9 contain all the points, and thus have a least squares sum of zero, although the lower-degree polynomials are actually better fits to the general trend of the data than are the higher-degree polynomials. The statements used to compute the points of the polynomials that are plotted in Figure 9.9 are the following:

```
x = [ 0, 1, 2, 3, 4, 5 ];
y = [ 0, 20, 60, 68, 77, 110 ];
newx = 0:0.05:5;
for n=1:9;
    f(:,n) = polyval(polyfit(x,y,n),newx)';
end
```

The corresponding plot statements are then of the following form:

```
plot(newx,f(:,2),x,y,'o')
```

Additional statements are necessary to set the axis limits, define the subplot, and label each plot.

SUMMARY

In this chapter we explained the difference between interpolation and least squares curve fitting. Two types of interpolation were presented: linear interpolation and cubic spline interpolation. After presenting the MATLAB commands for performing

these types of interpolations, we then turned to least squares curve fitting using polynomials. This discussion included determining the best fit to a set of data using a
polynomial with a specified degree, and then using the best fit polynomial to generate new values of the function.

MATLAB SUMMARY

This MATLAB summary lists all the special symbols, commands, and functions that
were defined in this chapter. A brief description is also included for each one.

Commands and Functions

polyfit	computes a least squares polynomial
polyval	evaluates a polynomial
spline	computes cubic spline interpolation
table1	computes one-dimensional linear interpolation
table2	computes two-dimensional linear interpolation

PROBLEMS

Problems 1 to 7 relate to the engineering applications presented in this chapter.
Problems 8 to 11 relate to new engineering applications.

Robot Arm Manipulator   These problems relate to the robot arm manipulator
problem in this chapter. Example coordinate files are included in the diskette that
accompanies this text, and have filenames such as msn1.dat, msn2.dat, and so
on, which stand for various missions assigned to the robot arm manipulator.

1. Write a program to perform the initial testing of the data values in
    points.dat to be sure that the x-coordinates in each of the three paths are
    strictly increasing or strictly decreasing. If errors are detected, identify the
    path in which they occurred and list the values in that path.

2. Assume that the data file points.dat can include more that three paths. For
    example, the paths might describe moving several objects to new places in the
    cargo bay. Write a program that will count the number of individual paths,
    where a path ends with either grasping an object, releasing an object, or re-
    turning to the home position.

3. Modify the program written to perform the cubic spline interpolation so that it
    prints the interpolated data points for the complete set of paths to an output file
    named paths.dat. Remove any repeated points. Be sure to include the
    proper codes with the data points.

4. Write a program to read the file paths.dat described in problem 3. Plot the
    entire set of paths and then insert circles at point at which the robot arm stops.

(The robot arm stops to grasp an object, or to release an object, or when it reaches the home position.)

5. Modify the program written in problem 1 so that it assumes that the file might contain multiple paths as described in problem 2.

6. Instead of assuming that an error occurs if the x values are not strictly increasing or strictly decreasing within a path, assume that when this occurs, the path is to be separated into subpaths. For example, if the x coordinates from the home position to the grasping position are [0,1,3,6,3,7], then the subpaths would have x coordinates of [0,1,3,6], [6,3], and [3,7]. Write a program to read the file input.dat that contains the x and y coordinates along with the codes for the paths. Your program should generate a new file points.dat that contain the coordinates and codes to accompany the new set of paths.

7. In problem 6, we were able to separate a path with x-coordinates such as [0,1,3,6,3,7] into subpaths. Suppose the path that we are considering has the following set of coordinates: (0,0), (3,2), (3,6), (6,9). The x coordinates have positions [0,3,3,6]. Breaking this path into subpaths as in problem 3 does not solve the problem, because the path is moving in a vertical direction. To solve this problem, we need to insert a new point between the points that are vertical with each other. Assume that the new point is to be midway between the points with an x coordinate that is 5% greater than the x coordinate of the vertical points. Write a program that reads a file points.dat and uses this method to convert it into subpaths where necessary, and store the complete new set of paths in a file named final.dat.

Oil Well Production   Assume that we are trying to determine how the production of an operating oil well is related to temperature. Therefore, we have collected a set of data that contains the average oil well production in barrels of oil per day along with the average temperature for the day. This set of data is stored in an ASCII file named oil.dat that is contained in the diskette at the back of this text.

8. Since the data are not ordered by oil production or by temperature, they will need to be reordered. Write a program to read the data and generate two new data files. The file oiltmp.dat should contain the data with the oil in ascending order, with the corresponding temperatures. The file tmpoil.dat should contain the data with the temperatures in ascending order, with the corresponding oil productions. If there are points with the same x coordinates, the output file should contain only one point with the x coordinate; the corresponding y coordinate should be the average of the y coordinates with the same x value.

9. Write a program to plot the data from the file tmpoil.dat along with second-degree and third-degree polynomial approximations to the data. Print the polynomial models and the least squares errors.

10. Write a program to plot the data from the file oiltmp.dat along with second-degree and third-degree polynomial approximation to the data. Print the polynomial models and the least squares errors.

11. Assume that a third-degree polynomial approximation is going to be used to model the oil well production in terms of the temperature. Write a program that will allow the user to enter a temperature, and the program will print the predicted production in barrels per day.

10

GRAND CHALLENGE: Weather Prediction

Weather balloons are used to collect data from the upper atmosphere to use in developing weather models. These balloons are filled with helium and rise to an equilibrium point where the difference between the densities of the helium inside the balloon and the air outside the balloon is just enough to support the weight of the balloon. During the day, the sun warms the balloon, causing it to rise to a new equilibrium point; in the evening, the balloon cools, and it descends to a lower altitude. The balloon can be used to measure the temperature, pressure, humidity, chemical concentrations, or other properties of the air around the balloon. A weather balloon may stay aloft for only a few hours or as long as several years collecting environmental data. The balloon falls back to earth as the helium leaks out or is released.

Polynomial Analysis

Introduction

This chapter presents a number of MATLAB functions for analyzing polynomials. We first discuss ways to evaluate polynomials and how to perform computations using polynomials. An application is then developed that uses polynomials to model the altitude and velocity of a weather balloon. Following the application, we define roots of a polynomial and illustrate the functions for determining the roots of a polynomial and for generating the polynomial given its roots.

10.1   COMPUTATIONS WITH POLYNOMIALS

A polynomial is a function of a single variable that can be expressed in the following general form:

general form

$$f(x) = a_0x^N + a_1x^{N-1} + a_2x^{N-2} + \cdots + a_{N-2}x^2 + a_{N-1}x + a_N$$

degree

where the variable is x and the coefficients are represented by the values of a_0, a_1, and so on. The degree of a polynomial is equal to the largest value used as an exponent. Therefore, the general form for a cubic (degree 3) polynomial is

$$g(x) = a_0x^3 + a_1x^2 + a_2x + a_3$$

and a specific example of a cubic polynomial is

$$h(x) = x^3 - 2x^2 + 0.5x - 6.5.$$

Note that the sum of the coefficient subscript and the variable exponent is equal to the polynomial degree.

Polynomials commonly occur in engineering and science applications because they are often good models to represent physical systems. In this section we discuss polynomial evaluation and polynomial computations. If you are interested in modeling a set of data using a polynomial model, refer to Chapter 9, which discusses curve fitting.

Evaluation

There are several ways to evaluate a polynomial for a set of x values using MATLAB. For example, assume that we are interested in evaluating the following polynomial:

$$f(x) = 3x^4 - 0.5x^3 + x - 5.2$$

If we want to evaluate this function for a scalar value that is stored in x, we can use scalar operations as shown in this command:

```
f = 3*x^4 - 0.5*x^3 + x - 5.2;
```

If x is a vector or a matrix, then we need to specify array or element-by-element operations:

```
f = 3*x.^4 - 0.5*x.^3 + x - 5.2;
```

The size of the matrix f will be the same as the matrix x.

polyval function

Polynomials can also be evaluated using the `polyval` function, which has two arguments. The first argument contains the coefficients of the polynomial and the second argument contains the matrix of values for which we want to evaluate the polynomial. Thus, the following commands can be used to evaluate the polynomial discussed in the previous paragraph, with the vector a containing the coefficients of the polynomial:

```
a = [3,-0.5,0,1,-5.2];
f = polyval(a,x);
```

These commands could also be combined into one command:

```
f = polyval([3,-0.5,0,1,-5.2],x);
```

The size of f will be equal to x, which can be a scalar, a vector, or a matrix.

Suppose that we want to evaluate the following polynomial over the interval [0,5]:

$$g(x) = -x^5 + 3x^3 - 2.5x^2 - 2.5$$

The following polyval reference will generate 201 points of the polynomial over the desired interval:

```
x = 0:5/200:5;
a = [-1,0,3,-2.5,0,-2.5];
g = polyval(a,x);
```

polyvalm
function

When x is a vector or a scalar, polyval evaluates the polynomial using element-by-element operations as described in the previous example. A polynomial can also be evaluated as a matrix operation using polyvalm. Its usage is the same as polyval, as in:

```
f = polyvalm(a,x);
```

Since this function will be multiplying matrices (instead of performing element-by-element operations), the matrix x is required to be a square ($n \times n$) matrix. Also, the scalar term in the polynomial (such as the term -2.5 in the previous polynomial), will be evaluated by polyvalm as an identity matrix multiplied by the scalar value.

Arithmetic Operations

If we assume that the coefficients of two polynomials are stored in row vectors a and b, we can then perform polynomial computations using the vectors a and b. For example, to add polynomials, we add the coefficients of like terms. Therefore, the coefficients of the sum of two polynomials is the sum of the coefficients of the two polynomials. Note that the vectors containing the polynomial coefficients must be the same size in order to add them. To illustrate, suppose that we want to perform the following polynomial addition:

$$g(x) = x^4 - 3x^2 - x + 2.4$$
$$h(x) = 4x^3 - 2x^2 + 5x - 16$$
$$s(x) = g(x) + h(x)$$

The MATLAB statements to perform this polynomial addition are

```
g = [1,0,-3,-1,2.4];
h = [0,4,-2,5,-16];
s = g + h;
```

As expected, the value of s is $[1, 4, -5, 4, -13.6]$.

The coefficients of the polynomial that represents the difference between two polynomials can be computed similarly. The coefficient vector of the difference is computed by subtracting the two polynomial coefficient vectors. Again, the size of the two coefficient vectors would need to be the same.

A scalar multiple of a polynomial can be specified by multiplying the coefficient vector of the polynomial by the scalar. Thus, if we want to specify the following polynomial:

$$g(x) = 3f(x)$$

we can represent $g(x)$ by the coefficient matrix that is a scalar times the coefficient vector of $f(x)$. If $f(x) = 3x^2 - 6x + 1$, then the coefficient vector g for $g(x)$ can be computed as follows:

```
f = [3,-6,1];
g = 3*f;
```

The scalar can of course be positive or negative.

Multiplying two polynomials is more complicated than adding or subtracting two polynomials because a number of terms are generated and combined. Therefore, rather than develop the steps to do the multiplication and recombining of terms, we use the MATLAB function conv, which performs these steps for us. If a and b are row vectors containing coefficients of polynomials A and B, the coefficients of the product C of the two polynomials can be computed with the following statement:

conv function

```
c = conv(a,b);
```

There is not a requirement that the row vectors a and b be the same size for this function.

deconv function

Computing the quotient between two polynomials can also be a tedious and error-prone process, so we use the deconv function to perform polynomial division. The deconv function returns two vectors—one vector contains the coefficients of the quotient and the other vector contains the coefficients of the remainder polynomial. The conv and deconv functions are frequently used in digital signal processing, and are discussed further in Chapter 14.

To illustrate the use of the conv and deconv functions for polynomial multiplication and division, consider the following polynomial product:

$$g(x) = (3x^3 - 5x^2 + 6x - 2)(x^5 + 3x^4 - x^2 + 2.5)$$

We can multiply these polynomials using the conv function as shown below:

```
a = [3,-5,6,-2];
b = [1,3,0,-1,0,2.5];
g = conv(a,b);
```

The values in g are $[3,4,-9,13,-1,1.5,-10.5,15,-5]$, which represent the following polynomial:

$$g(x) = 3x^8 + 4x^7 - 9x^6 + 13x^5 - x^4 + 1.5x^3 - 10.5x^2 + 15x - 5$$

The deconv function performs polynomial division. We can illustrate this function using the previous polynomials:

$$h(x) = \frac{3x^8 + 4x^7 - 9x^6 + 13x^5 - x^4 + 1.5x^3 - 10.5x^2 + 15\,x - 5}{x^5 + 3x^4 - x^2 + 2.5}$$

This polynomial division is specified by these commands:

```
n = [3,4,-9,13,-1,1.5,-10.5,15,-5];
d = [1,3,0,-1,0,2.5];
[q,r] = deconv(n,d);
```

As expected, the quotient coefficient vector is $[3,-5,6,-2]$, which represents a quotient polynomial of $3x^3 - 5x^2 + 6x - 2$, and the remainder vector contains zeros.

Caution is necessary when using the deconv function to perform polynomial division. If the remainder polynomial coefficients are not all zero, then you should use polyval to evaluate the numerator and denominator polynomials, and then divide those results instead of using polyval to evaluate the quotient polynomial from the deconv function. To illustrate, in the example below, ans1 and ans2 will have the same value if the remainder vector r contains all zeros; otherwise, ans1 and ans2 do not represent the same polynomial.

```
[q,r] = deconv(n,d);
ans1 = polyval(q,x);
ans2 = polyval(n,x)./polyval(d,x);
```

A number of engineering applications require that the ratio or quotient of two polynomials be expressed as a sum of polynomial fractions. Techniques for partial-fraction expansions of a ratio of two polynomials is discussed in Chapter 14.

Practice!

Assume that the following polynomials have been given:

$$f_1(x) = x^3 - 3x^2 - x + 3$$
$$f_2(x) = x^3 - 6x^2 + 12x - 8$$
$$f_3(x) = x^3 - 8x^2 + 20x - 16$$
$$f_4(x) = x^3 - 5x^2 + 7x - 3$$
$$f_5(x) = x - 2$$

Plot each of the following functions over the interval [0,4]. Compare plots using element-by-element computations and then using MATLAB functions with polynomial coefficient vectors. Be careful not to lose accuracy in performing polynomial division.

1. $f_1(x)$

2. $f_2(x) - 2f_4(x)$

3. $3f_5(x) + f_2(x) - 2f_3(x)$

4. $f_1(x)*f_3(x)$

5. $f_4(x)/(x - 1)$

6. $f_1(x)*f_2(x)/f_5(x)$

7. $f_2(x)/(x - 1)$

PROBLEM SOLVING APPLIED: WEATHER BALLOONS

Weather balloons are used to gather temperature and pressure data at various altitudes in the atmosphere. The balloon rises because the density of the helium in the balloon is less than the density of the surrounding air outside the balloon. As the balloon rises, the surrounding air becomes less dense, and thus the balloon's ascent slows until it reaches a point of equilibrium. During the day, sunlight warms the helium trapped inside the balloon, which causes the helium to expand and become less dense and the balloon to rise higher. During the night, however, the helium in the balloon cools and becomes more dense, causing the balloon to descend to a lower altitude. The next day, the sun heats the air again and the balloon rises. Over time this process generates a set of altitude measurements that can be approximated with a polynomial equation.

Assume that the following polynomial represents the altitude in meters during the first 48 hours following the launch of a weather balloon:

$$h(t) = -0.12t^4 + 12t^3 - 380t^2 + 4100t + 220$$

where the units of t are hours. The corresponding polynomial model for the velocity in meters per hour of the weather balloon is the following:

$$v(t) = -0.48t^3 + 36t^2 - 760t + 4100$$

Generate plots of the altitude and the velocity for this weather balloon using units of meters and meters/sec. Also determine and print the peak altitude and its corresponding time.

 1. PROBLEM STATEMENT

Using the polynomials given, plot the altitude and velocity for a weather balloon. Also find the maximum altitude and its corresponding time.

 2. INPUT/OUTPUT DESCRIPTION

Figure 10.1 contains a diagram that shows that there is no external input to the program. The output consists of the two plots and the maximum altitude and corresponding time.

 3. HAND EXAMPLE

A hand example is not needed, since the program is only plotting data and then computing the maximum value from the data plotted. However, it is important to note that, since the units of the x axis are in hours, we will need to convert meters per second by replacing the time in hours with the time in seconds.

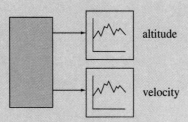

Fig. 10.1   I/O diagram.

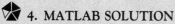 **4. MATLAB SOLUTION**

In the MATLAB solution, we use the `polyval` function to generate the data values to plot. The `max` function is then used to determine the maximum altitude and its corresponding position.

```
%
%      This program generates altitude and velocity
%      plots using polynomial models for the
%      altitude and velocity of a weather balloon.
%
t = 0:0.1:48;
alt_coef = [ -0.12 12 -380 4100 220 ];
vel_coef = [ -0.48 36 -760 4100 ];
alt = polyval(alt_coef,t);
vel = polyval(vel_coef,t)/3600;
%
subplot(211),plot(t,alt),...
title('Balloon Altitude'),...
xlabel('t, hours'),ylabel('meters'),grid,...
subplot(212),plot(t,vel),...
title('Balloon Velocity'),...
xlabel('t, hours'),ylabel('meters/sec'),grid, pause
%
[max_alt,k] = max(alt);
max_time = t(k);
fprintf('Maximum altitude: %8.2f  Time: %6.2f \n',...
   max_alt, max_time)
```

5. TESTING

The plots of the altitude and velocity are shown in Figure 10.2. The output from the program is shown below:

```
Maximum altitude: 17778.57  Time:   42.40
```

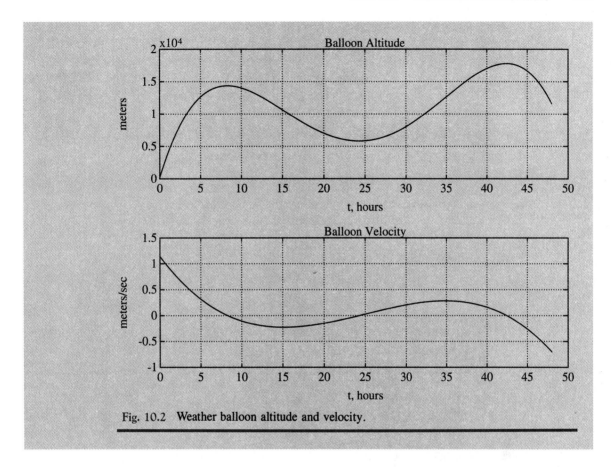

Fig. 10.2   Weather balloon altitude and velocity.

10.2   ROOTS OF POLYNOMIALS

The solution to many engineering problems involves finding the roots of an equation of the form

$$y = f(x)$$

roots

where the roots are the values of x for which y is equal to zero. Examples of applications in which we need to find roots of equations include designing the control system for a robot arm, designing springs and shock absorbers for an automobile, analyzing the response of a motor, and analyzing the stability of a digital filter.

multiple roots

If the function $f(x)$ is a polynomial of degree N, then $f(x)$ has exactly N roots. These N roots may contain multiple roots or complex roots, as will be shown in the following examples. If we assume that the coefficients $(a_0, a_1, \ldots)$ of the polyno-

complex
conjugate
roots

mial are real values, then any complex roots will always occur in complex conjugate pairs.

If a polynomial is factored into linear terms, it is easy to identify the roots of the polynomial by setting each term to zero. For example, consider the following equation:

$$f(x) = x^2 + x - 6$$
$$= (x - 2)(x + 3)$$

Then, if $f(x)$ is equal to zero, we have the following:

$$(x - 2)(x + 3) = 0$$

The roots of the equation, which are the values of x for which $f(x)$ is equal to zero, are then $x = 2$ and $x = -3$. The roots also correspond to the value of x where the polynomial crosses the x axis, as shown in Figure 10.3.

A cubic polynomial has the following general form:

$$f(x) = a_0 x^3 + a_1 x^2 + a_2 x + a_3$$

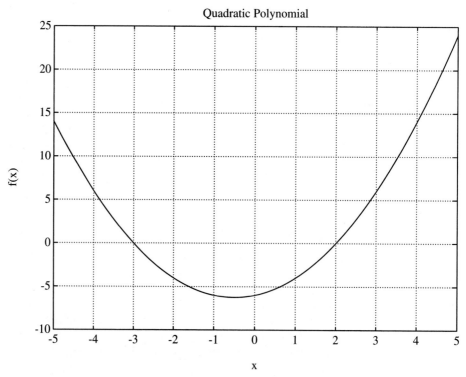

Fig. 10.3   Quadratic polynomial with 2 real roots.

Since the cubic polynomial has degree 3, it has exactly three roots. If we assume that the coefficients are real, then there are the following possibilities for the roots:

3 real distinct roots

3 real multiple roots

1 distinct real root and 2 multiple real roots

1 real root and a complex conjugate pair of roots

Examples of functions that illustrate these cases are

$$f_1(x) = (x - 3)(x + 1)(x - 1)$$
$$= x^3 - 3x^2 - x + 3$$
$$f_2(x) = (x - 2)^3$$
$$= x^3 - 6x^2 + 12x - 8$$
$$f_3(x) = (x + 4)(x - 2)^2$$
$$= x^3 - 12x + 16$$
$$f_4(x) = (x + 2)(x - (2+j))(x - (2-j))$$
$$= x^3 - 2x^2 - 3x + 10$$

Figure 10.4 contains plots of these functions. Note that the real roots correspond to the points where the function crosses the x axis.

It is relatively easy to determine the roots of polynomials of degree 1 or 2 by setting the polynomial equal to zero and then solving for x. If a second-degree polynomial cannot easily be factored, the quadratic equation can be used to solve for the two roots. For polynomials of degree 3 and higher, it can be difficult to determine the roots of the polynomials. A number of numerical techniques [4] exist for determining the roots of polynomials. Several techniques such as the incremental search, the bisection method, and the false-position technique identify the real roots by searching for points where the function changes sign because this indicates that the function has crossed the x axis. Additional techniques such as the Newton–Raphson method can be used to find complex roots.

roots
function

The MATLAB function for determining the roots of a polynomial is the roots function, which has as its input the coefficient row vector. The roots function then returns a column vector containing the roots of the polynomial; the number of roots is equal to the degree of the polynomial. To illustrate the use of this function, assume that we want to determine the roots of this polynomial:

$$f(x) = x^3 - 2x^2 - 3x + 10$$

The commands to compute and print the roots of this polynomial are

```
p = [1,-2,-3,10];
r = roots(p)
```

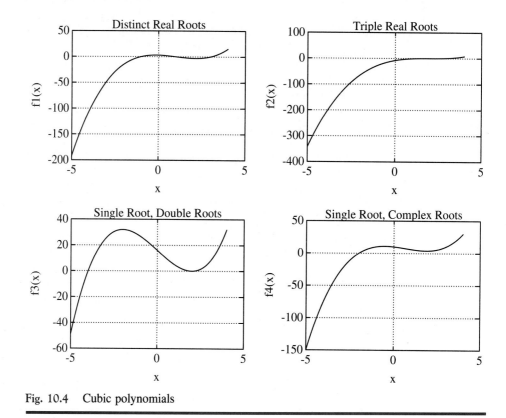

Fig. 10.4   Cubic polynomials

These two commands could also be combined into one command:

```
r = roots([1,-2,-3,10])
```

The values printed are $2 + i$, $2 - i$, and -2. We can verify that these values are roots by evaluating the polynomial at the roots and observing that the polynomial values are essentially zero:

```
polyval([1,-2,-3,10],r)
```

If we have the roots of a polynomial, and want to determine the coefficients of the polynomial when all the linear terms are multiplied, we can use the `poly` function. The function argument is a vector of the roots, and the output of the function is a row vector containing the polynomial coefficients. For example, we can compute the coefficients of the polynomial with roots -1, 1, 3 with the following statement:

```
a = poly([-1,1,3]);
```

The row vector a is equal to $[1, -3, -1, 3]$, as expected, since this is one of the example functions mentioned earlier in this section.

Practice!

Determine the roots for the following polynomials [4]. Then plot the polynomial over an appropriate interval to verify that the polynomial crosses the x axis at the root locations.

1. $g_1(x) = x^3 - 5x^2 + 2x + 8$

2. $g_2(x) = x^2 + 4x + 4$

3. $g_3(x) = x^2 - 2x + 2$

4. $g_4(x) = x^5 - 3x^4 - 11x^3 + 27x^2 + 10x - 24$

5. $g_5(x) = x^5 - 4x^4 - 9x^3 + 32x^2 + 28x - 48$

6. $g_6(x) = x^5 + 3x^4 - 4x^3 - 26x^2 - 40x - 24$

7. $g_7(x) = x^5 - 9x^4 + 35x^3 - 65x^2 + 64x - 26$

8. $g_8(x) = x^5 - 3x^4 + 4x^3 - 4x + 4$

SUMMARY

This chapter discussed polynomial analysis. We began with ways of evaluating polynomials, and then turned to evaluating expressions with polynomials. Computations with polynomials included scalar multiplication of a polynomial, addition and subtraction of polynomials, and then multiplication and division of polynomials. An important part of the analysis of a polynomial is the determination of the roots of the polynomial. We are also interested in determining the polynomial with a specified set of roots. Many of the analysis steps mentioned have special MATLAB functions to perform them, and these functions were discussed and contrasted with other solutions in MATLAB.

MATLAB SUMMARY

This MATLAB summary lists all the special symbols, commands, and functions that were defined in this chapter. A brief description is also included for each one.

Commands and Functions

conv	multiplies two polynomials
deconv	divides two polynomials
polyval	evaluates a polynomial
roots	determines the roots of a polynomial
poly	computes the coefficients of a polynomial

PROBLEMS

Problems 1 to 5 relate to the engineering application presented in this chapter. Problems 6 to 15 relate to general polynomial analysis.

Weather Balloons   These problems relate to the weather balloon problem given in this chapter.

1. Determine the time at which the weather balloon hits the ground based on the polynomial model for the altitude. (*Hint:* Consider the roots of the polynomial.)

2. Using the altitude data generated from the polynomial model, determine and print the periods of time during which the balloon is rising.

3. Using the altitude data generated from the polynomial model, determine and print the periods of time during which the balloon is falling.

4. The velocity of the weather balloon is equal to zero when it stops rising or stops falling. Determine the times at which the velocity of the weather balloon is zero using the polynomial model for the velocity.

5. In problem 4, we determined the points at which the velocity was zero, which corresponded to points where the balloon stops rising or stops falling. These points should correspond to the interval endpoints for periods of time during which the balloon was rising or falling. Compare the answers from problems 3 and 4 to the points at which the velocity was zero to see the relationship between the velocity and the rise and fall of the altitude.

Polynomial Analysis   Develop solutions using MATLAB for the following set of problems that analyze polynomials [7].

6. Assume that you have a row vector containing a set of polynomial coefficients. Determine and print the number of positive roots, the number of negative roots, and the number of complex roots.

7. Assume that you have a row vector containing a set of polynomial coefficients. Determine the polynomial coefficients for the polynomial that contains the same real roots, but none of the complex roots.

8. Assume that you have a row vector containing a set of polynomial coefficients. Determine the polynomials $A(x)$ and $B(x)$ that can be multiplied to give the original polynomial such that the roots of $A(x)$ are all real and the roots of $B(x)$ are all complex.

9. Find the value of k for which $x - 3$ is a factor of $kx^3 - 6x^2 + 2kx - 12$.

10. Find the value of k for which $x + 2$ is a factor of $3x^3 + 2kx^2 - 4x - 8$.

11. Generate a polynomial with integer coefficients having only the following set of roots. Plot the polynomial generated to demonstrate the root locations.
    (a) $1, 2, -3$
    (b) $2/3, -2, -1$
    (c) $\sqrt{5}, -\sqrt{5}, 4/3, 0$
    (d) $1/2, 2/3, 2i, -2i$

12. Generate a polynomial with a root of 3 plus the roots of the polynomial $2x^3 - 7x + 5$.

13. Given the polynomial coefficients for a polynomial $f(x)$, determine the polynomial coefficients for a polynomial $f(-x)$.

14. Given the polynomial coefficients for a polynomial $f(x)$, determine the number of variations in sign of the coefficients. For example, the number of variations in $2x^3 - 2x^2 + 2x - 3$ is three variations, while the number of variations in $x^2 + x - 2$ is one variation. (The number of variations is used in the solution to problem 15.)

15. Descartes' rule of signs states that if the polynomial coefficients are real, then the number of positive roots is not greater than the number of variations in sign and the number of negative roots is not greater than the number of variations in sign of the polynomial $f(-x)$. Given the polynomial coefficients for a polynomial, print the maximum number of positive roots and the maximum number of negative roots. (See problems 13 and 14.)

11

Courtesy of Chevron Corporation

GRAND CHALLENGE: Enhanced Oil and Gas Recovery

The design and construction of the Alaska pipeline presented numerous engineering challenges. One of the most important problems that had to be addressed was protecting the permafrost (the perennially frozen subsoil in arctic or subarctic regions) from the heat of the pipeline itself. The oil flowing in the pipeline is warmed by pumping stations and by friction from the walls of the pipe, enough so that the supports holding the pipeline have to be insulated or even cooled to keep them from melting the permafrost at their bases.

Numerical Integration
and Differentiation

Introduction

Integration and differentiation are the key concepts that are presented in calculus classes. These concepts are fundamental to solving a large number of engineering and science problems. While many of these problem solutions use analytical solutions derived using integration and differentiation, there are many problems that cannot be solved analytically and require numerical integration or numerical differentiation techniques to solve them. In this chapter we discuss numerical solutions to integration and differentiation, and then present MATLAB functions for computing these numerical solutions.

11.1   NUMERICAL INTEGRATION

integral

The integral of a function $f(x)$ over the interval $[a,b]$ is defined to be the area under the curve of $f(x)$ between a and b, as shown in Figure 11.1. If the value of this integral is K, then the notation to represent this integral of $f(x)$ between a and b is

$$K = \int_a^b f(x)\, dx$$

For many functions, this integral can be computed analytically. However, for a number of functions, this integral cannot be computed analytically and thus requires a numerical technique to estimate its value. The numerical evaluation of an integral

quadrature

is also called quadrature, which comes from an ancient geometrical problem.

The numerical integration techniques estimate the function $f(x)$ by another function $g(x)$, where $g(x)$ is chosen such that we can easily compute the area under $g(x)$. Then the better the estimate of $g(x)$ to $f(x)$, the better will be the estimate of the integral of $f(x)$. The two most common numerical integration techniques [4,8] estimate $f(x)$ with a set of piecewise linear functions or with a set of piecewise parabolic functions. If we estimate the function with piecewise linear functions, we can then compute the area of the trapezoids that compose the area under the piece-

trapezoidal rule

wise linear functions; this technique is called the trapezoidal rule. If we estimate the function with piecewise quadratic functions, we can then compute and add the areas

Simpson's rule

of these components; this technique is called Simpson's rule.

Trapezoidal Rule and Simpson's Rule

If the area under a curve is represented by trapezoids, and if the interval $[a,b]$ is divided into n equal sections, then the area can be approximated by the following formula (trapezoidal rule) [9]:

$$K_T = \frac{b - a}{2n}\left(f(x_0) + 2f(x_1) + 2f(x_2) + \cdots + 2f(x_{n-1}) + f(x_n)\right)$$

where the x_i values represent the end points of the trapezoids and where $x_0 = a$ and $x_n = b$.

If the area under a curve is represented by areas under quadratic sections of a curve, and if the interval $[a,b]$ is divided into $2n$ equal sections, then the area can be

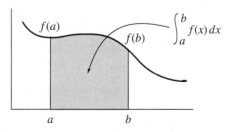

Fig. 11.1   Integral of $f(x)$ from a to b.

approximated by the following formula (Simpson's rule) [9]:

$$K_S = \frac{h}{3}(f(x_0) + 4f(x_1) + 2f(x_2) + 4f(x_3) + \cdots + 2f(x_{2n-2})$$

$$+ 4f(x_{2n-1}) + f(x_{2n}))$$

where the x_i values represent the end points of the sections and where $x_0 = a$ and $x_{2n} = b$, and $h = (b - a)/(2n)$.

where the x_i values represent the end points of the sections and where $x_0 = a$ and $x_{2n} = b$.

   If the piecewise components of the approximating function are higher-degree functions (trapezoidal rule uses linear functions and Simpson's rule uses quadratic functions), the integration techniques are referred to as Newton–Cotes integration techniques [9].

   The estimate of an integral improves as we use more components (such as trapezoids) to approximate the area under a curve. If we attempt to integrate a function with a singularity (a point at which the function or its derivatives are infinity or are not defined), we may not be able to get a satisfactory answer with a numerical integration technique.

singularity

MATLAB Quadrature Functions

quad, quad8
functions

MATLAB has two quadature functions for performing numerical function integration. The quad function uses an adaptive form of Simpson's rule, while quad8 uses an adaptive Newton–Cotes 8-panel rule. The quad8 function is better at handling functions with certain types of singularities, such as $\int_0^1 \sqrt{x}\, dx$. Both functions print a warning message if they detect a singularity, but an estimate of the integral is still returned. The quad and quad8 functions have the same types of arguments, so we will limit the following discussion to the quad function but recognize that the same argument options are available in quad8.

   The simplest form of the quad function requires three arguments. The first argument is the name (in quote marks) of the MATLAB function that returns a vector of values of $f(x)$ when given a vector of input values x; this function name can be the name of another MATLAB function such as sin, or it can be the name of a user-written MATLAB function. The second and third arguments are the integral limits a and b.

   To illustrate the quad function, we assume that we want to determine the integral of the square root function for nonnegative values of a and b:

$$K_Q = \int_a^b \sqrt{x}\, dx$$

The square root function $f(x) = \sqrt{x}$ is plotted in Figure 11.2 for the interval [0, 5]; the function values are complex for $x < 0$. This function can be integrated analytically to yield the following for nonnegative values of a and b:

$$K = \frac{2}{3}(b^{3/2} - a^{3/2})$$

To compare the results of the quad function with the analytical results for a user-specified interval, we use the following program:

```
%
%       This program compares the quad function with the
%       analytical results for the integration of the
%       square root of x over an interval [a,b], where
%       a and b are non-negative.
%
a = input('Enter left endpoint ≥ 0  ');
b = input('Enter right endpoint ≥ 0  ');
k = 2/3*(b^(1.5) - a^(1.5));
kq = quad('sqrt',a,b);
kq8 = quad('sqrt',a,b);
fprintf('Analytical: %f \n Numerical: %f %f ',k,kq,kq8)
```

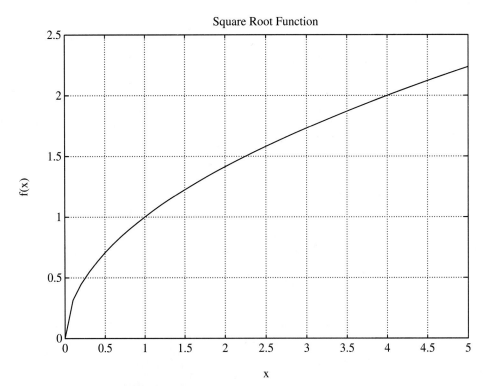

Fig. 11.2   Square root function.

This program was tested using several intervals, giving the following results:

Interval [0.5,0.6]

	Analytical:	0.074136	
	Numerical:	0.074136	0.074136

Interval [0,0.5]

	Analytical:	0.235702	
	Numerical:	0.235701	0.235702

Interval [0,1]

	Analytical:	0.666667	
	Numerical:	0.666663	0.666667

tolerance

The quad and quad8 functions can also include a fourth argument that represents a tolerance. The integration function continues to refine its estimate for the integration until the relative error is less than the tolerance:

$$\frac{\text{previous estimate} - \text{current estimate}}{\text{previous estimate}} < \text{tolerance}$$

If the tolerance is omitted, a default value of 0.001 is assumed. If an optional fifth argument is nonzero, a graph or trace is plotted containing a point plot of the function values used in the integration computations.

These integration techniques can handle some singularities that occur at one or the other interval end points, but they cannot handle singularities that occur within the interval. For these cases, you should consider dividing the interval into subintervals, and providing estimates of the singularities using other results such as L'Hopital's rule [10].

Practice!

Sketch the function $f(x) = |x|$, and indicate the areas specified by the following integrals. Then compute the integrals by hand, and compare your results to those generated by the quad function.

1. $\int_{0.5}^{0.6} |x|\, dx$
2. $\int_{.0}^{1} |x|\, dx$
3. $\int_{-1}^{-0.5} |x|\, dx$
4. $\int_{-1}^{0} |x|\, dx$
5. $\int_{-0.5}^{0.5} |x|\, dx$

PROBLEM SOLVING APPLIED: PIPELINE FLOW ANALYSIS

In this application, we address the flow of oil in a pipeline. However, the analysis of the flow of a liquid in a circular pipe has application in many different systems, including the veins and arteries in a body, the water system of city, the irrigation system for a farm, the piping system that transports fluids in a factor, the hydraulic lines of an aircraft, and the ink jet of a computer's printer [11].

The friction in a circular pipeline causes a "velocity profile" to develop in the flowing oil. Oil that is in contact with the walls of the pipe is not moving at all, while the oil at the center of the flow is moving the fastest. The diagram in Figure 11.3 shows how the velocity of the oil varies across the diameter of the pipe and defines the variables used in this analysis. The following equation describes this velocity profile:

$$v(r) = v_{max}\left(1 - \frac{r}{r_0}\right)^{1/n}$$

The variable n is an integer between 5 and 10 that defines the shape of the forward flow of the oil; in this case, the value of n for the diagram in Figure 11.3 is equal to 8. The average flow velocity of the pipe can be computed by integrating the velocity profile from zero to the pipe radius, r_0. Thus, we have

$$v_{ave} = \frac{\int_0^{r_0} v(r)2\pi r \, dr}{\pi r_0^2}$$

$$= \frac{2v_{max}}{r_0^2} \int_0^{r_0} r\left(1 - \frac{r}{r_0}\right)^{1/n} dr$$

The values of v_{max} and n can be measured experimentally, and the value of r_0 is the radius of the pipe. Write a MATLAB program to integrate the velocity profile to determine the average flow velocity of the pipe

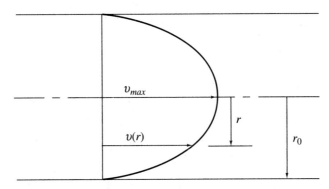

Fig. 11.3 Velocity profile in flowing oil.

1. PROBLEM STATEMENT
Compute the average flow velocity for a pipeline.

2. INPUT/OUTPUT DESCRIPTION
Figure 11.4 contains a diagram that shows that the output of the program is the value of the average flow velocity of the pipeline. The values of the maximum velocity v_{max}, the radius of the pipe r_0, and the value of n are specified as constants in the program.

3. HAND EXAMPLE
If we assume that the value of r_0 is $0.5m$ and that the value of n is 8, we can plot the function $r(1 - r/r_0)^{1/n}$, as shown in Figure 11.5. We can also compute an estimate to the integral of this function by summing the areas of the triangle and the rectangle shown in Figure 11.6. This estimate of the area is then

$$\text{area} = 0.5(0.4)(0.35) + (0.1)(0.35)$$
$$= 0.105$$

This area is then multiplied by the factor $2v_{max}/r_0^2$ to give the average flow velocity of the pipe. If we assume that v_{max} is $1.5\ m$, the average flow velocity is then approximately 1.260.

4. MATLAB SOLUTION
In the MATLAB solution, we use the quad function to evaluate the integral. One of the quad function parameters is the name of the function that computes values of the function to be numerically integrated, so we must also write an M-file that computes values of the function inside the integral. We will specify the values of v_{max}, r_0, and n as constants in both the program and

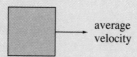

Fig. 11.4   I/O diagram.

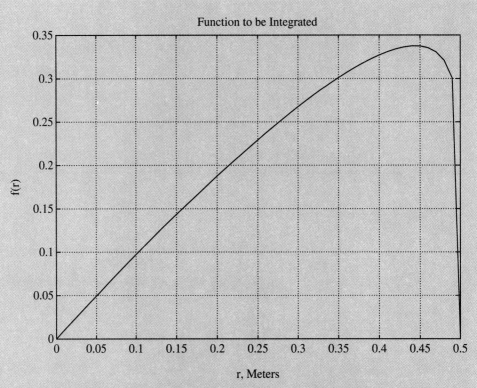

Fig. 11.5   Function related to average flow velocity.

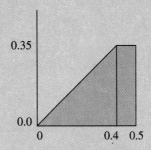

Fig. 11.6   Integral approximation.

the function. In the following function, we print the value of the integral so that we can compare it to the value determined in the hand calculation.

```
%
%  This program computes the value of the
%  average flow velocity for a pipeline
%  using numerical integration.
%
vmax = 1.5;
r0 = 0.5;
%
integral = quad('velocity',0,0.5)
%
ave_velocity = (2*vmax/(r0^2))*integral
```

The M-file that defines the function to be integrated in order to compute the average velocity is the following:

```
function v = velocity(r)
% VELOCITY   This function is related to the
%                  average flow velocity of the pipe.
%
r0 = 0.5;
n = 8;
%
v = r.*(1-r/r0).^(1/n);
```

5. TESTING

The output of the program is the following:

```
integral =
    0.1046
ave_velocity =
    1.2548
```

The value estimated in the hand example for the integral was 0.105, and the corresponding value estimated for the average velocity was 1.260.

11.2 NUMERICAL DIFFERENTIATION

derivative

The derivative of a function $f(x)$ is defined to be a function $f'(x)$ that is equal the rate of change of $f(x)$ with respect to x. The derivative can be expressed as a ratio, with the change in $f(x)$ indicated by $df(x)$ and the change in x indicated by dx, giving

$$f'(x) = \frac{df(x)}{dx}.$$

There are many physical processes in which we want to measure the rate of change of a variable. For example, velocity is the rate of change of position (as in meters per second) and acceleration is the rate of change of velocity (as in meters per second squared). It can also be shown that the integral of acceleration is velocity, and that the integral of velocity is position. Hence, integration and differentiation have a special relationship, in that they can be considered to be inverses of each other—the integral of a derivative returns the original function, and the derivative of an integral returns the original function, to within a constant value.

tangent line

The derivative $f'(x)$ can be described graphically as the slope of the function $f(x)$, where the slope of $f(x)$ is defined to be the slope of the tangent line to the function at the specified point. Thus, the value of $f'(x)$ at the point a is $f'(a)$, and it is equal to the slope of the tangent line at the point a, as shown in Figure 11.7.

critical points

Since the derivative of a function at a point is the slope of the tangent line at the point, a value of zero for the derivative of a function at the point x_k indicates that the line is horizontal at that point. Points with derivatives of zero are called critical points and can represent either a horizontal region of the function or a local maximum or a local minimum of the function. (The point may also be the global maximum or global minimum as shown in Figure 11.8, but more analysis of the entire function would be needed to determine this.) If we evaluate the derivative of a function at several points in an interval, and we observe that the sign of the derivative changes, then a local maximum or a local minimum occurs in the interval. The second derivative (the derivative of $f'(x)$) can be used to determine whether or not the critical points represent local maxima or local minima. More specifically, if the second derivative of a critical point is positive, then the function value at the point is a

local maximum,
local minimum

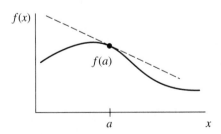

Fig. 11.7   Derivative of $f(x)$ at $x = a$.

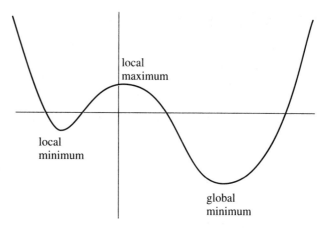

Fig. 11.8 Example of function with critical points.

local minimum; if the second derivative of a critical point is negative, then the function value at the point is a local maximum.

Difference Expresssions

Numerical differentiation techniques estimate the derivative of a function at a point x_k by approximating the slope of the tangent line at x_k using values of the function at points near x_k. The approximation of the slope of the tangent line can be done in several ways, as shown in Figure 11.9. Figure 11.9(a) assumes that the derivative at x_k is estimated by computing the slope of the line between $f(x_{k-1})$ and $f(x_k)$, as in

backward difference

$$f'(x_k) = \frac{f(x_k) - f(x_{k-1})}{x_k - x_{k-1}}$$

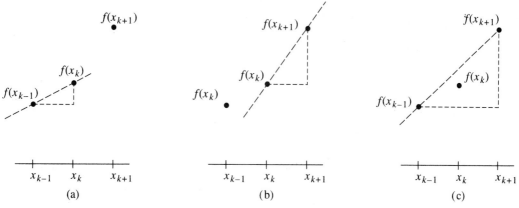

Fig. 11.9 Techniques for computing $f'(x_k)$.

This type of derivative approximation is called a backward difference approximation. Figure 11.9(b) assumes that the derivative at x_k is estimated by computing the slope of the line between $f(x_k)$ and $f(x_{k+1})$, as in

forward difference

$$f'(x_k) = \frac{f(x_{k+1}) - f(x_k)}{x_{k+1} - x_k}$$

This type of derivative approximation is called a forward difference approximation. Figure 11.9(c) assumes that the derivative at x_k is estimated by computing the slope of the line between $f(x_{k-1})$ and $f(x_{k+1})$, as in

central difference

$$f'(x_k) = \frac{f(x_{k+1}) - f(x_{k-1})}{x_{k+1} - x_{k-1}}$$

This type of derivative approximation is called a central difference approximation, and we usually assume that x_k is halfway between x_{k-1} and x_{k+1}. The quality of all of these types of derivative computations depends on the distance between the points used to estimate the derivative; the estimate of the derivative improves as the distance between the two points decreases.

second derivative

The second derivative of a function $f(x)$ is the derivative of the first derivative of the function

$$f''(x) = \frac{df'(x)}{dx}$$

This function can be evaluated using slopes of the first derivative. Thus, if we use backward differences, we have

$$f''(x_k) = \frac{f'(x_k) - f'(x_{k-1})}{x_k - x_{k-1}}$$

Similar expressions can be derived for computing estimates of higher derivatives.

DIFF Function

diff function

The `diff` function computes differences between adjacent values in a vector, generating a new vector with one less value. (If the `diff` function is applied to a matrix, it operates on the columns of the matrix as if each column were a vector.) For example, assume that the vector x contains the values [0,1,2,3,4,5] and that the vector y contains the values [2,3,1,5,8,10]. Then the vector generated by `diff (x)` is [1,1,1,1,1] and the vector generated by `diff (y)` is [1,−2,4,3,2]. The derivative dy is computed with `diff (y) . /diff (x)`. Note that these values of dy are correct for both the forward difference equation and the backward difference equation. The distinction between the two methods for computing the derivative is determined by the values of x, which correspond to the derivative dy. If the corresponding values of x are [1,2,3,4,5], then dy computes a backward difference; if the corresponding values of x are [0,1,2,3,4], then dy computes a forward difference.

Suppose that we have a function given by the following polynomial:

$$f(x) = x^5 - 3x^4 - 11x^3 + 27x^2 + 10x - 24$$

A plot of this function is shown in Figure 11.10. Assume that we want to compute the derivative of this function over the interval $[-4,5]$, using a backward difference equation. We can perform this operation using the diff function as shown in these equations, where df represents $f'(x)$ and xd represents the x values corresponding to the derivative

```
x = -4:0.1:5;
f = x.^5 - 3*x.^4 - 11*x.^3 + 27*x.^2 + 10*x - 24;
df = diff(f)./diff(x);
xd = x(2:length(x));
```

Figure 11.11 contains a plot of this derivative. Note that the zeros of the derivative correspond to the points of local minima or local maxima of this function; this function does not have a global minimum or a global maximum, because the function

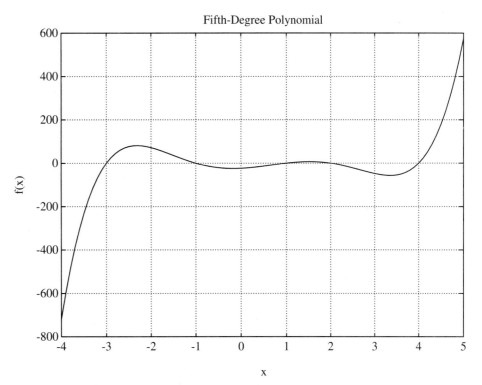

Fig. 11.10   Plot of a polynomial.

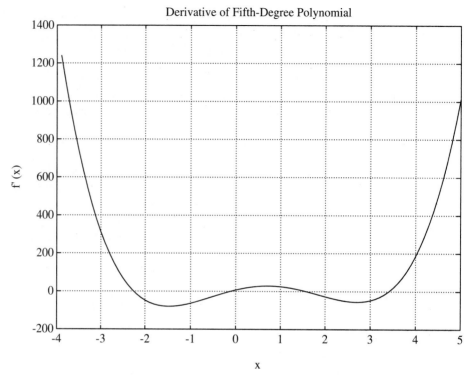

Fig. 11.11   Plot of the derivative of a polynomial.

ranges from $-\infty$ to $+\infty$. We can print the locations of the maxima and minima (which occur at -2.3, -0.2, 1.5, and 3.4) for this function with the following statements:

```
product = df(1:length(df)-1).*df(2:length(df));
local = xd(find(product < 0))
```

In this expression, the find function determines the indices k of the locations in product for which df(k) is equal to 0; these indices are then used with the xd vector to print the approximation to the locations of the maxima and minima.

To compute a central difference derivative using the x and f vectors, we could use the following statements:

```
numerator = f(3:length(f)) - f(1:length(f)-2);
denominator = x(3:length(x)) - x(1:length(x)-2);
dy = numerator./denominator;
xd = x(2:length(x)-1);
```

In the example discussed in this section, we assumed that we had the equation of the function to be differentiated, and thus we could generate points of the function. In many engineering problems, the data to be differentiated are collected from experiments, and thus we cannot choose the points to be close together to get a more accurate measure of the derivative. In these cases, it might be a good solution to use the techniques from Chapter 9 that allow us to determine an equation that fits a set of data and then compute points from the equation to use in computing values of the derivative.

Practice!

For each of the following, plot the function, its first derivative, and its second derivative over the interval $[-10,10]$. Then use MATLAB commands to print the locations of the local minima, followed on a separate line by the locations of the local maxima.

1. $g_1(x) = x^3 - 5x^2 + 2x + 8$

2. $g_2(x) = x^2 + 4x + 4$

3. $g_3(x) = x^2 - 2x + 2$

4. $g_4(x) = 10x - 24$

5. $g_5(x) = x^5 - 4x^4 - 9x^3 + 32x^2 + 28x - 48$

6. $g_6(x) = x^5 + 3x^4 - 4x^3 - 26x^2 - 40x - 24$

7. $g_7(x) = x^5 - 9x^4 + 35x^3 - 65x^2 + 64x - 26$

8. $g_8(x) = x^5 - 3x^4 + 4x^3 - 4x + 4$

SUMMARY

This chapter presented techniques for numerical integration and differentiation. Numerical integration techniques approximate the area under a curve, and numerical differentiation techniques approximate the slope of a curve. The MATLAB functions for integration are quad and quad8, which perform an iterative form of Simpson's rule and an iterative Newton–Cotes technique, respectively. Both functions require that the function to be integrated be a MATLAB function, which can be one of MATLAB's functions or a user-written function. The MATLAB function that can be used to compute the derivative of a function is the diff function, which computes differences between adjacent elements of a vector. To compute the derivative of a function f with respect to x, two references to the `diff` function are required, as in `diff(f)./diff(x)`.

MATLAB SUMMARY

This MATLAB summary lists all the special symbols, commands, and functions that were defined in this chapter. A brief description is also included for each one.

Commands and Functions

diff	computes the differences between adjacent values
quad	computes the integral under a curve (Simpson)
quad8	computes the integral under a curve (Newton–Cote)

PROBLEMS

Problems 1 to 4 relate to the engineering application presented in this chapter. Problems 5 to 11 relate to new applications.

Pipeline Flow Analysis These problems relate to the pipeline flow analysis problem given in this chapter.

1. Plot the velocity profile equation using the parameter values specified in this problem.

2. Generate a table showing the average flow velocity for a pipeline using the integer values of n from 5 to 10.

3. Generate a table showing the average flow velocity for pipelines with radii of 0.5, 1.0, 1.5, and 2.0 m. Assume that the other parameters are not changed from the values specified in this original problem.

4. Modify the program developed in this program so that the user can enter the value of v_{max} as the program is executing.

Trajectory Data Analysis Assume that an ASCII data file named altitude.dat contains a set of time and altitude data values from the flight of a new type of sounding rocket. Use these data to solve the following problems.

5. Compute and plot the velocity data for the rocket using backward differences.

6. Compute and plot the acceleration data for this rocket using backward differences.

7. Determine the number of rocket stages for this sounding rocket. (*Hint:* Consider critical points.)

8. Plot the velocity data on the same plot using all three difference equations.

9. Start with the acceleration data for this rocket that was computed in problem 6. Integrate the data to obtain velocity values. (You won't be able to use the quad functions, because you have only data points. Use either the trapezoidal rule or Simpson's rule.)

10. Start with the velocity data for this rocket that were computed in problem 5. Integrate the data to obtain altitude values. (You won't be able to use the quad functions, because you have only data points. Use either the trapezoidal rule or Simpson's rule.)

Function Analysis   The following problem relates to numerical integration.

11. Let the function f be defined by the following equation:

$$f(x) = 4e^{-x}$$

Plot the function over the interval [0,1]. Use numerical integration techniques to estimate the integral of $f(x)$ over [0,0.5] and over [0,1].

12

Courtesy of National Aeronautics and Space Administration

GRAND CHALLENGE: Vehicle Performance

One of the promising new propulsion technologies being developed for future transport aircraft is an advanced turboprop engine called the unducted fan (UDF). The UDF engine employs significant advancements in propeller technology. New materials, blade shapes, and higher rotation speeds enable UDF-powered aircraft to fly almost as fast as fanjets, and with greater fuel efficiency. The photograph shows a 2-foot model of a UDF being tested in a supersonic wind tunnel at NASA's Lewis Research Center. It uses sets of blades that rotate in opposite directions. A laser is being used to determine the shape of the blades when they are rotating at high speeds. Notice how the shape of the UDF blades differs from more traditional propeller blades.

Ordinary Differential Equations

Introduction

In this section we present a group of first-order differential equations and their analytical solutions. After describing the Runge-Kutta methods for integrating first-order differential equations, we then compare the numerical solutions for the group of first-order differential equations to the analytical solutions. An application problem that requires the solution of a differential equation is then discussed, and solved using the MATLAB function that implements a second- and third-order Runge-Kutta method. The chapter closes with a discussion of converting higher-order differential equations to first-order differential equations in order to solve them using the techniques discussed in this chapter.

12.1   FIRST-ORDER ORDINARY DIFFERENTIAL EQUATIONS

ODE

A first-order ordinary differential equation (ODE) is an equation that can be written in the following form:

$$y' = \frac{dy}{dx} = g(x,y)$$

where x is the independent variable and y is a function of x. The following equations are examples of first-order ODEs:

Equation 1 $y' = g_1(x, y) = 3x^2$
Equation 2 $y' = g_2(x, y) = -0.131y$
Equation 3 $y' = g_3(x, y) = 3.4444E{-}05 - 0.0015y$
Equation 4 $y' = g_4(x, y) = 2 \cdot x \cdot \cos^2(y)$
Equation 5 $y' = g_5(x, y) = 3y + e^{2x}$

Observe that y' is given as a function of x in Equation 1; y' is a function of y in Equations 2 and 3; y' is a function of both x and y in Equations 4 and 5.

integrating differential equations

A solution to a first-order ODE is a function $y = f(x)$ such that $f'(x) = g(x,y)$. Computing the solution of a differential equation involves integration in order to obtain y from y', and thus the techniques for solving differential equations are often referred to as techniques for integrating differential equations. The solution to a differential equation is generally a family of functions. An initial condition or boundary condition is usually needed in order to specify a unique solution. The ana-

initial conditions

lytical solutions to the ODEs presented at the beginning of this section were determined using certain initial conditions and are listed below:

Equation 1 Solution: $y = x^3 - 7.5$
Equation 2 Solution: $y = 4\,e^{-0.131\,x}$
Equation 3 Solution: $y = 0.022963 - 0.020763e^{-0.0015x}$
Equation 4 Solution: $y = \tan^{-1}(x^2 + 1)$
Equation 5 Solution: $y = 4e^{3x} - e^{2x}$

We do not present the analytical techniques for solving these ODEs here; refer to a differential equation text for the details on determining analytical solutions.

While an analytical solution to a differential equation is the preferred solution, many differential equations have complicated analytical solutions or no analytical solutions at all. For these cases, a numerical technique is needed to solve the differential equation. The most common numerical techniques for solving ordinary differential equations are Euler's method and the Runge-Kutta method.

Taylor's series

Both Euler's method and the Runge-Kutta methods approximate a function using its Taylor's series expansion [10]. Recall that a Taylor's series is an expansion that can be used to approximate a function whose derivatives exist on an interval containing a and b. The Taylor's series expansion for $f(b)$ is:

$$f(b) = f(a) + (b - a) f'(a) + \frac{(b - a)^2}{2!} f''(a) + \cdots + \frac{(b - a)^n}{n!} f^{(n)}(a) + \cdots$$

first-order

A first-order Taylor's series approximation uses the terms involving the function and its first derivative:

$$f(b) \approx f(a) + (b - a) f'(a)$$

second-order

A second-order approximation uses the terms involving the function, its first derivative, and its second derivative:

$$f(b) \approx f(a) + (b - a) f'(a) + \frac{(b - a)^2}{2!} f''(a)$$

The more terms of the Taylor's series that are used to approximate a function, the more accurate is the approximation. The MATLAB functions discussed in the next section use approximations of order 2, 3, 4, and 5 to approximate the function value $f(b)$.

12.2 RUNGE-KUTTA METHODS

Runge-Kutta method

Euler's method

The most popular methods for integrating a first-order differential equation are Runge-Kutta methods. These methods are based on approximating a function using its Taylor's series expansion, and thus a first-order Runge-Kutta method uses a first-order Taylor's series expansion, a second-order Runge-Kutta method uses a second-order Taylor's series expansion, and so on. (Euler's method is equivalent to a first-order Runge-Kutta method.) The MATLAB functions presented later in this section use approximations of order 2, 3, 4 and 5 to approximate values of an unknown function f using a differential equation.

The Taylor's series for evaluating $f(b)$ is given by the following expansion:

$$f(b) = f(a) + (b - a) f'(a) + \frac{(b - a)^2}{2!} f''(a) + \cdots + \frac{(b - a)^n}{n!} f^{(n)}(a) + \cdots$$

If we assume that the term $(b - a)$ represents a step size h, we can then rewrite the Taylor's series in this form:

$$f(b) = f(a) + h f'(a) + \frac{h^2}{2!} f''(a) + \cdots + \frac{h^n}{n!} f^{(n)}(a) + \cdots$$

Since $y = f(x)$, we can simplify the notation further if we assume that $y_b = f(b)$, $y_a = f(a)$, $y_a' = f'(a)$, and so on:

$$y_b = y_a + h y_a' + \frac{h^2}{2!} y_a'' + \cdots + \frac{h^n}{n!} y_a^{(n)} + \cdots$$

First-Order Approximation (Euler's Method)

A first-order Runge-Kutta integration equation is the following:

$$y_b = y_a + hy_a'$$

This equation estimates the function value y_b using a straight line that is tangent to the function at y_a, as shown in Figure 12.1. To compute the value of y_b (which is assumed to be on the tangent line), we use the step size h (which is equal to $b - a$) and a starting point y_a; the differential equation is used to compute the value of y_a'. Once we have determined the value of y_b, we can estimate the next value of the function $f(c)$ using the following:

$$y_c = y_b + hy_b'$$

This equation uses the tangent line at y_b to estimate y_c, as shown in Figure 12.2. Since an initial value or boundary value is needed to start the process of estimating

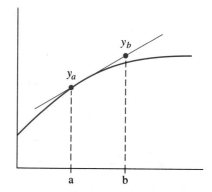

Fig. 12.1   Computation of y_b using
first-order Runge-Kutta.

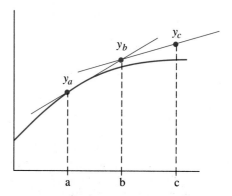

Fig. 12.2   Computation of y_c using
first-order Runge-Kutta.

boundary value
solutions

other points of the function $f(x)$, the Runge-Kutta methods (and Euler's method) are also called initial value solutions or boundary value solutions.

The first-order Runge-Kutta integration equation is simple to apply, but since it approximates the function with a series of short straight line segments, it may not be very accurate if the step size is large or if the slope of the function changes rapidly. Therefore, higher-order Runge-Kutta integration equations are often used to approximate the unknown function. These higher-order techniques average several tangent line approximations to the function, and thus obtain more accurate results. For example, a fourth-order Runge-Kutta integration equation uses terms in the Taylor's series expansion that include the first, second, third, and fourth derivatives, and computes the function estimate using four tangent line estimates.

MATLAB ODE Functions

ode23
ode45

MATLAB contains two functions for computing numerical solutions to ordinary differential equations—ode23 and ode45. The ode23 function uses second-order and third-order Runge-Kutta integration equations; the ode45 function uses fourth-order and fifth-order Runge-Kutta integration equations. The ode23 and ode45 functions have the same types of arguments, so we limit the following discussion to the ode23 function, but the same argument options are available for the ode45 function.

The simplest form of the ode23 function requires four arguments. The first argument is the name (in quotation marks) of a MATLAB function that returns the value of the differential equation $y' = g(x, y)$ when it receives values for x and y. The second and third arguments represent the endpoints of the interval over which we want to evaluate the function $y = f(x)$. The fourth argument contains the initial condition or boundary point that is needed to determine a unique solution to the ODE; it is assumed that this argument represents the function value at the left endpoint of the interval specified by the second and third arguments. The ode23 function has two outputs—a set of x coordinates and the corresponding set of y coordinates, which represent points of the function $y = f(x)$.

To illustrate the ode23 function, we present the steps to numerically compute the solutions to the differential equations given in Section 12.1. Since we know the analytical solutions to these ODEs, we also evaluate the analytical solutions and plot the analytical solution as a series of points while the numerical solution is plotted as a line graph. The MATLAB statements below define the functions required to evaluate the differential equations, assuming scalar inputs for x and y:

```
function dy = g1(x,y)
dy = 3*x^2;

function dy = g2(x,y)
dy = -0.131*y;

function dy = g3(x,y)
dy = 3.4444E-05 - 0.0015*y;
```

```
function dy = g4(x,y)
dy = 2*x*cos(y)^2;

function dy = g5(x,y)
dy = 3*y + exp(2*x);
```

We now present the commands to numerically compute the solutions to the differential equations using given initial conditions. The numerical solution (x, num_y) is plotted along with points from the analytical solution (x, anl_y) to demonstrate the accuracy of the numerical solutions.

Equation 1 The following statements solve $g_1(x, y)$ over the interval [2, 4], assuming that the initial condition $y = f(2)$ is equal to 0.5.

```
[x,num_y] = ode23('g1',2,4,0.5);
anl_y = x.^3 - 7.5;
subplot(211),plot(x,num_y,x,anl_y,'o'),...
title('Solution to Equation 1'),...
xlabel('x'),ylabel('y=f(x)'),grid
```

Figure 12.3 contains the comparison of the numerical solution to the analytical solution over the interval [2,4].

Equation 2 The following statements solve $g_2(x, y)$ over the interval [0, 5], assuming that the initial condition $y = f(0)$ is equal to 4:

```
[x,num_y] = ode23('g2',0,5,4);
anl_y = 4*exp(-0.131*x);
subplot(211),plot(x,num_y,x,anl_y,'o'),...
title('Solution to Equation 2'),...
xlabel('x'),ylabel('y=f(x)'),grid
```

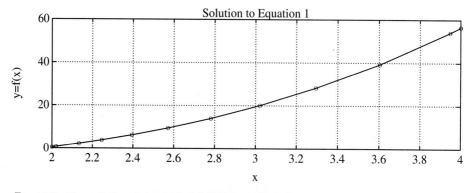

Fig. 12.3 Numerical and Analytical Solutions to Equation 1.

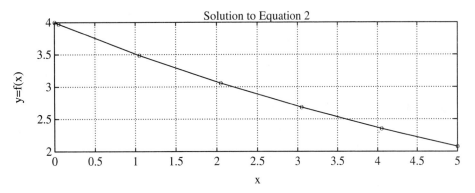

Fig. 12.4   Numerical and Analytical Solutions to Equation 2.

Figure 12.4 contains the comparison of the numerical solution to the analytical solution over the interval [0, 5].

Equation 3   The following statements solve $g_3(x, y)$ over the interval [0, 120], assuming that the initial condition $y = f(0)$ is equal to 0.0022:

```
[x,num_y] = ode23('g3',0,120,0.0022);
anl_y = 0.022963 - 0.020763*exp(-0.0015*x);
subplot(211),plot(x,num_y,x,anl_y,'o'),...
title('Solution to Equation 3'),...
xlabel('x'),ylabel('y=f(x)'),grid
```

Figure 12.5 contains the comparison of the numerical solution to the analytical solution over the interval [0, 120].

Equation 4   The following statements solve $g_4(x, y)$ over the interval [0, 2], assuming that the initial condition $y = f(0)$ is equal to $\pi/4$.

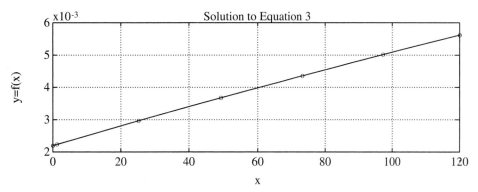

Fig. 12.5   Numerical and Analytical Solutions to Equation 3.

```
[x,num_y] = ode23('g4',0,2,pi/4);
anl_y = atan(x*x+1);
subplot(211),plot(x,num_y,x,anl_y,'o'),...
title('Solution to Equation 4'),...
xlabel('x'),ylabel('y=f(x)'),grid
```

Figure 12.6 contains the comparison of the numerical solution to the analytical solution over the interval [0, 2].

Equation 5   The following statements solve $g_5(x, y)$ over the interval [0, 3], assuming that the initial condition $y = f(0)$ is equal to 3:

```
[x,num_y] = ode23('g5',0,3,3);
anl_y = 4*exp(3*x) - exp(2*x);
subplot(211),plot(x,num_y,x,anl_y,'o'),...
title('Solution to Equation 5'),...
xlabel('x'),ylabel('y=f(x)'),grid
```

Figure 12.7 contains the comparison of the numerical solution to the analytical solution over the interval [0, 3].

The number of points computed for the function $y = f(x)$ by the ode23 and ode45 functions is determined by the MATLAB functions, and is not an input parameter. To compute more points of the function $f(x)$, an interpolation method can be used with the points returned by ode23 and ode45. For example, the cubic spline interpolation technique presented in Chapter 9 would be a good candidate for an interpolation method to give a smoother plot of the function $f(x)$.

The ode23 and ode45 functions can also be used with two additional parameters. A fifth parameter can be used to specify a tolerance that is related to the step size; the default tolerances are 0.001 for ode23 and 0.000001 for ode45. A sixth parameter can be used to request that the function print intermediate results (called a "trace"); the default value of 0 specifies no trace of the results.

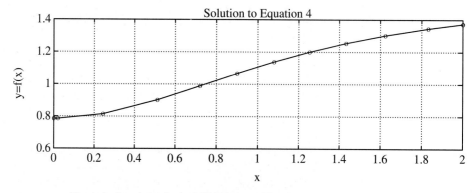

Fig. 12.6   Numerical and Analytical Solutions to Equation 4.

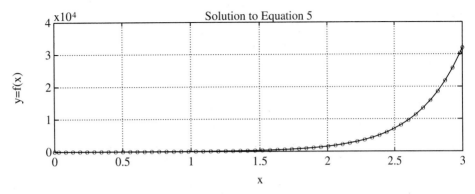

Fig. 12.7   Numerical and Analytical Solutions to Equation 5.

Practice!

Two ordinary differential equations are listed below:

$$y' = g_a(x, y) = -y$$

$$y' = g_b(x, y) = \frac{-x - e^x}{3y^2}$$

1. Write MATLAB functions to evaluate these differential equations given scalar values for x and y.

2. Assume that an initial condition of $f(0) = -3.0$ is given for the first differential equation. Use MATLAB to solve this differential equation over the interval $[0, 2]$. Plot the corresponding values of y.

3. The analytical solution to the first differential equation is:

$$y = -3\, e^{-x}$$

   Replot your solution in problem 2 and add points represented by this analytical solution in order to compare the numerical solution to the analytical solution.

4. Assume that an initial condition of $f(0) = 3.0$ is given for the second differential equation. Use MATLAB to solve this differential equation over the interval $[0, 2]$. Plot the corresponding values of y.

5. The analytical solution to the second differential equation is:

$$y = \sqrt[3]{28 - 0.5x^2 - e^x}$$

   Replot your solution in problem 4 and add points represented by this analytical solution in order to compare the numerical solution to the analytical solution.

PROBLEM SOLVING APPLIED: ACCELERATION OF UDF-POWERED AIRCRAFT

unducted fan (UDF)

An advanced turboprop engine called the unducted fan (UDF) is one of the promising new propulsion technologies being developed for future transport aircraft. Turboprop engines, which have been in use for decades, combine the power and reliability of jet engines with the efficiency of propellers. They are a significant improvement over earlier piston-powered propeller engines. Their application has been limited to smaller commuter-type aircraft, however, because they are not as fast or powerful as the fanjet engines used on larger airliners. The UDF engine employs significant advancements in propeller technology, which narrow the performance gap between turboprops and fanjets. New materials, blade shapes, and higher rotation speeds enable UDF-powered aircraft to fly almost as fast as fanjets, and with greater fuel efficiency. The UDF is also significantly quieter than the conventional turboprop.

During a test flight of a UDF-powered aircraft, the test pilot has set the engine power level at 40,000 Newtons, which causes the 20,000-kg aircraft to attain a cruise speed of 180 m/s (meters per second). The engine throttles are then set to a power level of 60,000 Newtons, and the aircraft begins to accelerate. As the speed of the plane increases, the aerodynamic drag increases in proportion to the square of the airspeed. Eventually, the aircraft reaches a new cruise speed where the thrust from the UDF engines is just offset by the drag. The differential equation that determines the acceleration of the aircraft is:

$$a = \frac{T}{m} - 0.000062\, v^2$$

where

$$a = \frac{dv}{dt}$$

T = thrust level in Newtons

m = mass in kg

v = velocity in m/s

Write a MATLAB program to determine the new cruise speed after the change in power level of the engines by plotting the solution to the differential equation.

 1. PROBLEM STATEMENT

Compute the new cruise speed of the aircraft after a change in power level.

 2. INPUT/OUTPUT DESCRIPTION

Figure 12.8 contains a diagram that shows that the output of the program is a plot from which the new cruise speed can be obtained.

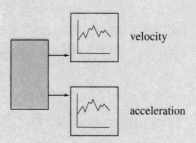

Fig. 12.8   I/O diagram.

 3. HAND EXAMPLE

The differential equation that is being solved is the following:

$$\frac{dv}{dt} = g(t,v) = \frac{T}{m} - 0.000062\,v^2$$

Thus, for the specified plane mass and the specified thrust, the differential equation is:

$$v' = 3.0 - 0.000062\,v^2$$

where $v = f(t)$. The velocity at the time that the higher thrust was applied was 180 m/s, and this velocity represents the initial condition $v = f(0)$. We can use the ode23 function to determine the velocity over a specified period of time that begins with the application of the higher thrust. We expect the velocity to increase initially, and then level off to a new higher cruise speed. Since the acceleration is equal to v', we can use the velocity values computed by the ode23 function to determine the acceleration over the specified period of time. We would expect that the acceleration should decrease after the initial new thrust, and return to a value of zero as the velocity (cruise speed) becomes constant.

 4. MATLAB SOLUTION

In the MATLAB solution, we use the ode23 function to evaluate the differential equation. The solution to the differential equation will give us values of velocity, which can then be used to determine values of acceleration. We will then plot both the velocity and the acceleration over an interval of 4 minutes to observe their changes. The velocity should increase and then stabilize at a new cruise speed while the acceleration should decrease to zero.

```
%
%       This program computes the velocity and accelera-
%       tion of an aircraft after a new thrust is applied.
%
```

```
initial_vel = 180;
seconds = 240;
[t,num_v] = ode23('g',0,seconds,initial_vel);
acc = 3 - 0.000062*num_v.^2;
clg
subplot(211),plot(t,num_v),title('Velocity'),...
xlabel('Time, s'),ylabel('m/s'),grid,...
subplot(212),plot(t,acc),title('Acceleration'),...
xlabel('Time, s'),ylabel('m/s^2'),grid
```

The M-file that defines the function used to compute values of the differential equation is the following:

```
function dv = g(t,v)
% g      This function of time and velocity computes a
%        value of a differential equation given velocity
%        values.
%
dv = 3 - 0.000062*v.^2;
```

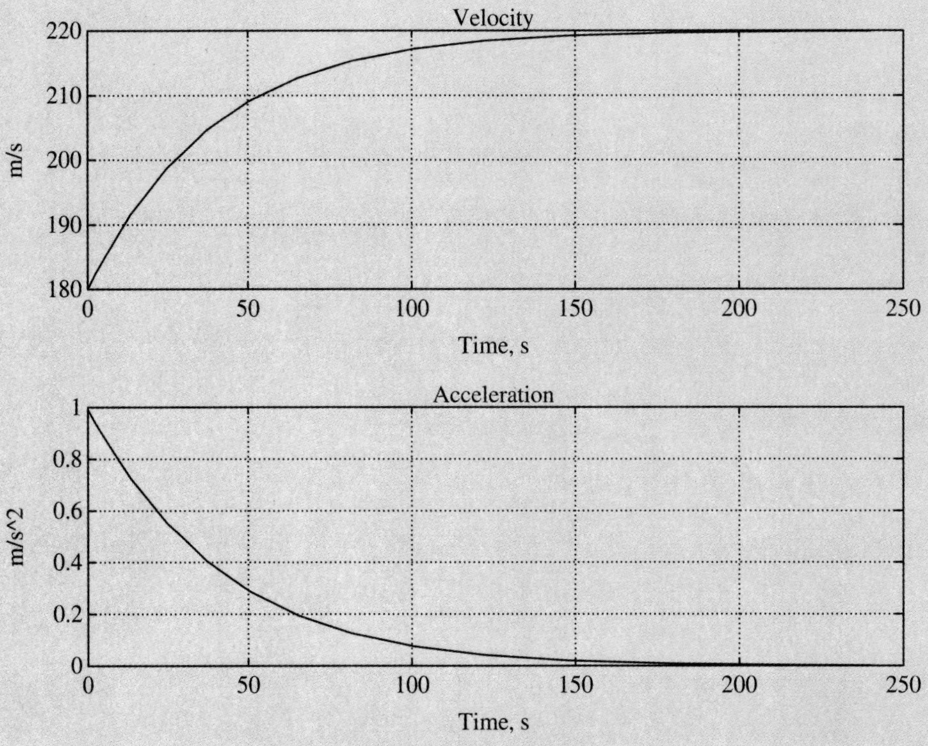

Fig. 12.9   Velocity and acceleration after new power thrust.

5. TESTING
The plots generated by this program are shown in Figure 12.9. The new cruise speed of the aircraft is approximately 220 m/s. As expected, the acceleration approaches zero as the new cruise speed is reached.

12.3   HIGHER-ORDER DIFFERENTIAL EQUATIONS

A higher-order differential equation can be written as a system of coupled first-order differential equations using a change of variables. For example, consider the following nth-order differential equation:

$$y^{(n)} = g(x, y, y', y'', \ldots, y^{(n-1)})$$

First, define n new unknown functions with these equations:

$$
\begin{aligned}
u_1(x) &= y^{(n-1)} \\
u_2(x) &= y^{(n-2)} \\
&\cdots \\
u_{n-2}(x) &= y'' \\
u_{n-1}(x) &= y' \\
u_n(x) &= y
\end{aligned}
$$

Then, the following system of n first-order equations is equivalent to the nth-order differential equation given above:

$$
\begin{aligned}
u_1' &= y^{(n)} = g(x, u_n, u_{n-1}, \ldots, u_1) \\
u_2' &= u_1 \\
&\cdots \\
u_{n-2}' &= u_{n-3} \\
u_{n-1}' &= u_{n-2}
\end{aligned}
$$

To demonstrate this process, consider this second-order linear differential equation [10]:

$$y'' = g(x, y, y') = y'(1 - y^2) - y$$

We first define two new functions:

$$
\begin{aligned}
u_1(x) &= y' \\
u_2(x) &= y
\end{aligned}
$$

We then obtain this system of coupled first-order differential equations:

$$
\begin{aligned}
u_1' &= y'' = g(x, u_2, u_1) = u_1(1 - u_2^2) - u_2 \\
u_2' &= u_1
\end{aligned}
$$

A system of coupled first-order differential equations can be solved using MAT-LAB's ode functions for solving first-order differential equations. However, the function that is used to evaluate the differential equation must compute the values of the coupled first-order differential equations in a vector. The initial condition must also be a vector that contains an initial condition for $y^{(n-1)}$, $y^{(n-2)}$, . . . , y', y. MAT-LAB's ode functions return solutions for each of the first-order differential equations, which in turn represent $y^{(n-1)}$, $y^{(n-2)}$, . . . , y', y.

To solve the set of two coupled equations developed in the previous example, we first define a function to compute values of the first-order differential equations:

```
function u_prime = eqns2(x,u)
u_prime(1) = u(1)*(1 - u(2)^2) - u(2);
u_prime(2) = u(1);
```

Then, to solve the system of first-order differential equations over the interval [0, 20] using initial conditions of $y'(0) = 0.0$ and $y(0) = 0.25$, we use these MATLAB statements:

```
initial = [0 0.25];
[x,num_y] = ode23('eqns2',0,20,initial);
subplot(211),plot(x,num_y(:,1)),...
title('1st Derivative of y'),xlabel('x'),grid,...
subplot(212),plot(x,num_y(:,2)),...
title('y'),xlabel('x'),grid
```

The plots generated by these statements are shown in Figure 12.10.

SUMMARY

This chapter described the Runge-Kutta methods for integrating first-order differential equations. The Runge-Kutta methods approximate the desired function using its Taylor series expansion; a first-order Runge-Kutta method uses a first-order Taylor series approximation, a second-order Runge-Kutta method uses a second-order Taylor series approximation, and so on. MATLAB contains two functions for integrating first-order differential equations. The function ode23 implements second- and third-order Runge-Kutta techniques and the function ode45 implements fourth- and fifth-order Runge-Kutta techniques. Higher-order differential equations can be written as a system of coupled first-order differential equations that can then be solved using the ode23 and ode45 functions.

MATLAB SUMMARY

This MATLAB summary lists all the special symbols, commands, and functions that were defined in this chapter. A brief description is also included for each one.

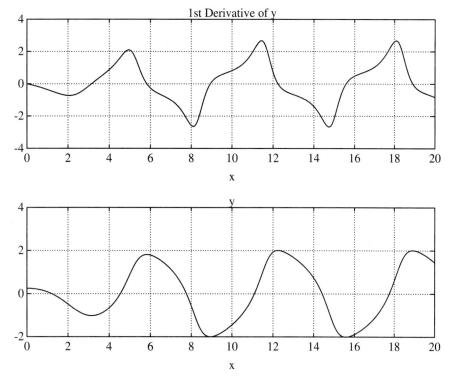

Fig. 12.10   Solution to 2nd-order differential equation.

Commands and Functions

ode23   second- and third-order Runge-Kutta solution
ode45   fourth- and fifth-order Runge-Kutta solution

PROBLEMS

Problems 1 to 4 relate to the engineering application presented in this chapter. Problems 5 to 18 relate to new applications.

Acceleration of UDF-Powered Aircraft   These problems relate to the aircraft acceleration problem given in this chapter.

1.   Modify the program to print the new cruise velocity. Assume that the new cruise velocity is achieved when three velocity values in a row are essentially the same values.

2.   Modify the program in problem 1 to also print the time (relative to the power thrust) at which the new cruise velocity was achieved.

3. Modify the program in problem 1 to assume that the new cruise velocity is achieved when three acceleration values in a row are essentially zero.

4. Modify the program such that the plots use units of miles/hour and ft/s^2.

Mixture Problems   The following problems use differential equations that are determined from considering the inflow and outflow of materials into a known solution.

5. The following differential equation describes the relationships between the volume of pollutants $x(t)$ in a lake and the time t (in years), using equal rates of inflow and outflow, and assuming an initial pollutant concentration [10]:

$$x' = 0.0175 - 0.3821x$$

Using an initial pollution volume at $t = 0.0$ of 0.2290, determine and plot the volume of pollutants over a period of 5 years.

6. Use data obtained from problem 5 to determine when the pollution volume in the lake will be reduced to 0.1.

7. The analytical solution to the differential equation presented in problem 5 is [10]:

$$x(t) = 0.0458 + 0.1832e^{-0.3821t}$$

Compare the analytical solution to the numerical solution determined in problem 5.

8. Using the analytical solution presented in problem 7, determine an analytical value for the answer to problem 6.

9. A 120-gallon tank contains 90 pounds of salt dissolved in 90 gallons of water. Brine containing 2 pounds of salt per gallon is flowing into the tank at a rate of 4 gallons per minute. The mixture flows out of the tank at the rate of 3 gallons per minute. The differential equation that specifies the amount of salt $x(t)$ in pounds in the tank at time t in minutes is [10]:

$$x' = 8 - \frac{3}{90 + t} \cdot x$$

The tank is full after 30 minutes. Determine and plot the amount of salt in the tank from time $= 0$ until the tank is full.

10. Using the data from problem 9, determine the amount of time required for the tank to contain 150 pounds of salt.

11. The analytical solution to the differential equation given in problem 9 is [10]:

$$x(t) = 2(90 + t) - \frac{90^4}{(90 + t)^3}$$

Compare the numerical solution to the analytical solution.

12. Use the analytical solution given in problem 11 to compute the amount of time required for the tank to contain 150 pounds of salt. Compare this answer to the answer determined in problem 10.

Bungee Jump   A bungee jumper is preparing to make a high-altitude jump from a hot air balloon using a 150-meter bungee line. He wants to estimate his peak acceleration, velocity, and drop distance so that he can be sure that the arresting force of the bungee is not too great, and that the balloon is high enough so that he will not hit the ground. The equation that he uses for his analysis is Newton's Second Law:

$$F = ma$$

where F is the sum of the gravitational, aerodynamic drag, and bungee forces acting on him, m is his mass, which is 70 kg, and a is his acceleration. He begins by defining the distance he falls as the variable x (which is a function of time, $x(t)$). His velocity and acceleration are then represented as x' and x'', respectively. He then rearranges Newton's equation to solve for the acceleration:

$$x'' = F/m$$

Next, he determines the forces making up F. The gravitational force will be his weight, which is

$$\begin{aligned} W &= m \cdot g \\ &= (70 \text{ kg}) \cdot (9.8 \text{ m/s}^2) \\ &= 686 \ N \end{aligned}$$

He knows that the aerodynamic drag, D, will be proportional to the square of his velocity, $D = c(x')^2$, but he does not know c, the constant of proportionality. However, he does know from his experience as a skydiver that his terminal velocity in a free-fall is about 55 m/s. At that speed, the aerodynamic drag is equal to his weight so he determines c using:

$$\begin{aligned} c &= D/(x')^2 \\ &= (686 \ N)/(55 \text{ m/s})^2 \\ &= 0.227 \text{ kg/m} \end{aligned}$$

Finally, after he has fallen 150 m, the slack in the bungee will be eliminated, and it will begin to exert an arresting force, B, of 10 N for every meter that it is stretched beyond 150 m. Thus, there will be two regions for computing the acceleration. The first will be used when the distance x is less than or equal to 150 m:

$$\begin{aligned} x'' &= F/m \\ &= (W - D)/m \\ &= (686 - 0.227(x')^2)/70 \\ &= 9.8 - 0.00324(x')^2 \text{ m/s}^2 \qquad \text{(Equation 1)} \end{aligned}$$

The second equation will be used when x is greater than 150 m:

$$
\begin{aligned}
x'' &= F/m \\
&= (W - D - B)/m \\
&= (686 - 0.227(x')^2 - 10(x - 150))/70 \\
&= 31.23 - 0.00324(x')^2 - 0.143x \text{ m/s}^2 \qquad \text{(Equation 2)}
\end{aligned}
$$

The following set of problems refer to this bungee jump problem.

13. Integrate Equation 1 for the interval beginning at 0 seconds to find the velocity and distance as a function of time from the beginning of the jump (which is assumed to occur at $t = 0.0$). From the results, determine the velocity and the time when $x = 150$ (this is the point at which the slack in the bungee is eliminated). You may need to experiment with the time interval in order to choose an interval that will give you the velocity and time desired.

14. Modify the program used for problem 13 so that it will use Equation 2 instead of Equation 1 to find the velocity and distance after the bungee becomes taut. Also compute the acceleration. What are the peak values of acceleration, velocity, and distance? The bungee jumper does not want the maximum acceleration to exceed 2 g's (1 $g = 9.8$ m/sec^2). Is the estimate of the peak acceleration higher or lower? How close does he come to reaching the estimated terminal velocity of 55 m/sec? How many seconds does he fall? How high should the balloon be to ensure a factor of safety of 4?

15. Assume that the bungee also has a viscous friction force, R, once it begins to stretch, that is given by:

$$
R = -7.0 \, x'
$$

Modify the equation used for problem 14 to include this force and determine the new results. How many seconds does it take for the bungee jumper to almost come to rest (or for the oscillations to almost stop)? What is the final stretched length of the bungee? Does this make sense?

16. For problem 15, determine the length of the bungee that will cause the peak upward acceleration to be close to 0.8 g's. How far below the balloon does the bungee jumper fall before he starts back up?

17. For problem 15, determine the arresting force per meter that will cause a peak upward acceleration of approximately 1.5 g's.

18.   From his experience as a skydiver, the bungee jumper knows that if he dives straight down such that he is streamlined into the wind, he could reach a speed of about 120 m/s. Determine the new value for the aerodynamic drag constant of proportionality c which corresponds to this situation and recompute the results of problem 15. Next, assume that the bungee is 300 meters long and determine the jumper's peak velocity, the maximum g level, and how far below the balloon the jumper falls if he dives such that he is streamlined into the wind. Does he reach the 2 g limit? Plot the net force acting on the bungee jumper as a function of time. Can you explain the appearance of the graph?

13

Courtesy of National Aeronautics and Space Administration

GRAND CHALLENGE: Speech Recognition

Communication satellites receive and retransmit large amounts of information. This information can represent many types of information, from images to speech signals. The receiving and transmitting processes can introduce errors or "noise" into a signal, and this noise makes it more difficult to design algorithms for operations such as automatic speech recognition. Therefore, before attempting to separate speech signals into words for further analysis, it is important to remove as much of the noise as possible. Adaptive digital signal processing techniques are commonly used to remove noise that has contaminated a signal. The performance of the adaptive digital signal processing techniques is related to the characteristics of the signals and the noise.

Matrix Decomposition and Factorization

Introduction

This chapter contains some of the more advanced features related to matrices that are useful in solving certain types of engineering problems. The first topic, eigenvalues and eigenvectors, is one that arises in a number of applications. After defining eigenvalues and eigenvectors and illustrating their properties with a simple example, the `eig` function is presented for computing both eigenvalues and eigenvectors. An application is then presented to demonstrate how eigenvalues and eigenvectors are used in analyzing the performance of adaptive noise-canceling algorithms. The rest of the chapter deals with decompositions and factorizations that can be applied to a matrix A.

13.1   EIGENVALUES AND EIGENVECTORS

Assume that A is an $n \times n$ square matrix. Let X be a column vector with n rows, and let λ be a scalar. Consider the following equation:

$$A X = \lambda X \tag{13.1}$$

Both sides of this equation are equal to a column vector with n rows. If X is filled with zeros, then this equation is true for any value of λ, but this is a trivial solution.

eigenvalues
eigenvectors

The values of λ for which X are non-zero are called the eigenvalues of the matrix A, and the corresponding values of X are called the eigenvectors of the matrix A.

Equation (13.1) can also be used to determine the following equation:

$$(A - \lambda I) X = 0 \tag{13.2}$$

homogeneous

where I is an $n \times n$ identity matrix. This equation represents a set of homogeneous equations, since the right side of the equations are zero. A set of homogeneous equations has nontrivial solutions if and only if the determinant is equal to zero:

$$\det (A - \lambda I) = 0 \tag{13.3}$$

characteristic
equation

Equation (13.3) represents an equation that is referred to as the characteristic equation of the matrix A. The solutions to the equation are also the eigenvalues of the matrix A.

In many applications, it is desirable to select eigenvectors such that $QQ^T = I$, where Q is the matrix whose columns are eigenvectors. This set of eigenvectors represents an orthonormal set, which means that they are both normalized and that they are mutually orthogonal. (A set of vectors is orthonormal if the dot product of a vector with itself is equal to unity, and the dot product of a vector with another vector in the set is zero.)

orthonormal

To illustrate these relationships between a matrix A and its eigenvalues and eigenvectors, consider the following matrix A:

$$A = \begin{bmatrix} 0.50 & 0.25 \\ 0.25 & 0.50 \end{bmatrix}$$

The eigenvalues can be computed using the characteristic equation:

$$\det (A - \lambda I) = \det \begin{bmatrix} 0.5 - \lambda & 0.25 \\ 0.25 & 0.5 - \lambda \end{bmatrix}$$
$$= \lambda^2 - \lambda + 0.1875$$
$$= 0$$

This equation can be easily solved using the quadratic equation, yielding $\lambda_0 = 0.25$ and $\lambda_1 = 0.75$. (For a matrix A with more than two rows and two columns, determining the eigenvalues by hand can be a formidable task.) The eigenvectors can be determined using the eigenvalues and Equation (13.2), as shown here with the eigenvalue 0.25:

$$\begin{bmatrix} 0.5 - 0.25 & 0.25 \\ 0.25 & 0.5 - 0.25 \end{bmatrix} \begin{bmatrix} x_1 \\ x_2 \end{bmatrix} = \begin{bmatrix} 0 \\ 0 \end{bmatrix}$$

or

$$\begin{bmatrix} 0.25 & 0.25 \\ 0.25 & 0.25 \end{bmatrix} \begin{bmatrix} x_1 \\ x_2 \end{bmatrix} = \begin{bmatrix} 0 \\ 0 \end{bmatrix}$$

But this pair of equations yields the following equations:

$$x_1 = -x_2$$

Therefore, there are an infinite number of eigenvectors that are associated with the eigenvalue 0.25. Some of these eigenvectors are shown below:

$$\begin{bmatrix} 1 \\ -1 \end{bmatrix} \quad \begin{bmatrix} 5 \\ -5 \end{bmatrix} \quad \begin{bmatrix} 0.2 \\ -0.2 \end{bmatrix}$$

Similarly, it can be shown that the eigenvectors associated with the eigenvalue 0.75 have the following relationship:

$$x_1 = x_2$$

Again, an infinite number of eigenvectors are associated with this eigenvalue, such as:

$$\begin{bmatrix} 1.5 \\ 1.5 \end{bmatrix} \quad \begin{bmatrix} -5 \\ -5 \end{bmatrix} \quad \begin{bmatrix} 0.2 \\ -0.2 \end{bmatrix}$$

To determine an orthonormal set of eigenvectors for the simple example that we have been using, recall that we want to select the eigenvectors such that $QQ^T = I$. Therefore, consider the following:

$$QQ^T = \begin{bmatrix} c_1 & c_2 \\ -c_1 & c_2 \end{bmatrix} \begin{bmatrix} c_1 & -c_1 \\ c_2 & c_2 \end{bmatrix}$$

$$= \begin{bmatrix} c_1^2 + c_2^2 & -c_1^2 + c_2^2 \\ -c_1^2 + c_2^2 & c_1^2 + c_2^2 \end{bmatrix}$$

$$= \begin{bmatrix} 1 & 0 \\ 0 & 1 \end{bmatrix}$$

Solving this set of equations gives:

$$c_1^2 = c_2^2 = 0.5$$

Thus, c_1 can be either $\dfrac{1}{\sqrt{2}}$ or $-\dfrac{1}{\sqrt{2}}$; similarly, c_2 can be either $\dfrac{1}{\sqrt{2}}$ or $-\dfrac{1}{\sqrt{2}}$. Thus, there are several variations of the same values that can be used to determine the set of orthonormal eigenvectors. We shall choose the following:

$$Q = \begin{bmatrix} \dfrac{1}{\sqrt{2}} & \dfrac{1}{\sqrt{2}} \\ -\dfrac{1}{\sqrt{2}} & \dfrac{1}{\sqrt{2}} \end{bmatrix}$$

The computations to obtain the eigenvalues and an associated set of orthonormal eigenvectors have been relatively simple for a matrix A with two rows and two columns. However, it should be evident that the computations become quite difficult as the size of the matrix A increases. Therefore, it is very convenient to be able to use MATLAB to determine both the eigenvectors and eigenvalues for a matrix A. The eig function has one argument—the matrix A. This function can be used to return a column vector containing only the eigenvalues, as in:

$$lambda = eig(A);$$

The function can also be used with a double assignment statement. In this case it returns two square matrices: one containing eigenvectors (X) as columns, and another containing the eigenvalues (λ) on the diagonal:

$$[Q,d] = eig(A);$$

The values of Q and d are such that $QQ^T = I$ and $AQ = Qd$.

We can illustrate the eig function with the example developed in this section as shown in the following statements:

$$A = [0.50, 0.25; 0.25, 0.50];$$
$$[Q,d] = eig(A);$$

The values of Q and d are the following:

$$Q = \begin{bmatrix} 0.7071 & 0.7071 \\ -0.7071 & 0.7071 \end{bmatrix}$$

$$d = \begin{bmatrix} 0.25 & 0.00 \\ 0.00 & 0.75 \end{bmatrix}$$

These values match the values that we computed by hand for this example. Using matrix multiplication, we can also easily verify that $QQ^T = I$ and $AQ = Qd$.

Practice!

Let A be the following matrix:

$$\begin{bmatrix} 4 & 3 & 0 \\ 3 & 6 & 2 \\ 0 & 2 & 4 \end{bmatrix}$$

Use MATLAB to answer the following questions:

1. Determine λ_1, λ_2, λ_3, the three eigenvalues of A.

2. Determine a set of orthonormal eigenvectors, X_1, X_2, X_3 such that X_1 is associated with λ_1, and so on.

3. Compute $\det(A - \lambda I)$ and verify that it is equal to zero for each eigenvalue.

4. Show that $AQ = Qd$ where Q is the matrix containing the eigenvectors as columns and d is the matrix containing the corresponding eigenvalues on the main diagonal and zeros elsewhere.

 PROBLEM SOLVING APPLIED: ADAPTIVE NOISE CANCELING

Adaptive noise canceling is used in many applications to reduce the effect of interfering noise in a signal. For example, assume that a microphone is used to record the speech signals of a speaker at the front of a large auditorium. Another microphone is used at the rear of the auditorium to collect primarily auditorium noise signals. Using adaptive noise-canceling techniques, the characteristics of the noise signal can be determined using the signals from the two microphones, and portions of the noise can be eliminated from the speech signal collected at the front of the auditorium, giving a clearer speech signal. This same adaptive noise-canceling process is also used in noisy jet cockpits. Adaptive signal processing is used to reduce the background noise in the speech signals that are then relayed to the control tower. This processing results in clearer speech signals and better communications.

The adaptive noise-canceling algorithms are beyond the scope of this text, but the performance and speed of the algorithms depend on characteristics of the two input signals. These characteristics determine a multidimensional quadratic surface for which we must obtain a minimum. This minimum is determined by tracking algorithms that track from a starting point in the direction of the unique minimum. If the quadratic surface has circular contours (similar to a circular bowl), the tracking algorithms are more accurate and find the minimum quicker. If the quadratic surface has elliptical contours (similar to an oblong bowl), the tracking algorithms are not as accurate. A matrix R can be computed from the input signals, and the eigenvalues of the matrix R will determine the type of contours of the surface. If the eigenvalues are all equal, the contours are circular. The wider the variation in the eigenvalues, the more elliptical are the contours. The eigenvectors also represent the principal axes of the surface. Therefore, to determine the speed and performance of the adaptive noise-canceling algorithms for certain types of data, and to analyze the surface, we need to determine the eigenvalues and eigenvectors of the matrix R that is computed from data values.

Write a program to read the values of a matrix from a MAT-file named dataR, and then compute the associated eigenvalues and eigenvectors.

1. PROBLEM STATEMENT
Compute the eigenvalues and eigenvectors of a matrix.

 2. INPUT/OUTPUT DESCRIPTION
Figure 13.1 contains a diagram that shows that the input to the program is a file named dataR.mat, and that the output values are the eigenvalues and eigenvectors of the matrix.

 3. HAND EXAMPLE
Assume that the input matrix is the following:

$$R = \begin{bmatrix} 0.50 & 0.25 \\ 0.25 & 0.50 \end{bmatrix}$$

From the example that we worked in Section 13.1, we know that the eigenvalues of this matrix are 0.25 and 0.75, and that the eigenvectors, in corresponding order, are the following:

$$V_1 = \begin{bmatrix} \dfrac{1}{\sqrt{2}} \\ -\dfrac{1}{\sqrt{2}} \end{bmatrix} \qquad V_2 = \begin{bmatrix} \dfrac{1}{\sqrt{2}} \\ \dfrac{1}{\sqrt{2}} \end{bmatrix}$$

Since the eigenvalues are not equal, we know that the quadratic surface does not have circular contours, and thus the performance of the adaptive noise canceling will be slower and less accurate.

 4. MATLAB SOLUTION
In the program, we use a for loop to print the eigenvalues and eigenvectors.

```
%
%       This program reads the values of a matrix R from
%       a MAT-file. The eigenvalues and corresponding
%       eigenvectors of the matrix are printed.
%
```

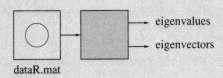

dataR.mat

Fig. 13.1 I/O diagram

```
load dataR;
[Q,d] = eig(R);
[m,n] = size(R);
for k=1:m
    fprintf('Eigenvalue %4.0f = %7.2f \n',k,d(k,k));
    disp('Corresponding Eigenvector')
    disp(Q(:,k)')
end
```

 5. TESTING

The output of this program using the data from the hand example is the following:

```
Eigenvalue    1 =    0.25
Corresponding Eigenvector
    0.7071    -0.7071

Eigenvalue    2 =    0.75
Corresponding Eigenvector
    0.7071     0.7071
```

While we have used a small matrix R in this example, it is not uncommon for adaptive noise-canceling applications to include configurations in which the matrix R may have hundreds of rows and columns. In fact, if the adaptive noise-canceling algorithm is being used with communication data from a satellite, the matrix R may have thousands of rows and columns in it.

13.2   DECOMPOSITIONS AND FACTORIZATIONS

In this section we present three decompositions or factorizations of matrices that can be useful in solving problems containing matrices. Each of these techniques decomposes a matrix A into a product of other matrices. The use of the factored product reduces the amount of calculations needed for many types of matrix computations, and thus many numerical techniques that use matrices convert the matrices into decomposed or factored forms.

Triangular Factorization

Triangular factorization expresses a square matrix as the product of two triangular matrices—a lower triangular matrix (or permuted lower triangular matrix) and an

upper triangular
lower triangular

upper triangular matrix. This factorization is often called an LU factorization (for lower-upper). The LU factorization is not a unique factorization.

Triangular factorization is often used to simplify computations involving matrices. It is one of the steps useful in computing the determinant of a large matrix, computing the inverse of a matrix, and solving simultaneous linear equations.

The factorization can be performed by starting with a square matrix A and an identity matrix of the same size. Row and column operations are performed on A to reduce it to an upper triangular form; the same operations are performed on the identity matrix. In the process of performing the row operations, we may find it necessary to interchange rows in order to produce the desired upper triangular form. These same row interchanges are performed on the identity matrix, and will result in the identity matrix being transformed into a permuted lower triangular matrix instead of a strict lower triangular matrix. To illustrate, let A and B be the following matrices:

permuted lower
triangular matrix

$$A = \begin{bmatrix} 1 & 2 & -1 \\ -2 & -5 & 3 \\ -1 & -3 & 0 \end{bmatrix}$$

$$B = \begin{bmatrix} 1 & 3 & 2 \\ -2 & -6 & 1 \\ 2 & 5 & 7 \end{bmatrix}$$

Using the process described above, it can be shown that A and B can be factored into the LU forms shown below:

$$A = \begin{bmatrix} 1 & 0 & 0 \\ -2 & 1 & 0 \\ -1 & 1 & 1 \end{bmatrix} \begin{bmatrix} 1 & 2 & -1 \\ 0 & -1 & 1 \\ 0 & 0 & -2 \end{bmatrix}$$

$$B = \begin{bmatrix} 1 & 0 & 0 \\ -2 & 0 & 1 \\ 2 & 1 & 0 \end{bmatrix} \begin{bmatrix} 1 & 3 & 2 \\ 0 & -1 & 3 \\ 0 & 0 & 5 \end{bmatrix}$$

Note that the factorization of B yields a permuted lower triangular form; if rows 2 and 3 are interchanged, the permuted lower triangular form becomes a strict lower triangular form.

lu

The lu function in MATLAB computes the LU factorization, and is specified as shown in the following statement:

$$[L, U] = lu(A);$$

The permuted lower triangular factor is stored in L; the upper triangular factor is stored in U, and the product of L and U is equal to A. To compute the LU factorization of the two matrices used in the previous example, we use the following statements:

$$A = [1, 2, -1; -2, -5, 3; -1, -3, 0];$$
$$[LA, UA] = lu(A);$$

$$B = [1,3,2; -2,-6,1; 2,5,7];$$
$$[LB, UB] = lu(B);$$

The LU factorization yields the following matrices:

$$LA = \begin{bmatrix} -0.5 & 1 & 0 \\ 1 & 0 & 0 \\ 0.5 & 1 & 1 \end{bmatrix} \qquad UA = \begin{bmatrix} -2 & -5 & 3 \\ 0 & -0.5 & 0.5 \\ 0 & 0 & -2 \end{bmatrix}$$

$$LB = \begin{bmatrix} -0.5 & 0 & 1 \\ 1 & 0 & 0 \\ -1 & 1 & 0 \end{bmatrix} \qquad UB = \begin{bmatrix} -2 & -6 & 1 \\ 0 & -1 & 8 \\ 0 & 0 & 2.5 \end{bmatrix}$$

It is easily verified that A = (LA)(UA) and B = (LB)(UB). It is also interesting to observe that neither factorization matches the one generated by hand earlier in this section; this is not a concern, since it was pointed out that the LU factorization is not a unique factorization. Also, note that both factorizations include a permuted lower triangular factor.

QR Factorization

The QR factorization technique factors a matrix A into the product of an orthonormal matrix and an upper triangular matrix. (Recall that a matrix Q is orthonormal if $QQ^T = I$.) It is not necessary that the matrix A be a square matrix in order to perform a QR factorization.

Gram–Schmidt   The QR factorization can be determined from performing the Gram–Schmidt process on the column vectors in A to obtain an orthonormal basis. The least squares solution of an overdetermined system AX = B is the solution of the square system $RX = Q^T B$.

The qr function is used to perform the QR factorization in MATLAB. The function is a double assignment function which returns the two matrices Q and R, as shown:

$$[Q, R] = qr(A);$$

For a matrix A of size $m \times n$, the size of Q is $n \times n$ and the size of R is $m \times n$.

Singular Value Decomposition

SVD decomposition   The singular value decomposition (SVD) is another orthogonal matrix factorization. SVD is the most reliable decomposition, but it can require as many as 10 times as many arithmetic operations as the QR factorization [9]. SVD decomposes a matrix A into a product of three matrix factors:

$$A = U S V$$

where U and V are orthogonal matrices and S is a diagonal matrix. The values on the diagonal matrix of S are called singular values, and thus give the decomposition technique its name. The number of nonzero singular values is equal to the rank of the matrix.

The three matrix factors of the SVD factorization can be obtained using the svd function in MATLAB, as shown in the following statement:

$$[U, S, V] = \text{svd}(A) \, ;$$

svd
When not used in the triple assignment above, the svd function returns just the diagonal elements of S, which are the singular values of A.

One of the main uses of the SVD factorization is in solving least squares problems. This application can also be extended to using the SVD for data compression.

SUMMARY

Techniques for advanced matrix computations were presented in this chapter. Determining the eigenvalues and eigenvectors of a matrix is a problem that occurs in a number of engineering applications, and the eig function will determine both sets of values. There are also applications that require that a matrix be decomposed into a product of factors in order to efficiently and accurately perform certain computations. In this chapter we presented three types of decompositions. The LU decomposition factors a matrix into a product of a permuted lower triangular matrix and an upper triangular matrix; the QR decomposition factors a matrix into a product of an orthonormal matrix and an upper triangular matrix; the singular value decomposition factors a matrix into a product of three matrices—two orthogonal matrices and a diagonal matrix.

MATLAB SUMMARY

This MATLAB summary lists all the special symbols, commands, and functions that were defined in this chapter. A brief description is also included for each one.

Commands and Functions

eig	computes eigenvalues and eigenvectors of a matrix
lu	computes an LU decomposition of a matrix
qr	computes an orthonormal decomposition of a matrix
svd	computes the SVD decomposition of a matrix

PROBLEMS

Problems 1 to 7 relate to the engineering application presented in this chapter. Problems 8 to 18 relate to new applications.

Adaptive Noise Canceling   These problems relate to the adaptive noise-canceling problem given in this chapter.

1. Modify the eigenvalue program to print the largest and smallest eigenvalues, λ_{max} and λ_{min}.

2. Modify the eigenvalue program to print the range of values represented by the eigenvalues. (A large range of eigenvalues indicates a surface that may present difficulties for the tracking algorithm.)

3. Modify the eigenvalue program to determine and print the average eigenvalue, λ_{ave}. (This value is also used in analyzing adaptive algorithms.)

4. Some of the computations that analyze the performance of an adaptive algorithm use the reciprocals of the maximum and minimum eigenvalues. Determine these values, $\dfrac{1}{\lambda_{max}}$ and $\dfrac{1}{\lambda_{min}}$.

5. Some of the computations that analyze the performance of an adaptive algorithm use the average reciprocal eigenvalue. Determine this value, $\left(\dfrac{1}{\lambda}\right)_{ave}$.

6. When there is a large number of eigenvalues, it is sometimes assumed that they are uniformly distributed between the maximum eigenvalue and the minimum eigenvalue. In this case the average value reciprocal can be computed using the following equation:

$$\left(\frac{1}{\lambda}\right)_{ave} = \frac{\ln(\lambda_{max}/\lambda_{min})}{\lambda_{max} - \lambda_{min}}$$

Determine $\left(\dfrac{1}{\lambda}\right)_{ave}$ using this equation and compare it to the answer in problem 5.

7. It can be shown that the maximum eigenvalue is less than the trace of the matrix R, where the trace is the sum of the diagonal elements of R. Print the value of the maximum eigenvalue and of the trace of R. (Use the MATLAB function trace, whose argument is a square matrix.)

Properties of Eigenvalues and Eigenvectors   The following problems relate to properties of eigenvalues and eigenvectors:

8. If λ is an eigenvalue of A with V as a corresponding eigenvector, then λ^k is an eigenvalue of A^k, again with V as a corresponding eigenvector, for any positive integer k. Write a program to demonstrate this property for $k = 1$ to 5, using the following matrix A:

$$A = \begin{bmatrix} 1 & 0 & 0 \\ -8 & 4 & -6 \\ 8 & 1 & 9 \end{bmatrix}$$

9. If λ is an eigenvalue of a matrix A (for which an inverse exists) with V as a corresponding eigenvector, then $\dfrac{1}{\lambda}$ is an eigenvalue of A^{-1}, again with V as a cor-

responding eigenvector. Write a program to demonstrate this property, using the following matrix A:

$$A = \begin{bmatrix} 1 & 0 & 0 \\ -8 & 4 & -6 \\ 8 & 1 & 9 \end{bmatrix}$$

10. If A has n distinct eigenvalues and Q is an $n \times n$ matrix having as its jth column vector an eigenvector corresponding to an eigenvalue λ_j, then $Q^{-1}AQ$ is the diagonal matrix having λ_j on the diagonal in the jth column. Write a program to demonstrate this property, using the following matrix A:

$$A = \begin{bmatrix} 1 & 0 & 0 \\ -8 & 4 & -6 \\ 8 & 1 & 9 \end{bmatrix}$$

Orthogonal Properties   The following problems relate to the orthogonality properties of matrices. An orthogonal matrix has rows (columns) that are mutually orthogonal, and that have magnitude one. A matrix with these properties is sometimes called an orthonormal matrix.

11. An $n \times n$ matrix A is orthogonal if $A^{-1} = A^T$. Write a program to determine if a matrix A is orthogonal.

12. Write a program to compute the length or magnitude of a vector V (with n elements) where the magnitude is defined as shown:

$$\|V\| = \sqrt{v_1^2 + v_2^2 + \dots + v_n^2}$$

13. Write a program to demonstrate that the magnitudes of the rows and columns of an orthogonal matrix are all equal to one.

14. The angle between two vectors V and W is equal to:

$$\theta = \cos^{-1}\left(\frac{V \cdot W}{\|V\|\,\|W\|}\right)$$

where $V \cdot W$ is the dot product of V and W. Write a program to determine if a set of eigenvectors are perpendicular.

Determinants   The following problems define cofactors and minors of a square matrix, and then use them to evaluate a determinant.

15. The minor of an element $a_{i,j}$ in a matrix A is the determinant of the matrix obtained by removing the row and column to which the given element $a_{i,j}$ belongs. Thus, if the original matrix has n rows and columns, the minor is the determinant of a matrix with n-1 rows and columns. Write a MATLAB function to compute the minor of a square matrix. The input arguments should be the matrix A and values of i and j.

16. A cofactor $A_{i,j}$ of a matrix A is the product of the minor of $a_{i,j}$ and the factor $(-1)^{i+j}$. Write a MATLAB function to compute a cofactor of a square matrix. The input arguments should be the matrix A and the values of i and j. (Hint: You may want to reference the function defined in problem 15.)

17. The determinant of a square matrix A can be computed in the following way:
    (a)   Select any column.
    (b)   Multiply each element in the column by its cofactor.
    (c)   Add the products obtained in step (b).
    Write a function det_col to compute the determinant using this technique. Compare the values computed by this function to those computed by the function det that is included in MATLAB. (Hint: you may want to reference the function defined in problem 16.)

18. The determinant of a square matrix A can be computed in the following way:
    (a)   Select any row.
    (b)   Multiply each element in the row by its cofactor.
    (c)   Add the products obtained in step (b).
    Write a function det_row to compute the determinant using this technique. Compare the values computed by this function to those computed by the function det that is included in MATLAB. (Hint: you may want to reference the function defined in problem 16.)

14

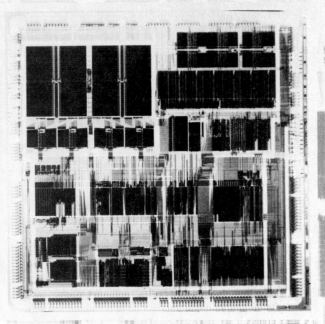

GRAND CHALLENGE: Machine Vision

Machine vision requires the rapid processing and identification of images. These steps require not only fast algorithms but also techniques for storing and accessing extremely large amounts of data. Many of the processes and algorithms for image processing contain steps that can be performed in parallel, and thus machine vision is an application that will benefit from parallel computing. Many of the operations performed in machine vision algorithms are digital filtering operations. The microprocessor shown in this photograph is Texas Instrument's TMS320C40 processor. It is the world's first parallel processing digital signal processor, and contains 884,000 on-chip transistors.

Signal Processing

Introduction

The student edition of MATLAB contains a number of functions that have been selected from the Signal Processing Toolbox and the Control Systems Toolbox, which are optional toolboxes that can be purchased with the professional version of MATLAB. These selected functions have been combined in a Signals and Systems Toolbox. This chapter discusses a number of these functions that are related to signal processing; the next chapter discusses a number of the remaining functions that are related to control systems. The functions discussed in this chapter have been divided into four categories—frequency domain analysis, filter analysis, filter implementation, and filter design. In these sections, we assume that the reader is already familiar with the signal processing concepts of time domain, frequency domain, transfer functions, and filters. Since the notation varies in the signal processing literature, we define the notation that is used in these discussions on the signal processing functions.

14.1   FREQUENCY DOMAIN ANALYSIS

DSP

analog signal

While this chapter discusses both analog signal processing and digital signal processing, the focus is on digital signal processing, or DSP. Recall that an analog signal is a continuous function [usually of time, as in $f(t)$] that represents information, such as a speech signal, a blood pressure signal, or a seismic signal. In order to process this information with the computer, an analog signal can be sampled every T seconds, thus generating a digital signal that is a sequence of values from the original analog signal. We represent a digital signal that has been sampled from a continuous signal $f(t)$ using the following notation:

digital signal

$$f_k = f(kT)$$

The digital signal is the sequence of samples represented by $[f_k]$.

sampling

The time that we begin collecting the digital signal is usually assumed to be zero, and thus the first sample of a digital signal is usually referred to as f_0. Hence, if a signal is sampled at 100 Hz (100 times per second), the first three values of the digital signal correspond to the following analog signal values:

$$f_0 = f(0T) = f(0.0)$$
$$f_1 = f(1T) = f(0.01)$$
$$f_2 = f(2T) = f(0.02)$$

Figure 14.1 compares an analog signal to its corresponding digital signal. In this figure, we show the digital signal as a sequence of points or samples, but in general we plot a digital signal with the points connected with line segments. The y-axis will be labeled as $[f(k)]$ or $f(kT)$ to indicate that it is a digital signal.

This is probably a good time to point out a source of confusion when performing signal processing with MATLAB. The subscripts of a MATLAB vector always start with 1, as in $x(1)$, $x(2)$, and so on. The subscripts of a signal usually start with zero, as in g_0, g_1, and so on. However, the subscripts of a signal might start with any integer value, even a negative value, as in h_{-2}, h_{-1}, h_0, and so on. Since many of the equations that relate to signal processing contain subscripts with these various possible subscripts, we would like to be able to use the equations without rewriting them to adjust the subscripts. This can often be accomplished by associating two vectors with a signal. One vector contains the subscripts that are associated with those values. Thus, if the signals g and h mentioned earlier in this paragraph contain 10 values, then the corresponding vectors g and h also contain 10 values. We can then use two additional vectors, such as kg and kh, to represent the subscripts that correspond to the 10 values in g and h. Thus, the vector kg would contain the values 0 to 9; the vector kh would contain the values -2 to 7. While the advantages of using this extra vector to represent subscripts will become clearer as we present examples using MATLAB and signals, the main advantage will be that we can use equations from signal processing without having to adjust the subscripts.

time domain

frequency domain

A signal is often analyzed in two domains—the time domain and the frequency domain. The time domain signal is represented by the data values; the frequency domain signal can be represented by complex values that represent the sinusoids (in

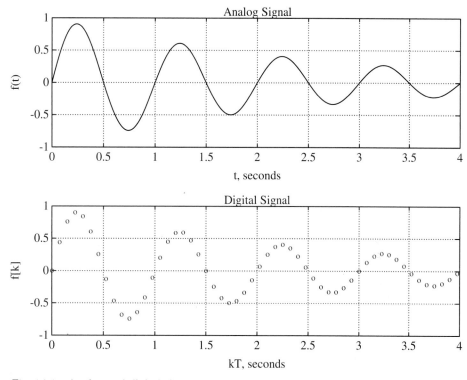

Fig. 14.1   Analog and digital signals.

cosine form) that compose the signal. Some types of information are most evident from the time domain representation of a signal. For example, by looking at a time domain plot, we can usually determine if the signal is periodic, or if it is a random signal. From the time domain values, we can also easily compute additional values such as mean, standard deviation, variance, and power. Other types of information, such as the frequency content of the signal, are not usually evident from the time domain, and thus we may need to compute the frequency content of the signal in order to determine if it is band-limited or if it contains certain frequencies. The frequency content of a signal is also called the frequency spectrum of the signal.

frequency spectrum

DFT algorithm

The discrete Fourier transform (DFT) algorithm [14] is used to convert a digital signal in the time domain into a set of points in the frequency domain. The input to the DFT algorithm is a set of N time values $[f_k]$; the algorithm then computes a set of N complex values $[F_k]$ that represent the frequency domain information. The DFT algorithm is in general a very computationally intensive algorithm, and may require a considerable amount of computer time if N is large. However, if the number of points is a power of two ($N = 2^M$), then a special algorithm called the fast Fourier transform (FFT) [14] can be used; the FFT algorithm greatly reduces the number of computations needed to convert the time signal into the frequency domain.

FFT algorithm

Since a digital signal is sampled every T seconds, there are $\frac{1}{T}$ samples per second; thus the sampling rate is $\frac{1}{T}$ samples per second, or $\frac{1}{T}$ Hz. The selection of the

aliasing

sampling rate for generating a digital signal must be done carefully to avoid aliasing, a problem that is caused by sampling too slowly. It can be shown that to avoid aliasing, a signal must be sampled at a sampling rate that is greater than twice the frequency of any sinusoid in the signal. Thus, if we are sampling a signal composed of the sum of two sinusoids, one with a frequency of 10 Hz and the other with a frequency of 35 Hz, then we must sample the signal with a sampling frequency greater

Nyquist frequency   than 70 Hz to avoid aliasing. The Nyquist frequency is equal to half the sampling frequency, and represents the upper limit of the frequencies that should be contained in the digital signal.

The MATLAB function for computing the frequency content of a signal is the

fft function   fft function. This function can be used with one or two input arguments. If a single input argument is used, then it is a vector containing a time signal, and the output of the function will be another vector of the same size containing complex values representing the frequency content of the input signal. If the number of values in the time signal is a power of 2, the function uses an FFT algorithm to compute the output values; if the number of values in the time signal is not a power of 2, the function uses a DFT algorithm to compute the output values. If two input arguments are used, the first argument is a vector containing the time signal and the second argument is an integer L specifying the number of points for the output vector. If L is a power of 2, then an FFT algorithm with L values is used; if L is not a power of 2, then a DFT with L values is used. If $L > N$, then $L-N$ zeros will be appended to the end of the time signal before the frequency domain values are computed; if $L < N$, then the first L values of the time signal will be used to compute the corresponding frequency domain values. In the remaining discussion, we assume N values are used in the FFT computation unless otherwise specified.

The frequency domain values computed by the fft function correspond to the

frequency
separation   frequencies separated by $\frac{1}{NT}$ Hz. Thus, if we have 32 samples of a time signal that was sampled at 1,000 Hz, the frequency values computed by the fft algorithm correspond to 0 Hz, $\frac{1}{0.032}$ Hz, $\frac{2}{0.032}$ Hz, and so on. These values are also equal to 0 Hz, 31.25 Hz, 62.5 Hz, and so on. The Nyquist frequency is equal to $\frac{1}{2T}$, and thus will correspond to the F_{16}. Since the discrete Fourier transform is a periodic function, the values above the Nyquist frequency do not represent new information, and thus only the first half of the output of the fft function is usually printed.

Consider the following set of MATLAB statements, which generate a time signal containing 64 samples:

```
N = 64;
T = 1/128;
```

```
k = 0:N-1;
f = sin(2*pi*20*k*T);
```

The signal f represents values of a 20-Hz sinusoid sampled every $\dfrac{1}{128}$ seconds, which is equivalent to a sampling rate of 128 Hz. (The sinusoid has a frequency of 20 Hz, so it should be sampled at a rate higher than 40 Hz. The specified sampling rate is 128 Hz, so we are assured that aliasing does not occur.) Figure 14.2 contains a plot of the digital signal f. (Note that the vector k gives the subscript values that correspond to the signal f.)

Since the signal f is a single sinusoid, we expect that the frequency content should be zero everywhere except at the point in the frequency domain that corresponds to 20 Hz. To determine the F_k that corresponds to 20 Hz, we need to compute the increment in Hz between points in the frequency domain, which is $\dfrac{1}{NT}$, or 2 Hz. Therefore, a 20-Hz component should appear as F_{10}, as we can see in Figure 14.3 which was generated with these additional statements:

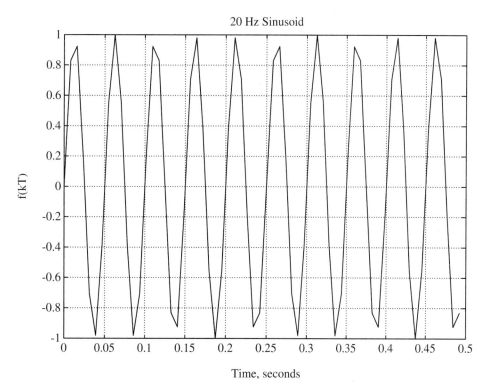

Fig. 14.2   Sinusoidal digital signal, f_k.

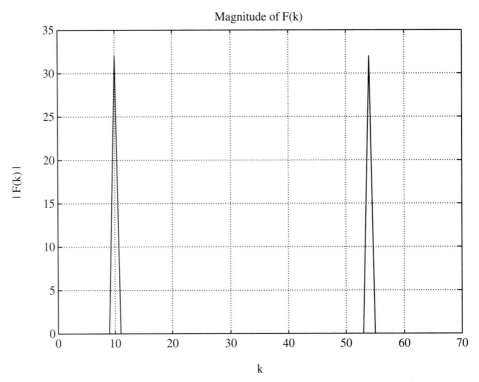

Fig. 14.3   Magnitude of F_k.

```
F = fft(f);
magF = abs(F);
plot(k,magF),title('Magnitude of F(k)'),...
xlabel('k'),ylabel('| F(k) |'),grid
```

In this plot, you can also see the symmetry that is caused by the periodicity of the DFT; the 20-Hz component also shows up as F_{54}. Also, note that the vector k contains the subscripts that correspond to the signals f and F. In general, we recommend that you plot only the first half of the magnitude values or the phase values to avoid the confusion that can occur when the symmetric components appear; the symmetry points can be misinterpreted as indicating that additional sinusoids are contained in the signal f. It is also convenient to plot the magnitude of F_k using an x-axis scale in Hz instead of the index k. The preferred plot of the magnitude of F_k is thus computed by the following statements, and shown in Figure 14.4.

```
hertz = k*(1/(N*T));
plot(hertz(1:N/2),magF(1:N/2)),...
title('Magnitude of F(k)'),...
xlabel('Hz'),ylabel('| F(k) |'),grid
```

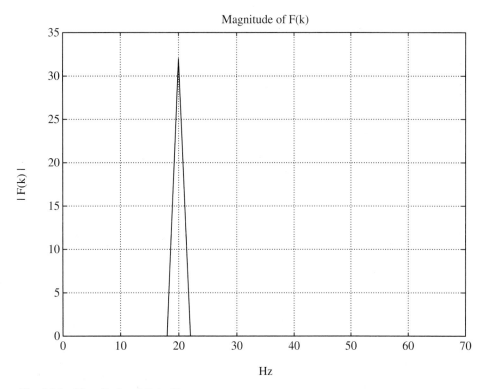

Fig. 14.4 Magnitude of F_k in Hz.

Suppose that the frequency of the sinusoid used in this example had been 19 Hz instead of 20 Hz. Since the increment in Hz between values of F_k for this example is 2 Hz, this sinusoid would appear at F_k where $k = 9.5$. However, the values of k are integers, so there is no value $F_{9.5}$. In this situation, the sinusoid appears at values of F near the computed index. For this example, the sinusoid appears primarily at values F_9 and F_{10}, which correspond to 18 and 20 Hz, as shown in Figure 14.5, which was generated with the following statements:

```
N = 64;
T = 1/128;
k = 0:N-1;
f = sin(2*pi*19*k*T);
magF = abs(fft(f));
hertz = k*(1/(N*T));
plot(hertz(1:N/2),magF(1:N/2)),...
title('Magnitude of F(k)'),...
xlabel('Hz'),ylabel('| F(k) |'),grid
```

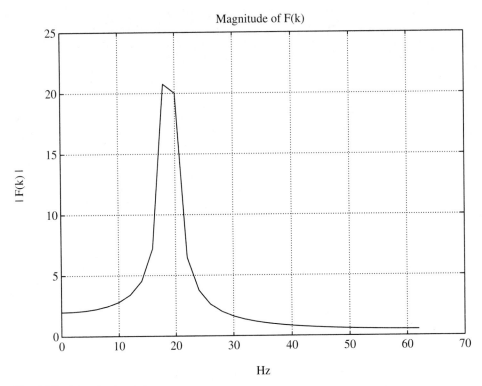

Fig. 14.5   Magnitude of signal with leakage.

Figures 14.4 and 14.5 both contain the frequency spectrum of a single sinusoid, but one sinusoid falls exactly on a point corresponding to an output point of the FFT algorithm, and the other does not. This is an example of "leakage," which occurs when a sinusoidal component does not fall exactly on one of the points in the FFT output. For more discussion on leakage, refer to a digital signal processing text.

leakage

inverse FFT algorithm

The ifft function uses the inverse FFT algorithm to compute the time domain signal $[f_k]$ from the complex values $[F_k]$. The following example computes the values of F_k, and then uses the ifft to compute the values of f_k from F_k. The final computation determines the sum of the differences between the original signal and the signal computed by the ifft function. The value printed was equal to zero.

```
N = 64;
T = 1/128;
k = 0:N-1;
f = sin(2*pi*19*k*T);
sum(f-ifft(fft(f)))
```

The FFT algorithm is an extremely powerful analysis tool for working with digital signals. Our discussion here has focused on the magnitude of the value F_k, but very important information is also obtained from the phase of F_k.

Practice!

Generate 128 points of the following signals. Plot the time domain signal. Then, using the FFT algorithm, generate and plot the frequency domain content. Plot only the first 64 points of the output of the FFT. Use a Hz scale on the x-axis. Assume a sampling rate of 1 kHz. Verify that the peaks occur where you expect them to occur in the frequency domain.

1. $f_k = 2 \sin(2\pi\, 50kT)$

2. $g_k = \cos(250\pi kT) - \sin(200\pi kT)$

3. $h_k = 5 - \cos(1000kT)$

4. $m_k = 4 \sin(250\pi kT - \pi/4)$

14.2   FILTER ANALYSIS

transfer function

The transfer function of an analog system is represented by a complex function $H(s)$, and the transfer function of a digital system is represented by a complex function $H(z)$. These transfer functions describe the effect of the system on an input signal, and thus also describe the filtering effect of the system. Both $H(s)$ and $H(z)$ are continuous functions of frequency, where $s = j\omega$ and $z = e^{j\omega T}$. (Recall that ω represents frequency in radians per second.) Thus, for a given frequency ω_0, assume that the transfer function magnitude is K and that the transfer function phase is ϕ. Then, if the input to the filter contains a sinusoid with frequency ω_0, the magnitude of the sinusoid will be multiplied by K and the phase will be incremented by ϕ. The effects of these changes are shown for both analog filters and digital filters in Figure 14.6.

lowpass
highpass

bandpass
bandstop

While the transfer function of a filter defines the effect of the filter in terms of frequencies, the transfer function can often be described in terms of the band of frequencies that it passes. For example, a lowpass filter will pass frequencies below a cutoff frequency and remove frequencies above the cutoff frequency. A highpass filter will pass frequencies above a cutoff frequency and remove frequencies below the cutoff frequency. A bandpass filter will pass frequencies within a specified band, and remove all other frequencies. A bandstop filter will remove frequencies within a specified band, and pass all other frequencies. Figure 14.7 contains example transfer functions for these four general types of filters. The transfer functions in Figure 14.7

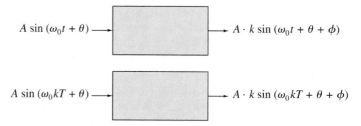

Fig. 14.6   Effect of filters on sinusoids.

Transfer Functions

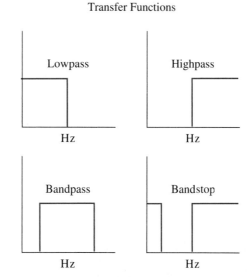

Fig. 14.7   Ideal transfer functions.

are ideal filters, with a frequency either passed or removed. We will see that it is not possible to design filters with exactly the same characteristics as these ideal filters.

Figure 14.8 contains an example of the magnitude of a typical lowpass filter that illustrates the types of characteristics of most lowpass filters. Instead of each frequency being passed or rejected, there are three regions—a passband, a transition band, and a stopband. These regions are defined by a cutoff frequency ω_c and a rejection frequency ω_r. Unless otherwise indicated, we will assume that the frequency that corresponds to a magnitude of 0.7 is the cutoff frequency, and the frequency that corresponds to a magnitude of 0.1 is the rejection frequency. With these definitions of the cutoff and rejection frequencies, we can then be more specific about the definition of the passband, the transition band, and the stopband. The passband contains frequencies with magnitudes above the magnitude of the cutoff frequency; the transition band contains frequencies with magnitudes between the magnitudes of the cutoff and rejection frequencies; the stopband contains frequencies with magnitudes less than the magnitude of the rejection frequency.

cutoff frequency
rejection
frequency

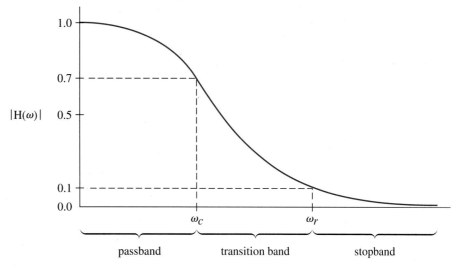

Fig. 14.8 Typical lowpass filter.

Since a transfer function is a complex function, the analysis of the corresponding filter usually includes plots of the magnitude and phase of the transfer function. The MATLAB functions abs, angle, and unwrap can be used to determine the magnitude and phase of the complex functions $H(s)$ and $H(z)$. In addition, the functions freqs and freqz can be used to compute the values of the functions $H(s)$ and $H(z)$ as will be shown in the following examples.

Analog Transfer Functions

An analog filter is defined by a transfer function $H(s)$ where $s = j\omega$. The general form of the transfer function $H(s)$ is the following:

$H(s)$

$$
\begin{aligned}
H(s) &= \frac{B(s)}{A(s)} \\
&= \frac{b_0 s^n + b_1 s^{n-1} + b_2 s^{n-2} + \ldots + b_n}{a_0 s^n + a_1 s^{n-1} + a_2 s^{n-2} + \ldots + a_n}
\end{aligned}
\tag{14.1}
$$

This transfer function corresponds to an nth order analog filter. Some examples of specific transfer functions are the following:

$$H_1(s) = \frac{0.5279}{s^2 + 1.0275s + 0.5279}$$

$$H_2(s) = \frac{s^2}{s^2 + 0.1117s + 0.0062}$$

$$H_3(s) = \frac{1.05s}{s^2 + 1.05s + 0.447}$$

$$H_4(s) = \frac{s^2 + 2.2359}{s^2 + 2.3511s + 2.2359}$$

331

freqs

In order to determine the characteristics of the systems with the transfer functions that are given, we need to plot the corresponding magnitude and phase functions. The MATLAB function `freqs` computes values of the complex function $H(s)$, using three input arguments. The first argument is a vector containing the coefficients of the polynomial $B(s)$ in Equation (14.1); the second argument is a vector containing the coefficients of the polynomial $A(s)$ in Equation (14.1); the third argument is a vector of frequency values in rps. The coefficient vectors come directly from the transfer function. It may require several trials to find an appropriate range of values for the frequency vector. In general, you want the frequency range to start at 0 and include all the critical information in the filter. Therefore, you want to be able to determine the filter type (lowpass, highpass, bandpass, bandstop), and you want to be able to determine the critical frequencies (cutoff, rejection).

The program below determines and plots the magnitudes of the four example transfer functions.

```
%
%       This program determines and plots the
%       magnitudes of four analog filters.
%
w1 = 0:0.05:5.0;
B1 = [0.5279];
A1 = [1,1.0275,0.5279];
H1s = freqs(B1,A1,w1);
%
w2 = 0:0.001:0.3;
B2 = [1,0,0];
A2 = [1,0.1117,0.0062];
H2s = freqs(B2,A2,w2);
%
w3 = 0:0.01:10;
B3 = [1.05,0];
A3 = [1,1.05,0.447];
H3s = freqs(B3,A3,w3);
%
w4 = 0:0.005:5;
B4 = [1,0,2.2359];
A4 = [1,2.3511,2.2359];
H4s = freqs(B4,A4,w4);
clg
subplot(221),plot(w1,abs(H1s)),title('Filter H1(s)'),...
xlabel('w, rps'),ylabel('Magnitude'),grid
subplot(222),plot(w2,abs(H2s)),title('Filter H2(s)'),...
xlabel('w, rps'),ylabel('Magnitude'),grid
subplot(223),plot(w3,abs(H3s)),title('Filter H3(s)'),...
xlabel('w, rps'),ylabel('Magnitude'),grid
```

```
subplot(224),plot(w4,abs(H4s)),title('Filter H4(s)'),...
xlabel('w, rps'),ylabel('Magnitude'),grid
```

Figure 14.9 contains the plots of these filter magnitudes. Note that there was no confusion due to the fact that the subscripts of the vectors A and B did not match the subscripts in Equation (14.1); the `freqs` function needed only the values in the vectors, and did not use the values of the subscripts.

angle unwrap

The phase of a filter can be plotted using the `angle` function or the `unwrap` function. Since the phase of a complex number is an angle in radians, the angle is only unique for a 2π interval. The output of the `angle` function will convert all angles to equivalent values between $-\pi$ and π, and thus has discontinuities at points where the phase exceeds these bounds. The `unwrap` function removes the 2π discontinuities introduced by the `angle` function when used in the reference `unwrap(angle(H))`.

Digital Transfer Functions

A digital filter is defined by a transfer function $H(z)$ where $z = e^{j\omega T}$. The variable z can be written as a function of frequency (ω), or it can written as a function of nor-

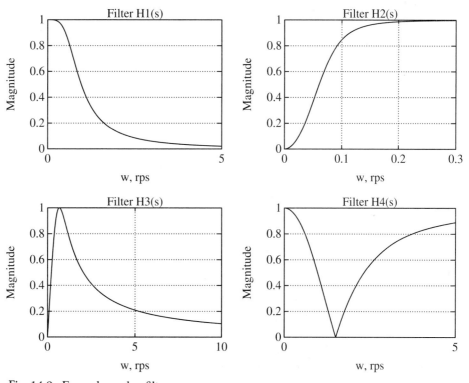

Fig. 14.9 Example analog filters.

malized frequency (ωT). If z is used as a function of frequency, then $H(z)$ is also a function of frequency. Since $H(z)$ is applied to input signals with a sampling time of T, then the appropriate range of frequencies is from 0 to the Nyquist frequency, which is $\dfrac{\pi}{T}$ rps or $\dfrac{1}{2T}$ Hz. If we assume that z is used as a function of normalized frequency, then $H(z)$ has a corresponding range of frequencies from 0 to π.

A general form of the transfer function $H(z)$ can be written in the following form:

H(z)

$$H(z) = \frac{B(z)}{A(z)}$$

$$= \frac{b_0 + b_1 z^{-1} + b_2 z^{-2} + \ldots + b_n z^{-n}}{a_0 + a_1 z^{-1} + a_2 z^{-2} + \ldots + a_n z^{-n}} \tag{14.2}$$

This transfer function corresponds to an nth order digital filter. Some examples of specific transfer functions are the following:

$$H_1(z) = \frac{0.2066 + 0.4131 z^{-1} + 0.2066 z^{-2}}{1 - 0.3695 z^{-1} + 0.1958 z^{-2}}$$

$$H_2(z) = \frac{0.894 - 1.789 z^{-1} + 0.894 z^{-2}}{1 - 1.778 z^{-1} + 0.799 z^{-2}}$$

$$H_3(z) = \frac{0.42 - 0.42 z^{-2}}{1 - 0.443 z^{-1} + 0.159 z^{-2}}$$

$$H_4(z) = \frac{0.5792 + 0.4425 z^{-1} + 0.5792 z^{-2}}{1 + 0.4425 z^{-1} + 0.1584 z^{-2}}$$

If a function $H(z)$ is written in terms of positive powers of z, the numerator and denominator can be divided by a power of z to convert them to the form of Equation (14.2). (It is also possible for a transfer function to include terms in the numerator with positive powers of z. We will discuss this case later in the chapter.)

difference equation

A digital filter can also be specified using a standard difference equation, SDE, which has this general form:

$$y_n = \sum_{k=-N_1}^{N_2} b_k x_{n-k} - \sum_{k=1}^{N_3} a_k y_{n-k} \tag{14.3}$$

There is a direct relationship between the difference equation and the transfer equation if we assume that N_1 in Equation (14.3) is equal to zero:

$$y_n = \sum_{k=0}^{N_2} b_k x_{n-k} - \sum_{k=1}^{N_3} a_k y_{n-k} \tag{14.4}$$

In this form, the coefficients $[b_k]$ and $[a_k]$ from Equation (14.4) are precisely the same coefficients $[b_k]$ and $[a_k]$ in Equation (14.2), with a_0 equal to 1. Thus, the difference equation that corresponds to the first example transfer function given after Equation (14.2) is the following:

$$y_n = 0.2066 x_n + 0.4131 x_{n-1} + 0.2066 x_{n-2} + 0.3695 y_{n-1} - 0.1958 y_{n-2}$$

If the coefficients $[a_k]$ are all equal to zero, with the exception of a_0, which is equal to 1, then the corresponding transfer function has a denominator polynomial equal to 1, as shown in the following example:

$$y_n = 0.5x_n - 1.2x_{n-1} + 0.25x_{n-3}$$
$$H(z) = 0.5 - 1.2z^{-1} + 0.25z^{-3}$$

FIR

IIR

If the denominator of the transfer function is equal to 1, the filter is an FIR (finite impulse response) filter; if the denominator of the transfer function is not equal to a constant, the filter is an IIR (infinite impulse response) filter. Both types of filters are commonly used in digital signal processing.

freqz

In order to determine the characteristics of a system with a given transfer function, we need to plot the magnitude and phase of the transfer function. The MATLAB function freqz computes values of the complex function $H(z)$ using three input arguments. The first argument is a vector containing the coefficients of the polynomial $B(z)$ in Equation (14.2); the second argument is a vector containing the coefficients of the polynomial $A(z)$ in Equation (14.2); the third argument is an integer that specifies the number of normalized frequency values to use over the interval $[0, \pi]$. The coefficient vectors come directly from the transfer function. The number of points used in computing the transfer function determines the resolution. The resolution should be fine enough that you can determine the filter type (lowpass, highpass, bandpass, bandstop), and the critical frequencies (cutoff, rejection).

normalized
frequency

This program determines and plots the magnitudes of the four example transfer functions given at the beginning of this discussion:

```
%
%      This program determines and plots the
%      magnitudes of four digital filters.
%
B1 = [0.2066,0.4131,0.2066];
A1 = [1,-0.3695,0.1958];
[H1z,w1T] = freqz(B1,A1,100);
%
B2 = [0.894,-1.789,0.894];
A2 = [1,-1.778,0.799];
[H2z,w2T] = freqz(B2,A2,100);
%
B3 = [0.42,0,-0.42];
A3 = [1,-0.443,0.159];
[H3z,w3T] = freqz(B3,A3,100);
%
B4 = [0.5792,0.4425,0.5792];
A4 = [1,0.4425,0.1584];
[H4z,w4T] = freqz(B4,A4,100);
clg
subplot(221),plot(w1T,abs(H1z)),title('Filter H1(z)'),...
```

```
xlabel('Normalized Frequency'),ylabel('Magnitude'),grid
subplot(222),plot(w2T,abs(H2z)),title('Filter H2(z)'),...
xlabel('Normalized Frequency'),ylabel('Magnitude'),grid
subplot(223),plot(w3T,abs(H3z)),title('Filter H3(z)'),...
xlabel('Normalized Frequency'),ylabel('Magnitude'),grid
subplot(224),plot(w4T,abs(H4z)),title('Filter H4(z)'),...
xlabel('Normalized Frequency'),ylabel('Magnitude'),grid
```

Figure 14.10 contains the plots of these filter magnitudes. Again, you can see that these four filters represent a lowpass filter, a highpass filter, a bandpass filter, and a bandstop filter.

group delay

The phase of a digital filter can be plotted using the `angle` function or the `unwrap` function. In addition, MATLAB also contains a function `grpdelay` that is used to determine the group delay of a digital filter. The group delay is a measure of the average delay of the filter as a function of frequency. It is defined as the negative first derivative of the filter's phase response. If $\theta(\omega)$ represents the phase re-

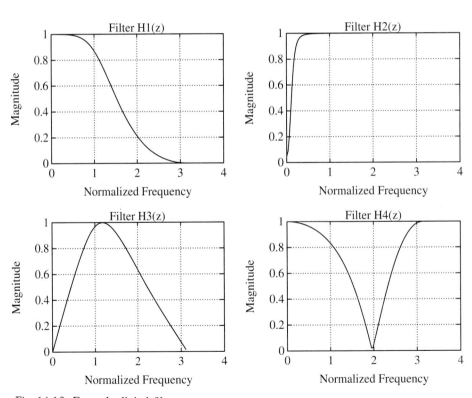

Fig. 14.10 Example digital filters.

sponse of the filter $H(z)$, then the group delay is

$$\tau(\omega) = -\frac{d\theta(\omega)}{d\omega}$$

The grpdelay function has three input arguments. The first two arguments are the coefficients of the $B(z)$ and $A(z)$ polynomials from Equation (14.2). The third argument is an integer that specifies the number of values of the group delay to determine over the range of normalized frequencies from zero to π. Since this function uses the fft function, it is desirable to select a value for the third argument that is a power of 2.

Partial Fraction Expansions

In analyzing both analog and digital filters, we often need to perform a partial fraction expansion of the transfer function $H(s)$ or $H(z)$. This partial fraction expansion may be used to express the filter in a cascaded structure or in a parallel structure of subfilters. A partial fraction expansion may also be used to perform an inverse transformation on a frequency domain function to obtain the corresponding time domain function. MATLAB includes a function residue that performs a partial fraction expansion (also called a residue computation) of the ratio of two polynomials B and A. Therefore, this function can be used with either $H(s)$ or $H(z)$, since both of these functions can be represented as the ratio of two polynomials. A precise definition of a partial fraction expansion is now needed in order to represent the most general case.

residue

Let G be a ratio of two polynomials in the variable v. Since G may represent an improper fraction, we can also express G as a mixed fraction as shown below:

$$G(v) = \frac{B(v)}{A(v)} \tag{14.5}$$

$$= \sum_{n=0}^{N} k_n v^n + \frac{N(v)}{D(v)} \tag{14.6}$$

If we then consider the proper fraction $\dfrac{N(v)}{D(v)}$, it can be written as a ratio of two polynomials in v. The denominator polynomial can be written as a product of linear factors with roots p_1, p_2, and so on. (The roots of the numerator are called the zeros of the function, and the roots of the denominator are called the poles of the function.) Each root of the denominator may represent a single root or a multiple root as shown below:

zeros
poles

$$\frac{N(v)}{D(v)} = \frac{b_1 v^{n-1} + b_2 v^{n-2} + \ldots + b_{n-1} v + b_n}{(v - p_1)^{m_1}(v - p_2)^{m_2} \ldots (v - p_r)^{m_r}}$$

This proper fraction can then be written as a sum of partial fractions. Single roots will correspond to one term in the partial fraction; a root with multiplicity k will correspond to k terms in the partial fraction expansion. This partial fraction expansion can then be written as:

$$\frac{N(v)}{D(v)} = \frac{C_{1,1}}{v - p_1} + \frac{C_{1,2}}{(v - p_1)^2} + \cdots \frac{C_{1,m_1}}{(v - p_1)^{m_1}}$$

$$+ \frac{C_{2,1}}{v - p_2} + \frac{C_{2,2}}{(v - p_2)^2} + \cdots \frac{C_{2,m_2}}{(v - p_2)^{m_2}}$$

$$+ \cdots$$

$$+ \frac{C_{r,1}}{v - p_r} + \frac{C_{r,2}}{(v - p_r)^2} + \cdots \frac{C_{r,m_r}}{(v - p_r)^{m_r}} \qquad (14.7)$$

The residue function has two inputs—the coefficients of the B polynomial and of the A polynomial from Equation (14.5). The residue function computes three output vectors—r, p, k. The r vector contains the coefficients $C_{i,j}$, the p vector contains the values of the poles p_n, and the k vector contains the values of k_n. Thus, to interpret the output of the residue function, it is necessary to understand the notation presented in Equations (14.5), (14.6), and (14.7) for factoring a polynomial. It is also important to recognize that a partial fraction expansion is generally not a unique expansion. There are often several different expansions that represent the same polynomial. Of course, the residue function will always give the same expansion for a given pair of numerator and denominator polynomials, but it may not be the only expansion for the polynomial.

To illustrate the use of the residue function, we now present several examples. First, consider the following polynomial:

$$F(z) = \frac{z^2}{z^2 - 1.5z + 0.5}$$

The MATLAB statement for computing the partial fraction expansion of this ratio of polynomials is the following:

```
B = [1, 0, 0];
A = [1, -1.5, 0.5];
[r, p, k] = residue (B, A)
```

The values of the three vectors computed by the residue function are the following:

$$r = \begin{bmatrix} 2.0 \\ -0.5 \end{bmatrix} \qquad p = \begin{bmatrix} 1.0 \\ 0.5 \end{bmatrix} \qquad k = [1]$$

Thus, the partial fraction expansion is the following:

$$F(z) = \frac{z^2}{z^2 - 1.5z + 0.5}$$

$$= 1 + \frac{2}{z - 1.0} - \frac{0.5}{z - 0.5}$$

To double-check this expansion, you can combine the three terms over a common denominator, or use the MATLAB functions for polynomial analysis presented in Chapter 10 to perform the combinations.

Consider the following function:

$$G(z) = \frac{z - 1}{z^2 + 4z + 4}$$

To use the residue function to factor this function, we use the following statements:

```
B = [1, -1];
A = [1, 4, 4];
[r, p, k] = residue (B, A)
```

The values of the three vectors computed by the residue function are the following:

$$r = \begin{bmatrix} 1 \\ -3 \end{bmatrix} \qquad p = \begin{bmatrix} -2 \\ -2 \end{bmatrix} \qquad k = [\,]$$

Thus, the partial fraction expansion is the following:

$$G(z) = \frac{z - 1}{z^2 + 4z + 4}$$

$$= \frac{1}{z - 2} - \frac{3}{(z - 2)^2}$$

As a final example, consider the following function:

$$H(z) = \frac{z^{-2}}{1 - 3.5z^{-1} + 1.5z^{-2}}$$

To fit the general form presented, we multiply numerator and denominator by z^2, giving the following equivalent form:

$$H(z) = \frac{1}{z^2 - 3.5z + 1.5}$$

To use the residue function to factor this function, we use the following statements:

```
B = [1];
A = [1, -3.5, 1.5];
[r, p, k] = residue (B, A)
```

The values of the three vectors computed by the residue function are the following:

$$r = \begin{bmatrix} 0.4 \\ -0.4 \end{bmatrix} \qquad p = \begin{bmatrix} 3.0 \\ 0.5 \end{bmatrix} \qquad k = [\,]$$

Thus, the partial fraction expansion is the following:

$$H(z) = \frac{1}{z^2 - 3.5z + 1.5}$$

$$= \frac{0.4}{z - 3} - \frac{0.4}{z - 0.5}$$

To write this using negative powers of z, we can multiply the numerator and denominator of each term by z^{-1}:

$$H(z) = \frac{0.4z^{-1}}{1 - 3z^{-1}} - \frac{0.4z^{-1}}{1 - 0.5z^{-1}}$$

Thus, the `residue` function can be used to determine terms with negative powers of z that can be useful in performing inverse z transformations.

Practice!

For each of the following transfer functions, plot the magnitude response. Determine the transition band or bands for these filters. Use normalized frequency on the x-axis for the digital filters.

1. $H(s) = \dfrac{s^2}{s^2 + \sqrt{2}\,s + 1}$

2. $H(z) = \dfrac{0.707z - 0.707}{z - 0.414}$

3. $H(z) = -0.163 - 0.058z^{-1} + 0.116z^{-2} + 0.2\,z^{-3}$
   $+ 0.116z^{-4} - 0.058z^{-5} - 0.163z^{-6}$

4. $H(s) = \dfrac{5s + 1}{s^2 + 0.4s + 1}$

14.3   DIGITAL FILTER IMPLEMENTATION

Analog filters are implemented in hardware using components such as resistors and capacitors. Digital filters are implemented in software, and thus this section refers specifically to digital filters. Recall from the previous section that a digital filter can be defined in terms of either a transfer function $H(z)$ or a standard difference equation. The input to the filter is a digital signal, and the output of the filter is another digital signal. The difference equation defines the steps involved in computing the output signal from the input signal. This process is shown in the diagram in Figure 14.11, with x_n as the input signal and y_n as the output signal.

Fig. 14.11   Digital filter input and output.

The relationship between the output signal y_n and the input signal x_n is described by the difference equation, as repeated in this general form:

$$y_n = \sum_{k=-N_1}^{N_2} b_k x_{n-k} - \sum_{k=1}^{N_3} a_k y_{n-k} \tag{14.8}$$

Examples of difference equations are the following:

$$y_n = 0.04x_{n-1} + 0.17x_{n-2} + 0.25x_{n-3} + 0.17x_{n-4} + 0.04x_{n-5}$$
$$y_n = 0.42x_n - 0.42x_{n-2} + 0.44y_{n-1} - 0.16y_{n-2}$$
$$y_n = 0.33x_{n+1} + 0.33x_n + 0.33x_{n-1}$$

The three difference equations described represent different filters with different characteristics. The output of the first filter depends only on past values of the input signal. For example, to compute y_{10}, we need values of x_9, x_8, x_7, x_6, and x_5. Then, using the difference equation, we can compute the value for y_{10}. This type of filter is an FIR filter (see the previous section), and the denominator of its transfer function $H(z)$ is equal to 1. The second filter requires values not only of the input signal, but past values also of the output signal in order to compute new output values. This type of filter is an IIR filter (also discussed in the previous section). The third filter is an FIR filter, since the output values depend only on the input signal. However, note that the subscripts in this third difference equation require that we be able to "look ahead" in the input signal. Thus, to compute y_5, we need values for x_6, x_5, and x_4. This look-ahead requirement is not a problem if we are computing the input signal values from a function or if they are stored in a file. However, it can be a problem if the input values are being generated in "real time" by an experiment.

real-time filter

filter

The simplest way to apply a digital filter to an input signal in MATLAB is with the filter function. The filter function assumes that the standard difference equation has the following form:

$$y_n = \sum_{k=0}^{N_2} b_k x_{n-k} - \sum_{k=1}^{N_3} a_k y_{n-k} \tag{14.9}$$

which also corresponds to the following transfer function form that was discussed in the previous section:

$$H(z) = \frac{B(z)}{A(z)}$$

$$= \frac{b_0 + b_1 z^{-1} + b_2 z^{-2} + \ldots + b_n z^{-n}}{a_0 + a_1 z^{-1} + a_2 z^{-2} + \ldots + a_n z^{-n}}$$

Equation (14.9) is different from Equation (14.8) in that the first summation begins with $k = 0$ instead of $k = -N_1$.

The first two arguments of the filter function are the vector of coefficients $[b_k]$ and the vector of coefficients $[a_k]$. The third argument is the input signal, as shown in these statements that apply the first filter example to a signal x.

```
B = [0.0,0.04,0.17,0.25,0.17,0.04];
A = [1];
y = filter(B,A,x);
```

To apply the second filter example to a signal x, we could use the following statements:

```
B = [0.42,0.0,-0.42];
A = [-0.44,0.16];
y = filter(B,A,x);
```

We cannot use the filter function to apply the third filter to a signal x_k, because the difference equation does not fit the general form used by the filter function, Equation (14.9). The difference equation of the third filter requires that the first summation start with $k = -1$, not $k = 0$. In this case, we implement the filter using vector arithmetic. Assuming that the input signal x is stored in the vector x, we can compute the corresponding output signal y using the following statements:

```
N = length(x);
y(1) = 0.33*x(1) + 0.33*x(2);
for n=2:N-1
    y(n) = 0.33*x(n+1) + 0.33*x(n) + 0.33*x(n-1);
end
y(N) = 0.33*x(N-1) + 0.33*x(N);
```

Note that we assume that the values of x for which we do not have a value $[x(-1)$ and $x(N + 1)]$ are equal to zero. Another way to compute the signal y is the following:

```
N = length(x);
y(1) = 0.33*x(1) + 0.33*x(2);
y(2:N-1) = 0.33*x(3:N) + 0.33*x(2:N-1) + 0.33*x(1:N-2);
y(N) = 0.33*x(N-1) + 0.33*x(N);
```

Any digital filter could be implemented using vector operations. However, the filter function generally provides a simpler solution.

The filter function can be used with two output assignments, as in:

```
[y,state] = filter(b,a,x);
```

The vector y contains the output of the filter when the vector x is the input signal, and the vector state contains the final set of values used by the filter. Thus, if we want to filter a vector $x2$ that is another segment of the signal x, we can specify that the initial conditions are the values in state, and then the output values will be computed as though x and $x2$ were one long vector instead of two separate vectors:

```
y2 = filter(b,a,x2,state);
```

conv

Finally, the conv function can be used to compute the output of an FIR filter. Since this function is restricted to only FIR filters, we prefer to use the conv function for polynomial multiplication and to use the filter function for implementing

deconv

filters. The deconv function can be used to compute the impulse response for an IIR filter, but is generally used to perform polynomial division. The conv and deconv functions were discussed in Chapter 12, which covers polynomial analysis.

Practice!

The following transfer function has been designed to pass frequencies between 500 Hz and 1500 Hz in a signal sampled at 5 kHz:

$$H(z) = \frac{0.42z^2 - 0.42}{z^2 - 0.443z + 0.159}$$

Use the following signals as input to the filter. Plot the input and output of the filter on the same plot, and determine the effect of the filter on the input signal.

1. $x_k = \sin(2\pi\,1000kT)$

2. $x_k = 2\cos(2\pi\,100kT)$

3. $x_k = -\sin(2\pi\,2000kT)$

4. $x_k = \cos(2\pi\,1600kT)$

14.4   DIGITAL FILTER DESIGN

In this section we present the MATLAB functions for designing digital filters. The discussion is separated into two techniques for designing IIR filters and one technique for designing FIR filters.

IIR Filter Design Using Analog Prototypes

Butterworth

MATLAB contains functions for designing four types of digital filters based on analog filter designs. Butterworth filters have maximally flat passbands and stop-

Chebyshev
elliptic

ripple

bands, Chebyshev Type I filters have ripple in the passband, Chebyshev Type II filters have ripple in the stopband, and elliptic filters have ripple in both the passband and the stopband. However, for a given filter order, elliptic filters have the sharpest transition (narrowest transition band) of all these filters. The Chebyshev filters have a sharper transition than a Butterworth filter with the same design specifications. Figure 14.12 illustrates the definitions of the passband ripple (Rp) and the stopband ripple (Rs). The values of Rp and Rs are specified in decibels (where x in decibels is equal to $-20 \log_{10} x$).

The functions for designing digital lowpass IIR filters using analog prototypes have the following format:

```
[B,A] = butter(N,Wn);
[B,A] = cheby1(N,Rp,Wn);
[B,A] = cheby2(N,Rs,Wn);
[B,A] = ellip(N,Rp,Rs,Wn);
```

The input arguments represent the order of the filter (N), the ripple (Rp and Rs), and the normalized cutoff frequency (Wn). The normalized frequency is based on a scale with the Nyquist frequency equal to 1.0. (Note that this differs from the normalized frequency scale used by the freqz function.) The output vectors B and A are the vectors in the general expression for IIR filters given in Equation (14.2). Remember that the vectors B and A can be used to determine the transfer function $H(z)$ or the standard difference equation. They can also be used in the freqz function and the filter function.

To design bandpass filters, the arguments appear to be the same as the arguments for designing lowpass filters. However, Wn should be a vector containing two elements that represent the normalized frequencies specifying the passband of frequencies from $Wn(1)$ to $Wn(2)$.

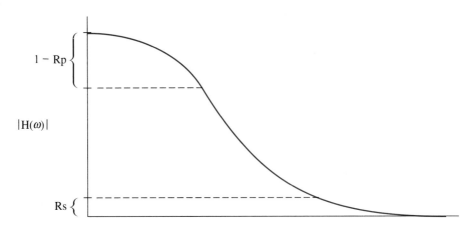

Fig. 14.12 Regions with ripple.

To design a highpass filter, an additional parameter with the literal 'high' should be added to the function reference, as shown in these references:

```
[B,A] = butter(N,Wn,'high');
[B,A] = cheby1(N,Rp,Wn,'high');
[B,A] = cheby2(N,Rs,Wn,'high');
[B,A] = ellip(N,Rp,Rs,Wn,'high');
```

To design bandstop filters, the arguments are the same as in highpass filters, but with the literal 'stop' instead of the literal 'high'. In addition, the argument Wn must be a vector containing two values that define the band of frequencies from $Wn(1)$ to $Wn(2)$ to be rejected.

To illustrate the use of these functions, suppose that we want to design a highpass Chebyshev Type II filter of order 6. We would also like to limit the passband ripple to 0.1, or 20 db. The filter is to be used with a signal sampled at 1 kHz, and thus the Nyquist frequency is 500 Hz. The cutoff is to be 300 Hz, and thus the normalized frequency is 300/500, or 0.6. The statements to design this filter, and to then plot its magnitude characteristics are the following:

```
[B,A] = cheby2(6,20,0.6,'high');
[H,wT] = freqz(B,A,100);
T = 0.001;
hertz = wT/(2*pi*T);
plot(hertz,abs(H)),title('Highpass Filter'),...
xlabel('Hz'),ylabel('Magnitude'),grid
```

The plot from these statements is shown in Figure 14.13. To apply this filter to a signal x, we could use the following statement:

```
y = filter(B,A,x);
```

The values of the vectors B and A can also be used to determine the filter's standard difference equation.

Direct IIR Filter Design

Yule–Walker designs

MATLAB contains a function for performing Yule–Walker filter designs. This design technique can be used to design an arbitrarily shaped, possibly multiband frequency response. The command to design a filter with this function is the following:

```
[B,A] = yulewalk(n,f,m);
```

The output vectors B and A contain the coefficients of the nth order IIR filter. The vectors f and m specify the frequency–magnitude characteristics of the filter over the frequency range from 0 to 1, which represents the range from 0 to the Nyquist

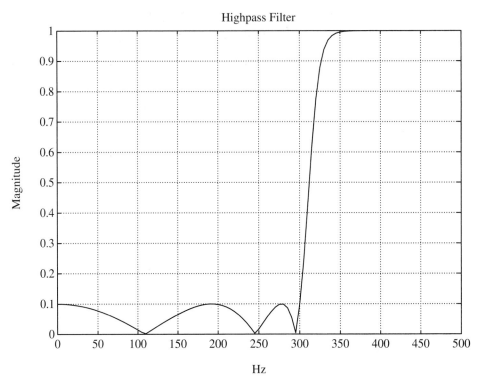

Fig. 14.13   Chebyshev Type II filter.

frequency. The frequencies in f must begin with 0, end with 1, and be increasing. The magnitudes in m must correspond to the frequencies in f, and represent the desired magnitude for each frequency. Thus, the following example designs a filter with two passbands, and then plots the magnitude response in normalized frequency:

```
m = [0 0 1 1 0 0 1 1 0 0];
f = [0 .1 .2 .3 .4 .5 .6 .7 .8 1];
[B,A] = yulewalk(12,f,m);
[H,wT] = freqz(B,A,100);
plot(f,m,wT/pi,abs(H)),...
title('IIR Filter with Two Passbands'),...
xlabel('Normalized Frequency'),...
ylabel('Magnitude'),grid
```

Figure 14.14 contains the plot from these statements.

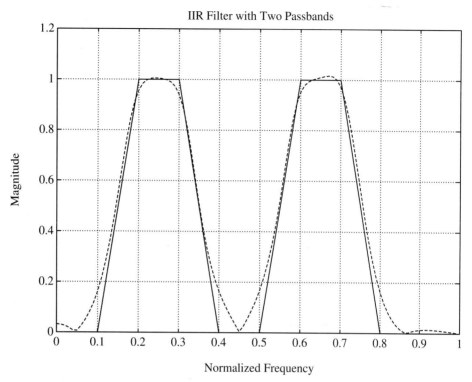

Fig. 14.14   Yule–Walker design method.

Direct FIR Filter Design

Parks–McClellan design

Remez exchange algorithm

FIR filters are designed in MATLAB using the Parks–McClellan filter design algorithm that uses a Remez exchange algorithm [15]. Recall that FIR filters require only a B vector because the denominator polynomial of $H(z)$ is equal to 1. Therefore, the MATLAB function remez computes only a single output vector, as shown in this statement:

```
B = remez(n,f,m);
```

The first input argument defines the order of the filter, while the constraints on the values of f and m are very similar to the constraints on similar input arguments for the Yule–Walker filter design technique. The vectors f and m specify the frequency-magnitude characteristics of the filter over the frequency range from 0 to 1, which represents the range from 0 to the Nyquist frequency. The frequencies in f must begin with 0, end with 1, and be in increasing orders. The magnitudes in m must correspond to the frequencies in f, and represent the desired magnitude for each frequency. In addition, the number of points of *f* and *m* must be an even number. To obtain desirable filter characteristics with an FIR filter, it is not unusual for the filter order to become large.

The following example designs a filter with two passbands, and then plots the magnitude response in normalized frequency:

```
m = [0 0 1 1 0 0 1 1 0 0];
f = [0 .1 .2 .3 .4 .5 .6 .7 .8 1];
B = remez(50,f,m);
[H,wT] = freqz(B,[1],100);
plot(f,m,wT/pi,abs(H)),...
title('FIR Filter with Two Passbands'),...
xlabel('Normalized Frequency'),...
ylabel('Magnitude'),grid
```

Figure 14.15 contains the plot from these statements.

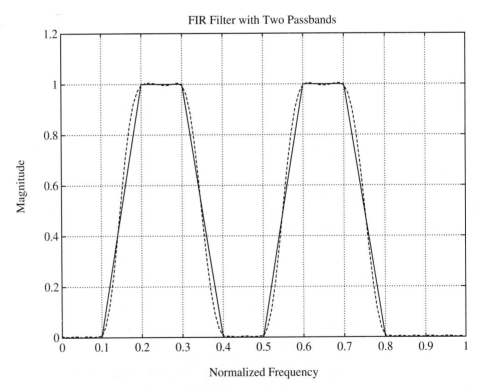

Fig. 14.15   Remez exchange design method.

weighting vector

Additional variations of the remez function allow it to be used to design Hilbert transformers or differentiators. A weighting vector can also be used to give a weighting or priority to the values in each band defined by f and m.

Practice!

Use the MATLAB functions described in this section to design the following filters. Plot the magnitude of the filter designed to confirm that it has the correct characteristics.

1. Lowpass IIR filter with a cutoff of 75 Hz when used with a sampling rate of 500 Hz. (Use an order 5 filter.)

2. Highpass IIR filter with a cutoff of 100 Hz when used with a sampling rate of 1 kHz. (Use an order 6 filter.)

3. Lowpass FIR filter with a cutoff of 75 Hz when used with a sampling rate of 500 Hz. (Use an order 40 filter.)

4. Bandpass FIR filter with a passband of 100 to 200 Hz when used with a sampling rate of 1 kHz. (Use an order 80 filter.)

 PROBLEM SOLVING APPLIED: CHANNEL SEPARATION FILTERS

Images collected from spacecraft sent into deep space or from satellites circling the earth are transmitted back to the earth in data streams. These data streams are converted into digitized signals that contain the information that can be reconstructed into the original images. Information collected by other sensors is also transmitted back to earth. The frequency content of the information in the sensor signals depends on the type of data being measured.

Modulation techniques can be used to move the frequency content of the data to specified frequency bands so that a signal can contain multiple signals at one time. For example, suppose that we want to send three signals in parallel. The first signal contains components from 0 to 100 Hz, the second signal contains components from 500 Hz to 1 kHz, and the third signal contains components from 2 kHz to 5 kHz. Assume that the signal containing these three components is sampled at 10 kHz. To separate these components after the signal is received, we need a lowpass filter with a cutoff at 100 Hz, a bandpass filter with cutoffs at 500 Hz and 1 kHz, and a highpass filter with a cutoff at 2 kHz. The order of the filters should be large enough to generate narrow transition bands so that frequencies from one component do not contaminate other components.

1. PROBLEM STATEMENT

Design three filters to be used with a signal sampled at 10 kHz. One filter is to be a lowpass filter with a cutoff of 100 Hz; another is to be a bandpass filter with a passband from 500 Hz to 1 kHz; another is to be a highpass filter with a cutoff of 2 kHz.

 2. INPUT/OUTPUT DESCRIPTION

There are not input values for this problem. The output values are the coefficient vectors that define the three filters $H_1(z)$, $H_2(z)$, and $H_3(z)$, as shown in Figure 14.16.

 3. HAND EXAMPLE

The sketch in Figure 14.17 shows the frequency range from 0 to the Nyquist frequency (5 kHz) with the three desired filters. We will use Butterworth filters in order to have filters with flat passbands and flat stopbands. We may need to experiment with the filter orders in order to be sure that the transition bands of the filters do not interfere with each other.

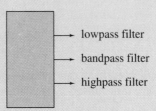

Fig. 14.16   I/O diagram.

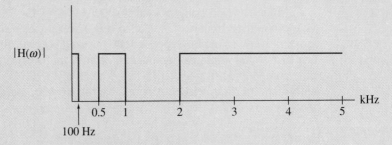

Fig. 14.17   Sketch of desired filters.

4. MATLAB SOLUTION

The following MATLAB program determines the normalized frequency values (between 0 and 1, where 1 represents the Nyquist frequency) for the cutoff frequencies in the butter function. After computing the coefficients for the filters, we use the freqz function to plot the filter characteristics. Recall that the freqz function normalizes the frequencies to values between 0 and π, where π represents the Nyquist frequency. We will use Hz as the units for the frequency axis in order to easily verify the characteristics of the filters designed.

```
%
%       This program designs three digital filters
%       for use in a channel separation problem.
%
fs = 10000;             % sampling frequency
T = 1/fs;               % sampling time
fn = fs/2;              % Nyquist frequency
f1n = 100/fn;           % normalized lowpass cutoff
f2n = 500/fn;           % normalized bandpass left cutoff
f3n = 1000/fn;          % normalized bandpass right cutoff
f4n = 2000/fn;          % normalized highpass cutoff
%
[B1,A1] = butter(8,f1n);
[B2,A2] = butter(7,[f2n,f3n]);
[B3,A3] = butter(10,f4n,'high');
%
[H1,wT] = freqz(B1,A1,200);
[H2,wT] = freqz(B2,A2,200);
[H3,wT] = freqz(B3,A3,200);
%
hertz = wT/(2*pi*T);
plot(hertz,abs(H1),'-',hertz,abs(H2),'-',hertz,abs(H3),'-'),...
title('Channel Separation Filters'),...
xlabel('Hz'),ylabel('Magnitude'),grid
```

 5. TESTING

The filter magnitudes for all three filters are shown on the same plot to verify that the filters do not overlap with each other, as shown in Figure 14.18.

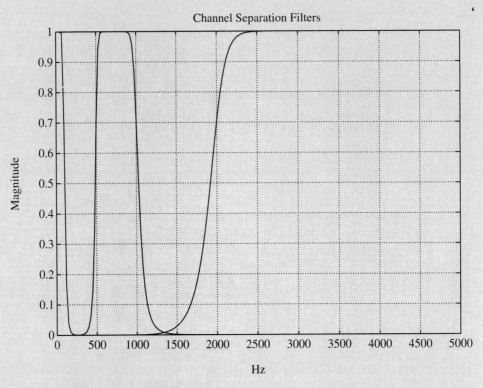

Fig. 14.18   Three channel separation filters.

SUMMARY

A number of MATLAB functions for performing signal processing operations were presented. The `fft` function was presented for analyzing the frequency content of a digital signal. The `freqs` and `freqz` functions were discussed for computing the frequency content of an analog or a digital filter from a transfer function. From the complex signal that represents the frequency content, it is then straightforward to compute and plot the magnitude and phase of the filter. The `filter` function can be used to implement either an IIR or an FIR filter. Finally, a number of functions were presented for designing IIR and FIR filters.

MATLAB SUMMARY

This MATLAB summary lists all the special symbols, commands, and functions that were defined in this chapter. A brief description is also included for each one.

Commands and Functions

`butter`	designs a Butterworth digital filter
`cheby1`	designs a Chebyshev Type I digital filter
`cheby2`	designs a Chebyshev Type II digital filter
`ellip`	designs an elliptic digital filter
`fft`	computes the frequency content of a signal
`filter`	applies a digital filter to an input signal
`freqs`	computes the analog frequency content
`freqz`	computes the digital frequency content
`grpdelay`	measures the group delay of a digital filter
`ifft`	computes the inverse FFT
`remez`	designs an optimal FIR digital filter
`residue`	performs a partial fraction expansion
`unwrap`	removes 2π discontinuities in a phase angle
`yulewalk`	designs an optimal IIR digital filter

PROBLEMS

Problems 1 to 6 relate to the engineering application presented in this chapter. Problems 7 to 23 relate to new engineering applications.

Channel Separation Filters These problems relate to the channel separation filter design problem. In these problems we develop a computer simulation of this system.

1. We first want to generate signals in the three bands described in this filter. We will do this using sums of sinusoids, all of which are sampled at 10 kHz. Signal 1 should contain a sum of sinusoids with frequencies at 25 Hz, 40 Hz, and 75 Hz. Signal 2 should contain a sum of sinusoids with frequencies at 500 Hz, 730 Hz, and 850 Hz. Signal 3 should contain a sum of sinusoids with frequencies at 3,500 Hz, 4,000 Hz, and 4,200 Hz. Choose a variety of amplitudes and phase shifts for the sinusoids. Plot 500 points of signal 1, signal 2, and signal 3 in separate graphs.

2. Compute and plot the magnitude and phase of each of the three signals generated in problem 1. Use Hz as the units for the x-axis in the plots. Make sure that the sinusoidal components appear where they would be expected.

3. Add the three time signals generated in problem 1. Plot the time signal. Also plot the magnitude of the time signal, using Hz as the units for the x-axis.

4. Apply the lowpass filter to the signal generated in problem 3. Plot the output of the filter (time domain), and plot the magnitude of the frequency content of the output of the filter. Compare these plots to those generated in problems 1 and 2. The time plot in this problem should be similar to the one generated in problem 1 for signal 1, with perhaps a phase shift. The magnitude plots should be very similar.

5. Repeat problem 4 using the bandpass filter. Compare these plots to those generated in problems 1 and 2. The time plot in this problem should be similar to the one generated in problem 1 for signal 2, with perhaps a phase shift. The magnitude plots should be very similar.

6. Repeat problem 4 using the highpass filter. Compare these plots to those generated in problems 1 and 2. The time plot in this problem should be similar to the one generated in problem 1 for signal 3, with perhaps a phase shift. The magnitude plots should be very similar.

Filter Characteristics   For each of the following filters, determine the passband(s), transition band(s), and stopband(s). Use 0.7 to determine cutoff frequencies and use 0.1 to determine rejection frequencies.

7. $H(s) = \dfrac{0.5279}{s^2 + 1.0275s + 0.5279}$

8. $H(s) = \dfrac{s^2}{s^2 + 0.1117s + 0.0062}$

9. $H(s) = \dfrac{1.05s}{s^2 + 1.05s + 0.447}$

10. $H(s) = \dfrac{s^2 + 2.2359}{s^2 + 2.3511s + 2.2359}$

11. $H(z) = \dfrac{0.2066 + 0.4131z^{-1} + 0.2066z^{-2}}{1 - 0.3695z^{-1} + 0.1958z^{-2}}$

12. $H(z) = \dfrac{0.894 - 1.789z^{-1} + 0.894z^{-2}}{1 - 1.778z^{-1} + 0.799z^{-2}}$

13. $H(z) = \dfrac{0.42 - 0.42z^{-2}}{1 - 0.443z^{-1} + 0.159z^{-2}}$

14. $H(z) = \dfrac{0.5792 + 0.4425z^{-1} + 0.5792z^{-2}}{1 + 0.4425z^{-1} + 0.1584z^{-2}}$

15. $y_n = 0.04x_{n-1} + 0.17x_{n-2} + 0.25x_{n-3} + 0.17x_{n-4} + 0.04x_{n-5}$

16. $y_n = 0.42x_n - 0.42x_{n-2} + 0.44y_{n-1} - 0.16y_{n-2}$

17. $y_n = 0.33x_{n+1} + 0.33x_n + 0.33x_{n-1}$

18. $y_n = 0.33x_n + 0.33x_{n-1} + 0.33x_{n-2}$

Filter Design   The following problems use the filter design functions discussed in this chapter. Use Hz as the units for the x-axis in all plots of magnitude or phase.

19. Design a lowpass filter with a cutoff frequency of 1 kHz when used with a sampling frequency of 8 kHz. Compare designs for the four standard IIR filter types with an order 8 filter by plotting the magnitudes of the four designs on the same plot.

20. Design a highpass filter with a cutoff frequency of 500 Hz when used with a sampling frequency of 1,500 Hz. Compare designs using an order 8 elliptic filter to an order 32 FIR filter by plotting the magnitudes of the two designs on the same plot.

21. Design a bandpass filter with a passband of 300 Hz to 4,000 Hz when used with a sampling frequency of 9.6 kHz. Compare designs using a Butterworth filter of order 8 to an FIR filter. Choose the order of the FIR filter so that the band of frequencies passed are similar. Plot the magnitudes of the two designs on the same plot.

22. Design a filter that removes frequencies from 500 Hz to 1,000 Hz in a signal that is sampled at 10 kHz. Compare an elliptic filter of order 12 to an order 12 Yule–Walker filter design. Plot the magnitudes of the two designs on the same plot.

23. Design a filter that will eliminate frequencies between 100 and 150 Hz and between 500 and 600 Hz in a signal that is sampled at 2.5 kHz. Compare designs using an FIR filter and an IIR filter. Plot the magnitudes of the two designs on the same plot.

15

Courtesy of Lawrence Livermore National Laboratory

GRAND CHALLENGE: Nuclear Fusion

Nova, the world's most powerful laser system, is being used by scientists and engineers at Lawrence Livermore National Laboratory to make energy the way the sun does. Nuclear fusion is the process that stars use to generate the massive amounts of heat and light that sustain life on earth. If the process can be successfully implemented on earth, the promise is great—clean, unlimited energy. Nova is attempting to develop nuclear fusion using a technique called inertial confinement fusion (ICF). Ten huge lasers, each longer than a football field, are tuned and brought together through large pipes to focus at the center of the spherical target chamber shown in the photograph. For an instant, 200 times more power than can be produced by all the electrical generating plants in the United States strikes a tiny fuel pellet that is located at the point where the laser beams meet. The lasers compress the pellet by a factor of 30, the equivalent of compressing a basketball to the size of a pea, and for about a billionth of a second, a tiny star is created.

Control Systems

Introduction

The student edition of MATLAB has an extensive set of functions that are very useful for linear system and control system design and analysis. These functions, which are combined in the Signals and Systems Toolbox, have been selected from the Signal Processing Toolbox and the Control Systems Toolbox, which are available for the professional version of MATLAB. Many of the design and analysis tasks associated with linear systems and control systems involve matrix operations, complex arithmetic, root determination, model conversions, and plotting of complicated functions. As we have seen earlier, MATLAB has been designed to make many of these operations easy to do. This chapter is divided into three topics—system modeling, model conversion functions, and analysis functions. Since linear system theory and control system theory are extensive fields of study, they cannot be addressed in depth in this chapter. Therefore, the information that follows assumes at least some familiarity with the topics.

15.1   SYSTEM MODELING

Linear and control system analysis and design begins with models of real systems. These models, which are mathematical representations of such things as chemical processes, machinery, and electrical circuits, are used to study the dynamic response of real systems. The mathematical techniques used by MATLAB to design and analyze these systems assume processes that are physically realizable, linear, and time-invariant (LTI). Thus, the models themselves are similarly constrained; nonlinear, time-varying systems either cannot be analyzed or must be approximated by LTI functions.

LTI system

MATLAB uses models in the form of transfer functions or state-space equations, thus allowing both "classical" and "modern" control system design and analysis techniques to be used. Either model form can be expressed in continuous-time (analog) or discrete-time (digital) forms. Transfer functions can be expressed as a polynomial, a ratio of polynomials, or one of two factored forms: zero-pole-gain or partial-fraction form. State-space system models are particularly well suited to MATLAB, since they are a matrix-based expression.

To demonstrate the various ways models can be formulated, we will use the classic example of a spring-mass-damper system shown in Figure 15.1. In this system, a mass m is acted on by three forces: a time-dependent input force $u(t)$, a spring with spring constant k, and a viscous damper with damping constant b. The position of the mass as a function of time is represented by $x(t)$. Attached to the mass is a measurement potentiometer p that provides an output voltage $y(t)$ proportional to $x(t)$. The equation of motion of the mass m is given by the second-order differential equation

$$mx'' + bx' + kx = u(t)$$

and the measurement equation for the potentiometer is

$$y(t) = px(t)$$

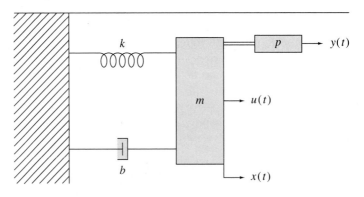

Fig. 15.1   Spring-mass-damper system.

The equation for the potentiometer is an example of a situation where the variable representing the dynamics of system (x in this case) is not the output variable (y in this case). Together, these two equations provide a mathematical model of the dynamic behavior of the system. By integrating the equation of motion, using the techniques discussed in Chapter 12, we can determine the motion of the mass as a function of time. Such an analysis would therefore be called a time-domain analysis.

Transfer Functions

The analysis of linear systems and control systems often involves determining certain dynamical properties, such as stability and frequency response, that cannot easily be determined using time-domain analyses. For these analyses, we often perform a Laplace transform on the time-domain equation so that we can analyze the system in the frequency domain. The Laplace transform of our spring-mass-damper differential equation is

Laplace transform

$$(ms^2 + bs + k)x(s) = u(s)$$

where s is a complex variable ($\sigma + j\omega$), called the Laplace variable. (The complex variable s used in Chapter 14 in defining the Fourier transform assumed that $\sigma = 0$.) This equation is easily rearranged to give a transfer function $H(s)$, which relates the output motion of the system $x(s)$ to the input force $u(s)$:

$$H(s) = \frac{x(s)}{u(s)}$$

$$= \frac{1}{ms^2 + bs + k}$$

The transfer function for the potentiometer is simply

$$\frac{y(s)}{x(s)} = p$$

block diagram

Block diagrams are frequently used to show how the transfer functions and the input and output variables of a system are related. Assuming for our spring-mass-damper example that $m = 1$, $b = 4$, $k = 3$, and $p = 10$, the block diagram in Figure 15.2 depicts the system. The first block represents the plant model, which is the part of the system that is controlled, and the second block represents the measurement model.

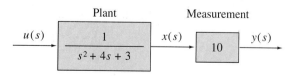

Fig. 15.2   Plant model and measurement model.

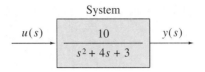

Fig. 15.3   System model.

We can also combine the blocks into a single system model block as shown in Figure 15.3. This transfer function is expressed as a ratio of two polynomials, where the numerator polynomial is simply a scalar. For systems having a single input and a single output (SISO), the form for this method of writing transfer functions is

continuous
polynomial form

$$H(s) = \frac{b_0 s^n + b_1 s^{n-1} + \ldots + b_{n-1}s + b_n}{a_0 s^m + a_1 s^{m-1} + \ldots + a_{m-1}s + a_m}$$

More generally, the numerator of this transfer function can be a three-dimensional matrix for multi-input/multi-output (MIMO) systems. Since MATLAB matrices are two-dimensional, we are limited to single-input/multiple-output (SIMO) system representations.

Very often, the numerator and denominator of a transfer function are factored into the zero-pole-gain form, which is

continuous
zero-pole-gain
form

$$H(s) = k \frac{(s - z_1)(s - z_2) \ldots (s - z_n)}{(s - p_1)(s - p_2) \ldots (s - p_m)}$$

For example, the zero-pole-gain form of the transfer function

$$H(s) = \frac{3s^2 + 18s + 24}{s^3 + 9s^2 + 23s + 15}$$

is

$$H(s) = 3 \frac{(s + 2)(s + 4)}{(s + 1)(s + 3)(s + 5)}$$

This form is particularly useful because it directly shows the roots of the numerator and denominator polynomials, which are called the zeros and the poles of the system, respectively.

Finally, transfer functions can also be written in the partial-fraction expansion or residue form, which is

continuous
partial-fraction
form

$$H(s) = \frac{r_1}{s - p_1} + \frac{r_2}{s - p_2} + \ldots + \frac{r_n}{s - p_n} + k(s)$$

This form is useful in determining the inverse Laplace transform and for designing certain types of filters. For more on the use of the residue form, see the section on filter analysis in Chapter 14.

State-Space Models

In Chapter 12, we showed how a second or higher-order differential equation could be expressed as a set of coupled first-order differential equations. This technique is also the basis of the matrix or state-space model form. Using our earlier spring-mass-damper example, whose equation of motion was

$$mx'' + bx' + kx = u(t)$$

we can define

$$x_1 = x$$
$$x_2 = x'$$

Next, we rewrite the second-order differential equation as a set of coupled first-order differential equations:

$$x_1' = x_2$$
$$x_2' = -\frac{k}{m}x_1 - \frac{b}{m}x_2 + \frac{u}{m}$$
$$= -3x_1 - 4x_2 + u$$

and the measurement equation as:

$$y = g(x, u)$$
$$= 10x_1$$

Using matrix notation, this system model can now be written as

continuous state-space form

$$x' = Ax + Bu$$
$$y = Cx + Du$$

which, for this example represents

$$\begin{bmatrix} x_1' \\ x_2' \end{bmatrix} = \begin{bmatrix} 0 & 1 \\ -3 & -4 \end{bmatrix} \begin{bmatrix} x_1 \\ x_2 \end{bmatrix} + \begin{bmatrix} 0 \\ 1 \end{bmatrix} u$$
$$y = \begin{bmatrix} 10 & 0 \end{bmatrix} \begin{bmatrix} x_1 \\ x_2 \end{bmatrix} + [0]\, u$$

Discrete-Time Systems

Many systems have variables that vary only at discrete times, or are available for measurement or use only at discrete times. Some of these systems are analog systems whose continuous variables are sampled at regular time intervals; others may be digital systems whose quantized variables are similarly sampled.

z-transform

Discrete-time systems are analyzed in a manner similar to continuous-time systems. The main difference is that they utilize the z-transform rather than the Laplace transform to derive transfer functions. The discrete-time variable z is mathematically related to the continuous-time Laplace variable s by the equation

$$z = e^{sT}$$

where T is the sampling time.

MATLAB represents discrete-time systems using the same forms that it does for continuous-time systems: as polynomial, zero-pole-gain, and partial-fraction transfer functions, and as state-space equations. For example, the discrete-time version of a transform that is expressed as the ratio of two polynomials is

discrete
polynomial form

$$H(z) = \frac{b_0 z^n + b_1 z^{n-1} + \ldots + b_{n-1} z + b_n}{a_0 z^m + a_1 z^{m-1} + \ldots + a_{m-1} z + a_m}$$

The zero-pole-gain and partial-fraction discrete-time transfer function forms are similarly related to the continuous-time forms.

The state-space equations for discrete-time systems are also very similar to those for continuous systems:

discrete
state-space form

$$x[n + 1] = Ax[n] + Bu[n]$$
$$y[n] = Cx[n] + Du[n]$$

where n indicates the current sample, and $n + 1$ indicates the next sample. Notice in the discrete-time form that $x[n + 1]$ replaces x' in the continuous-time form. This is because discrete-time systems use difference equations instead of differential equations. The equations for discrete-time systems compute the value of the state vector at the next sample time, rather than the derivative of the state vector at the current time, which continuous-time state-space equations compute.

In summary, there are several different forms of system models for both continuous-time and discrete-time systems. The next section presents MATLAB functions that can be used to convert from one model form to another.

15.2 MODEL CONVERSION

MATLAB has a number of functions that make it easy to convert from one model form to another and to convert continuous-time systems into discrete-time systems. These conversion functions and their uses are summarized in Table 15.1

TABLE 15.1 Model Conversion Functions

FUNCTION	PURPOSE
c2d	continuous state-space to discrete state-space
residue	partial fraction expansion
ss2tf	state-space to transfer function
ss2zp	state-space to zero-pole-gain
tf2ss	transfer function to state-space
tf2zp	transfer function to zero-pole-gain
zp2ss	zero-pole-gain to state-space
zp2tf	zero-pole-gain to transfer function

Following are descriptions of each conversion function and examples of how to use them.

c2d Function   The c2d function converts the continuous-time state-space equation

$$x' = Ax + Bu$$

to the discrete-time state-space equation

$$x[n + 1] = A_d x[n] + B_d u[n]$$

The function has two output matrices as shown in this statement:

```
[ad, bd] = c2d(a, b, Ts);
```

The input arguments a and b of the c2d function are the matrices A and B of the continuous-time state-space equation that is to be converted, and Ts is the desired sample period. The outputs ad and bd are the matrices A_d and B_d of the discrete-time equation.

For example, our earlier continuous-time state-space plant equation

$$x' = Ax + Bu$$

where

$$A = \begin{bmatrix} 0 & 1 \\ -3 & -4 \end{bmatrix}, \qquad B = \begin{bmatrix} 0 \\ 1 \end{bmatrix}$$

can be converted to a discrete-time state-space equation with a sampling period of 0.1 seconds using the statements

```
a = [0 1; -3 -4];
b = [0 1]';
[ad, bd] = c2d(a, b, 0.1);
```

The values of the matrices computed by the c2d function are the following:

$$ad = \begin{bmatrix} 0.9868 & 0.0820 \\ -0.2460 & 0.6588 \end{bmatrix}, \qquad bd = \begin{bmatrix} 0.0044 \\ 0.0820 \end{bmatrix}$$

Thus, the discrete-time state-space equation of the plant model is

$$x[n + 1] = A_d x[n] + B_d u[n]$$

which represents

$$\begin{bmatrix} x_1 \\ x_2 \end{bmatrix}_{n+1} = \begin{bmatrix} 0.9868 & 0.0820 \\ -0.2460 & 0.6588 \end{bmatrix} \begin{bmatrix} x_1 \\ x_2 \end{bmatrix}_n + \begin{bmatrix} 0.0044 \\ 0.0820 \end{bmatrix} u_n$$

residue Function   The residue function converts the polynomial transfer function:

$$H(s) = \frac{b_0 s^n + b_1 s^{n-1} + \ldots + b_{n-1} s + b_n}{a_0 s^m + a_1 s^{m-1} + \ldots + a_{m-1} s + a_m}$$

to the partial-fraction transfer function:

$$H(s) = \frac{r_1}{s - p_1} + \frac{r_2}{s - p_2} + \ldots + \frac{r_n}{s - p_n} + k(s)$$

The function has three output matrices as shown in this statement:

```
[r,p,k] = residue(b,a);
```

The input arguments b and a of the residue function are the coefficients of the numerator and denominator polynomials, respectively, of the polynomial transfer function that is to be converted. They must be entered as row vectors in descending powers of s. The outputs include r, a column vector containing the residue values; p, a column vector of the corresponding pole locations; and k, a row vector of the direct terms. For a more detailed discussion on the residue function that includes the consideration of multiple poles at the same value, see Chapter 14. The discussion in Chapter 14 also illustrates how to compute a partial fraction expansion of the discrete-time transfer function $H(z)$.

For example, the partial-fraction expansion of our system transfer function

$$\frac{y(s)}{u(s)} = \frac{10}{s^2 + 4s + 3}$$

can be computed with the statements

```
b = [10];
a = [1 4 3];
[r,p,k] = residue(b,a);
```

The values of the matrices computed by the residue function are the following:

$$r = \begin{bmatrix} -5 \\ 5 \end{bmatrix}, \qquad p = \begin{bmatrix} -3 \\ -1 \end{bmatrix}, \qquad k = [\,]$$

Thus, the partial-fraction expansion of our system polynomial transfer function is

$$H(s) = \frac{y(s)}{u(s)}$$

$$= \frac{-5}{s + 3} + \frac{5}{s + 1}$$

ss2tf Function   The ss2tf function converts the continuous-time, state-space equations

$$x' = Ax + Bu$$
$$y = Cx + Du$$

to the polynomial transfer function

$$H(s) = \frac{b_0 s^n + b_1 s^{n-1} + \ldots + b_{n-1} s + b_n}{a_0 s^m + a_1 s^{m-1} + \ldots + a_{m-1} s + a_m}$$

The function has two output matrices as shown in this statement:

```
[num, den] = ss2tf (a, b, c, d, iu);
```

Since multi-input systems can be converted for one input at a time by MATLAB, the input arguments a, b, c, and d of the ss2tf function are the matrices A, B, C, and D of the state-space equations corresponding to the iu'th input. In the case of a single-input system, iu is 1. The output vectors num and den contain the coefficients, in descending powers of s, of the numerator and denominator of the polynomial transfer function for the iu'th input.

For example, the state-space equations for our system

$$\begin{bmatrix} x_1' \\ x_2' \end{bmatrix} = \begin{bmatrix} 0 & 1 \\ -3 & -4 \end{bmatrix} \begin{bmatrix} x_1 \\ x_2 \end{bmatrix} + \begin{bmatrix} 0 \\ 1 \end{bmatrix} u$$

$$y = \begin{bmatrix} 10 & 0 \end{bmatrix} \begin{bmatrix} x_1 \\ x_2 \end{bmatrix} + \begin{bmatrix} 0 \end{bmatrix} u$$

can be converted to a polynomial transfer function using the statements

```
a = [0 1; -3 -4];
b = [0 1]';
c = [10 0];
d = 0;
iu = 1;
[num, den] = ss2tf (a, b, c, d, iu);
```

The values of the vectors computed by the ss2tf function are the following:

$$\text{num} = \begin{bmatrix} 0 & 0 & 10 \end{bmatrix}, \qquad \text{den} = \begin{bmatrix} 1 & 4 & 3 \end{bmatrix}$$

Thus, the transfer function is

$$\frac{y(s)}{u(s)} = \frac{10}{s^2 + 4s + 3}$$

The ss2tf function can also be used with discrete-time systems.

ss2zp Function   The ss2zp function converts the continuous-time, state-space equations

$$x' = Ax + Bu$$
$$y = Cx + Du$$

to the zero-pole-gain transfer function

$$H(s) = k \frac{(s - z_1)(s - z_2) \ldots (s - z_n)}{(s - p_1)(s - p_2) \ldots (s - p_m)}$$

The function has three ouput matrices as shown in this statement:

```
[z, p, k] = ss2zp (a, b, c, d, iu) ;
```

The input arguments a, b, c, and d of the ss2zp function are the matrices A, B, C, and D of the state-space equations corresponding to the iu'th input, where iu is the number of the input for a multi-input system. In the case of a single-input system, iu is 1. The output vectors z and p are the zeros and poles, respectively, of the zero-pole-gain transfer function for the iu'th input, and k is the associated gain.
    For example, the state-space equations of our system

$$\begin{bmatrix} x_1' \\ x_2' \end{bmatrix} = \begin{bmatrix} 0 & 1 \\ -3 & -4 \end{bmatrix} \begin{bmatrix} x_1 \\ x_2 \end{bmatrix} + \begin{bmatrix} 0 \\ 1 \end{bmatrix} u$$

$$y = \begin{bmatrix} 10 & 0 \end{bmatrix} \begin{bmatrix} x_1 \\ x_2 \end{bmatrix} + [0] u$$

can be converted to a zero-pole-gain transfer function using these statements:

```
a = [0 1; -3 -4];
b = [0 1]';
c = [10 0];
d = 0;
iu = 1;
[z, p, k] = ss2zp (a, b, c, d, iu) ;
```

The values of the matrices computed by the ss2zp function are the following:

$$z = [\,], \qquad p = \begin{bmatrix} -1 \\ -3 \end{bmatrix}, \qquad k = [10]$$

Thus, the zero-pole-gain transfer function is

$$\frac{y(s)}{u(s)} = \frac{10}{(s + 1)(s + 3)}$$

The ss2zp function can also be used with discrete-time systems.

tf2ss Function   The tf2ss function converts the polynomial transfer function

$$H(s) = \frac{b_0 s^n + b_1 s^{n-1} + \ldots + b_{n-1} s + b_n}{a_0 s^m + a_1 s^{m-1} + \ldots + a_{m-1} s + a_m}$$

to the controller canonical form state-space equations

$$x' = Ax + Bu$$
$$y = Cx + Du$$

The function has four output matrices as shown in this statement:

```
[a, b, c, d] = tf2ss (num, den) ;
```

The input arguments num and den of the tf2ss function are vectors that contain the coefficients, in descending powers of s, of the numerator and denominator polynomials of the transfer function that is to be converted. The outputs a, b, c, and d are the matrices of the controller canonical form state-space equations.

For example, the polynomial transfer function

$$\frac{y(s)}{u(s)} = \frac{10}{s^2 + 4s + 3}$$

can be converted to controller canonical form state-space equations using these statements:

```
num = 10;
den = [1 4 3];
[a, b, c, d] = tf2ss (num, den) ;
```

The values of the matrices computed by the tf2ss function are the following:

$$a = \begin{bmatrix} -4 & -3 \\ 1 & 0 \end{bmatrix}, \qquad b = \begin{bmatrix} 1 \\ 0 \end{bmatrix}, \qquad c = [0 \quad 10], \qquad d = [0]$$

Thus, the controller-canonical form state-space equations are

$$\begin{bmatrix} x_1' \\ x_2' \end{bmatrix} = \begin{bmatrix} -4 & -3 \\ 1 & 0 \end{bmatrix} \begin{bmatrix} x_1 \\ x_2 \end{bmatrix} + \begin{bmatrix} 1 \\ 0 \end{bmatrix} u$$
$$y = [0 \quad 10] \begin{bmatrix} x_1 \\ x_2 \end{bmatrix}$$

(The controller-canonical form is a variation of the state-space form presented earlier.) The tf2ss function can also be used with discrete-time systems.

tf2zp Function   The tf2zp function converts the polynomial transfer function

$$H(s) = \frac{b_0 s^n + b_1 s^{n-1} + \ldots + b_{n-1}s + b_n}{a_0 s^m + a_1 s^{m-1} + \ldots + a_{m-1}s + a_m}$$

to the zero-pole-gain transfer function

$$H(s) = k\frac{(s - z_1)(s - z_2) \ldots (s - z_n)}{(s - p_1)(s - p_2) \ldots (s - p_m)}$$

The function has three output matrices as shown in this statement:

```
[z,p,k] = tf2zp(num,den);
```

The input arguments num and den are vectors containing the coefficients, in descending powers of s, of the numerator and denominator of the polynomial transfer function that is to be converted. The output vectors z and p are the zeros and poles of the zero-pole-gain transfer function, and k is the associated gain.

For example, the polynomial transfer function

$$\frac{y(s)}{u(s)} = \frac{10}{s^2 + 4s + 3}$$

can be converted to a pole-zero-gain transfer function using these statements:

```
num = 10;
den = [1 4 3];
[z,p,k] = tf2zp(num,den);
```

The values of the matrices computed by the tf2zp function are the following:

$$z = [\,], \qquad p = \begin{bmatrix} -3 \\ -1 \end{bmatrix}, \qquad k = [10]$$

Thus, the zero-pole-gain transfer function is:

$$\frac{y(s)}{u(s)} = \frac{10}{(s + 3)(s + 1)}$$

The tf2zp function can also be used with discrete-time systems.

zp2ss Function The zp2ss function converts the zero-pole-gain transfer function

$$H(s) = k\frac{(s - z_1)(s - z_2) \ldots (s - z_n)}{(s - p_1)(s - p_2) \ldots (s - p_m)}$$

to the controller canonical form state-space equations

$$x' = Ax + Bu$$
$$y = Cx + Du$$

The function has four output matrices as shown in this statement:

```
[a,b,c,d] = zp2ss(z,p,k);
```

The input argument p is a column vector of the pole locations of the zero-pole-gain transfer function that we wish to convert to state-space equations. Input argument z is a matrix of the corresponding zero locations, having one column for each output of a multi-output system. In the case of a single-output system, z is a column vector of the zero locations corresponding to the pole locations of vector p. The third input argument k is the gain of the zero-pole-gain transfer function. The outputs a, b, c and d are the matrices of the controller canonical form state-space equations.

For example, the zero-pole-gain transfer function

$$\frac{y(s)}{u(s)} = \frac{10}{(s + 3)(s + 1)}$$

can be converted to the controller canonical state-space representation using the statements

```
z = [];
p = [-3 -1]';
k = 10;
[a,b,c,d] = zp2ss(z,p,k);
```

The values of the matrices computed by the zp2ss function are the following:

$$a = \begin{bmatrix} -4 & -1.7321 \\ 1.7321 & 0 \end{bmatrix}, \qquad b = \begin{bmatrix} 1 \\ 0 \end{bmatrix}, \qquad c = [0 \quad 5.7735], \qquad d = 0$$

Thus, the controller canonical form state-space equations are

$$\begin{bmatrix} x_1' \\ x_2' \end{bmatrix} = \begin{bmatrix} -4 & -1.7321 \\ 1.7321 & 0 \end{bmatrix} \begin{bmatrix} x_1 \\ x_2 \end{bmatrix} + \begin{bmatrix} 1 \\ 0 \end{bmatrix} u$$

$$y = [0 \quad 5.7735] \begin{bmatrix} x_1 \\ x_2 \end{bmatrix}$$

(This controller canonical form is equivalent to the one computed earlier using tf2ss)

zp2tf Function   The zp2tf function converts the zero-pole-gain transfer function

$$H(s) = k \frac{(s - z_1)(s - z_2) \ldots (s - z_n)}{(s - p_1)(s - p_2) \ldots (s - p_m)}$$

into the polynomial transfer function

$$H(s) = \frac{b_0 s^n + b_1 s^{n-1} + \ldots + b_{n-1} s + b_n}{a_0 s^m + a_1 s^{m-1} + \ldots + a_{m-1} s + a_m}$$

The function has two output matrices as shown in this statement:

```
[num,den] = zp2tf(z,p,k);
```

The input argument p is a column vector of the pole locations of the zero-pole-gain transfer function that we wish to convert to a polynomial transfer function. Input argument z is a matrix of the corresponding zero locations, having one column for each output of a multi-output system. In the case of a single-output system, z is a column vector of the zero locations corresponding to the pole locations of vector p. The third input argument k is the gain of the zero-pole-gain transfer function. The output vectors num and den contain the coefficients, in descending powers of s, of the numerator and denominator of the polynomial transfer function.

For example, this zero-pole-gain transfer function

$$\frac{y(s)}{u(s)} = \frac{10}{(s + 3)(s + 1)}$$

can be converted into a polynomial transfer function using the statements

```
z = [];
p = [-3 -1]';
k = 10;
[num,den] = zp2tf(z,p,k);
```

The values of the matrices computed by the zp2tf function are the following:

$$num = \begin{bmatrix} 0 & 0 & 10 \end{bmatrix}, \qquad den = \begin{bmatrix} 1 & 4 & 3 \end{bmatrix}$$

Thus, the polynomial transfer function is

$$\frac{y(s)}{u(s)} = \frac{10}{s^2 + 4s + 3}$$

Practice!

Consider the polynomial transfer function

$$H(s) = \frac{s + 3}{s^2 + 4s - 12}$$

1. Use the residue function to find the partial fraction expansion.

2. Use the tf2ss function to convert the transfer function to continuous-time state-space equations.

3. Use the c2d function to convert the continuous-time state-space equations of problem 2 to discrete-time equations, using a sample period of 0.5 second.

4. Use the tf2zp function to convert the polynomial transfer function to a zero-pole-gain tranfer function.

5.   Use the `zp2ss` function to convert the zero-pole-gain function in problem 4 to continuous-time state-space equations.

15.3   DESIGN AND ANALYSIS FUNCTIONS

MATLAB has several functions that are useful for designing and analyzing linear systems. These functions can be used to study the response of systems in both the time domain and frequency domain, and to design Kalman filters, which are used in optimal estimation and control problems. The functions that are described in this section are summarized in Table 15.2.

Bode Plots

This frequency-domain analysis tool consists of two separate plots of the amplitude ratio and the phase angle of a transfer function plotted versus the frequency of an input sinusoid. When plotted using the customary scales of amplitude ratio in decibels and phase angle in degrees, versus the $\log_{10}$ of frequency, these plots are referred to as Bode plots, after H. W. Bode [16], where the nondimensional amplitude ratio AR

decibels (dB)   in decibels (dB) is defined as

$$AR_{dB} = 20 \log_{10} AR$$

Nichols plots   Nichols plots can also be generated by plotting the log of the amplitude versus the phase angle. Both plots are extremely useful for designing and analyzing control systems.

The bode function calculates the magnitude and phase frequency responses of continuous-time linear-time-invariant systems for use in making Bode and Nichols plots. This function can be used in a variety of ways. If it is used without left-side arguments, it generates a Bode plot. It can also be used with output arguments that represent the magnitude and phase of the transfer function defined by the input to the function, as shown next:

TABLE 15.2   Design and Analysis Functions

FUNCTION	PURPOSE
bode	magnitude and phase frequency response plots
nyquist	Nyquist frequency response plot
rlocus	Evans root-locus plot
step	unit step time response
lqe	linear-quadratic estimator
lqr	linear-quadratic regulator

```
[mag,phase] = bode(num,den);
[mag,phase] = bode(num,den,w);
[mag,phase] = bode(a,b,c,d);
[mag,phase] = bode(a,b,c,d,iu);
[mag,phase] = bode(a,b,c,d,iu,w);
```

We now illustrate the use of some of these forms with examples.

The input arguments num and den are vectors containing the coefficients, in descending powers of s, of the numerator and denominator polynomials of the transfer function for which the Bode plot is needed. For example, the Bode plot for the second-order system transfer function

$$\frac{x(s)}{u(s)} = \frac{10}{s^2 + s + 3}$$

can be generated with these statements:

```
num = 10;
den = [1 1 3];
bode(num,den),...
title('Bode Plot for a Second-Order System')
```

The plot generated is shown in Figure 15.4.

A Bode plot using a user-specified input frequency vector w can also be generated. For example, the following statements produce the plots in Figure 15.5 for 100 logarithmically evenly spaced points between the frequencies 10^{-1} and 10^2 radians/second:

```
num = 10;
den = [1 1 3];
w = logspace(-1,2,100);
bode(num,den,w),...
title('Bode Plot for a Second-Order System')
```

A Bode plot can also be generated using input arguments that represent the continuous state-space system:

$$x' = Ax + Bu$$
$$y = Cx + Du$$

For example, the state-space equations for the previous example are

$$\begin{bmatrix} x_1' \\ x_2' \end{bmatrix} = \begin{bmatrix} 0 & 1 \\ -3 & -1 \end{bmatrix} \begin{bmatrix} x_1 \\ x_2 \end{bmatrix} + \begin{bmatrix} 0 \\ 1 \end{bmatrix} u$$

$$y = \begin{bmatrix} 10 & 0 \end{bmatrix} \begin{bmatrix} x_1 \\ x_2 \end{bmatrix} + 0$$

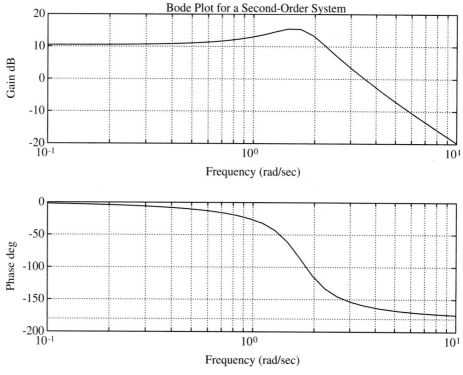

Fig. 15.4   Bode plot for second-order system.

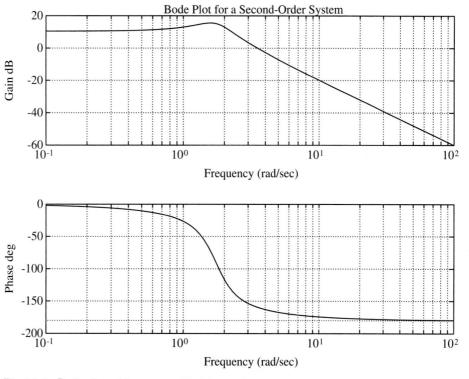

Fig. 15.5   Bode plot with user-specified frequencies.

Thus, these statements will produce the same Bode plot as shown in Figure 15.4:

```
a = [0 1; -3 -1];
b = [0 1]';
c = [10 0];
d = 0;
bode(a,b,c,d),...
title(`Bode Plot for a Second-Order System')
```

When used with multi-input systems, bode (a,b,c,d) produces a series of Bode plots—one for each input. A plot for a specific input iu of a multi-input system can be obtained using bode (a,b,c,d,iu). Including a vector of user-specified frequency values w in the argument list will cause the bode function to calculate the system response at those frequencies. Note, however, that if w is included in the argument list of a state-space system, iu must also be included, even for a single-input system.

Adding the left-side arguments in the bode function statement causes the function to calculate, but not plot the amplitude, phase, and frequency vectors. This makes it possible to plot the results in different formats. For example, the following statements plot the Bode plot and the Nichols plot of the previous state-space system, using 100 logarithmically evenly spaced frequency values:

```
w = logspace(-1,2,100);
[mag,phase,w] = bode(a,b,c,d,1,w);
%
subplot(211),semilogx(w,20*log10(mag)),...
title('Bode Plot'),...
xlabel('Frequency (rad/sec)'),ylabel('Gain (dB)'),grid,...
subplot(212),semilogx(w,phase),...
xlabel('Frequency (rad/sec)'),ylabel('Phase (deg)'),grid
%
subplot(111),axis([-180 180 -20 20]),...
axis('square'),plot(phase,20*log10(mag)),...
title('Nichols Plot'),...
xlabel('Phase (deg)'),ylabel('Gain (dB)'),grid
```

These statements produce the plots shown in Figures 15.6 and 15.7.

Nyquist Plots

The nyquist function is a frequency-domain analysis function that is similar to the bode function in that it uses exactly the same input arguments to produce frequency response plots. The difference between the two functions is the output. The Nyquist plot is a single plot, as opposed to two plots for the Bode plot. It plots the real component of the open-loop transfer function versus the imaginary component for differ-

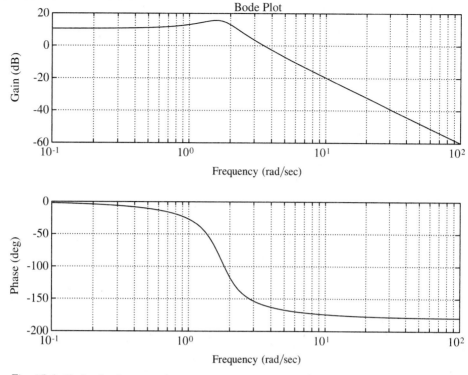

Fig. 15.6   Bode plot for a continuous state-space system.

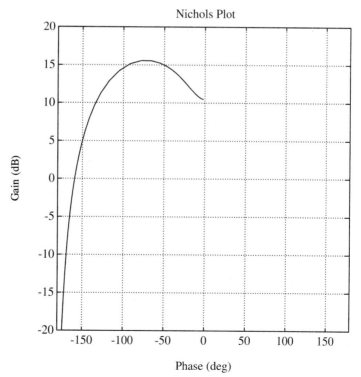

Fig. 15.7   Nichols plot for a continuous state-space system.

ent values of frequency. It is also frequently referred to as a Nyquist path or as a polar plot. The Nyquist plot is most often used for stability analysis.

The nyquist function can be used for continuous-time, linear-time-invariant systems. The various forms for its use are shown below:

```
[re, im, w] = nyquist (num, den);
[re, im, w] = nyquist (num, den, w);
[re, im, w] = nyquist (a, b, c, d);
[re, im, w] = nyquist (a, b, c, d, iu);
[re, im, w] = nyquist (a, b, c, d, iu, w);
```

The input arguments represent the same system information as described by the input arguments for the Bode function. The output matrices re and im are the real and imaginary components of the system response, respectively, and w is the vector of frequencies at which the response is computed.

We now illustrate the use of some of these forms with examples. As is the case with the bode function, a Nyquist plot can be generated without using the left-side arguments. For example, the Nyquist plot of the spring-mass-damper system model, which is shown again in Figure 15.8, can be produced using these statements:

```
num = 10;
den = [1 4 3];
nyquist (num, den), title ('Nyquist Plot')
```

which produce the plot shown in Figure 15.9. In this case, the function automatically selects the frequency range.

Similarly, the statement nyquist (num, den, w) produces a Nyquist plot using a user-specified input frequency vector w. For example, the statements

```
num = 10;
den = [1 4 3];
w = logspace (0, 1, 100);
nyquist (num, den, w), title ('Nyquist Plot')
```

result in the plot shown in Figure 15.10 for 100 logarithmically evenly spaced points between the frequencies of 1 and 10 radians/second.

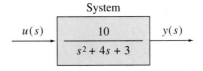

Fig. 15.8   System model.

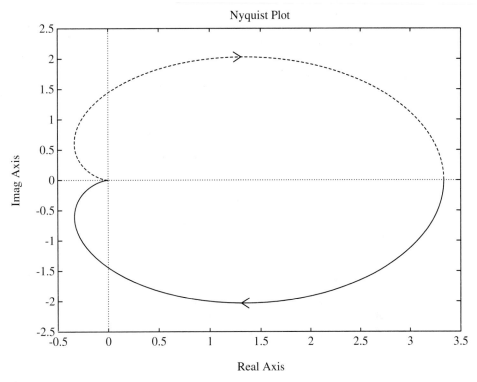

Fig. 15.9   Nyquist plot of spring-mass-damper system.

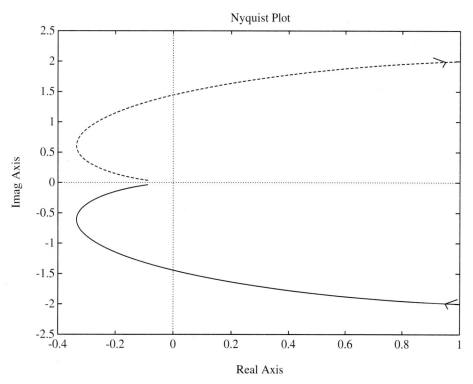

Fig. 15.10   Nyquist plot with user-specified frequencies.

A Nyquist plot can also be generated using arguments that represent the continuous state-space system:

$$x' = Ax + Bu$$
$$y = Cx + Du$$

For example, the state-space equations for the spring-mass-damper example are

$$a = \begin{bmatrix} 0 & 1 \\ -3 & -4 \end{bmatrix} \qquad b = \begin{bmatrix} 0 \\ 1 \end{bmatrix} \qquad c = \begin{bmatrix} 10 & 0 \end{bmatrix} \qquad d = \begin{bmatrix} 0 \end{bmatrix}$$

Thus, the statements below will produce the same Nyquist plot as shown in Figure 15.9:

```
a = [0 1; -3 -4];
b = [0 1]';
c = [10 0];
d = 0;
nyquist(a,b,c,d),title('Nyquist Plot')
```

When used with multi-input systems, nyquist(a, b, c, d) produces a series of plots—one for each input. A plot for a specific input iu of a multi-input system can be obtained using nyquist(a, b, c, d, iu). Including a vector of user-specified frequency values w in the argument list will cause the nyquist function to calculate the system response at those frequencies. Note, however, that if w is included in the argument list of a state-space system, iu must also be included, even for a single-input system.

Adding the left-side arguments in the nyquist function statement causes the function to calculate the output matrices, but does not generate a plot. This makes the real and imaginary values of the response, as well as the frequency values used in the calculations, available.

Root Locus Plots

The root locus is an extremely useful analysis tool for single-input/single-output systems. It is a graphical method that was developed by W. R. Evans [17] for assessing the stability and transient response of a system, and for determining qualitatively at least, ways to improve the system performance. The Evans root locus is a plot of the location of the roots of the characteristic equation of a system. The roots of the characteristic equation determine the stability of the system and, generally, how the system will respond to an input. By analyzing the root locus plot, the system designer can determine where to locate the roots and what changes may be needed in the transfer function to achieve the desired stability and response. When used in conjuction with frequency response tools, such as Bode plots, the overall dynamic performance of a system can be thoroughly evaluated.

Evans root locus

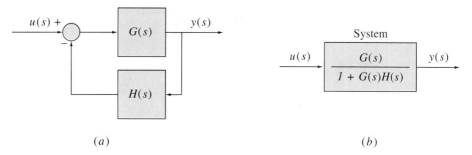

Fig. 15.11 Feedback control system.

feedback control

Figure 15.11(a) is a commonly used block diagram that depicts a feedback control system [16]. In this figure, $G(s)$ represents the forward path transfer function and $H(s)$ represents the feedback path transfer function. The system closed-loop transfer function, shown in Figure 15.11(b), is

$$\frac{y(s)}{u(s)} = \frac{G(s)}{1 + G(s)H(s)}$$

The Evans root locus is obtained by setting the denominator of the closed-loop transfer function equal to zero, which gives the characteristic equation of the system:

$$1 + G(s)H(s) = 0$$

or

$$G(s)H(s) = -1$$

The roots of the characteristic equation are then computed as some parameter, usually the forward-path gain, is varied from zero to infinity.

The rlocus function can be used to produce root locus plots for both continuous-time and discrete-time systems. The forms for using rlocus are

```
[r,k] = rlocus (num, den);
[r,k] = rlocus (num, den, k);
[r,k] = rlocus (a, b, c, d);
[r,k] = rlocus (a, b, c, d, k);
```

The input arguments num and den for the first forms of the rlocus function shown are vectors containing the coefficients, in descending powers of s, of the numerator and denominator polynomials of the open-loop transfer function $G(s)H(s)$. The output matrices r and k are the root locations and the corresponding gains, respectively.

We now demonstrate the use of some of these forms with examples. First, a root locus plot can be quickly generated by using only the the right-side arguments.

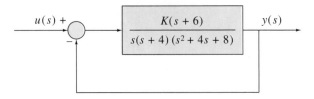

$$u(s) + \quad \xrightarrow{} \quad \boxed{\frac{K(s + 6)}{s(s + 4)\,(s^2 + 4s + 8)}} \quad \xrightarrow{} y(s)$$

Fig. 15.12   Fourth-order control system.

For example, the root locus plot of the unity feedback fourth-order control system [16] shown in the block diagram in Figure 15.12 can be produced using the statements

```
num = [1 6];
p1 = [1 4 0];
p2 = [1 4 8];
den = conv(p1,p2);
rlocus(num,den),title('Root Locus Plot')
```

The root locus plot is shown in Figure 15.13. In this case, the function automatically selects the values for the gain, K. Similarly, the statement rlocus (num, den, k) produces a root locus plot using a user-specified input gain vector k.

A root locus plot can also be generated using arguments that represent the continuous state-space system:

$$x' = Ax + Bu$$
$$y = Cx + Du$$

For example, the root locus of the previous example can be produced by first using the conversion function tf2ss to obtain the state-space matrices a, b, c, and d, and then using the state-space argument for the rlocus function:

```
[a,b,c,d] = tf2ss(num,den);
rlocus(a,b,c,d),title('Root Locus Plot')
```

These statements produce the same root locus plot as before.

Adding the left-side arguments in the rlocus function statement causes the function to calculate the output roots r and gains k, but does not generate a plot. This makes the pole locations and the corresponding gain values available. The plot of the individual pole locations for each gain value can then be generated with the statements

```
[r,k] = rlocus(a,b,c,d);
plot(r,'x'),title('Root Locus Plot'),...
xlabel('Real'),ylabel('Imag'),grid
```

which produce the plot shown in Figure 15.14.

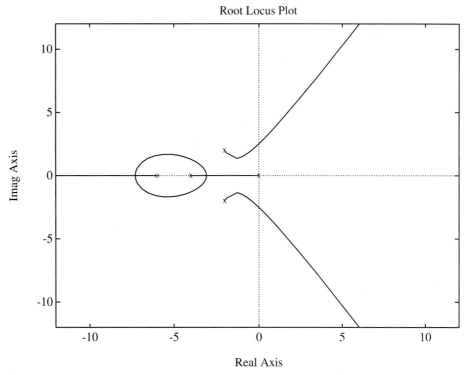

Fig. 15.13   Root locus plot.

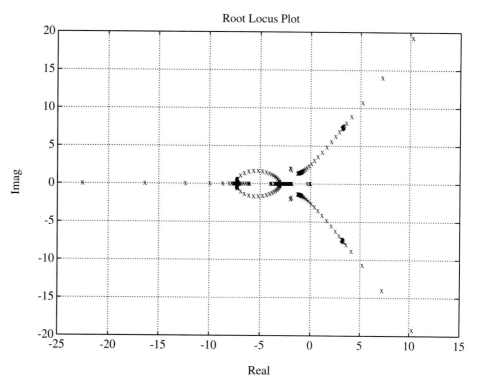

Fig. 15.14   Root locus plot with individual pole locations.

> ## Practice!
>
> The following problems refer to this transfer function:
>
> $$H(s) = \frac{s + 30}{s^2(s + 300)}$$
>
> 1.   Generate a Bode plot.
>
> 2.   Generate a root locus plot.
>
> 3.   Generate a Nyquist plot.

Step Response

The step response generally shows how the system will respond to an input. More specifically, it shows the time-domain transient behavior of a system when it is subjected to an instantaneous unit change in one of its inputs. It is a graphical method that is especially useful in making an initial evaluation of a control system design. Information on the stability, damping, and bandwidth of a system can be obtained from this one analysis tool.

The step function can be used to produce step response plots for each input and output combination of a continuous-time system. The algorithm used by step converts the continuous system to an equivalent discrete system, which is then propagated to produce the output matrices. The forms for using the step function are

```
[y,x,t]  =  step (num, den) ;
[y,x,t]  =  step (num, den, t) ;
[y,x,t]  =  step (a, b, c, d) ;
[y,x,t]  =  step (a, b, c, d, iu) ;
[y,x,t]  =  step (a, b, c, d, iu, t) ;
```

The input arguments num and den for the first forms of the step function shown are vectors containing the coefficients, in descending powers of s, of the numerator and denominator polynomials of the transfer function for which the step response plot is needed. The output matrix y is the output of the system. The output matrix x contains the values of the state vector of the equivalent discrete system, whose sampling period is the interval between the time values in output vector t. We now demonstrate the use of some of these forms with examples.

As is the case with several of the analysis functions discussed previously, a step response plot can be generated without using the left-side arguments. For example, the step response of the second-order transfer function

$$\frac{y(s)}{u(s)} = \frac{5(s + 3)}{s^2 + 3s + 15}$$

can be produced using the statements

```
num = 5*[1 3];
den = [1 3 15];
step(num,den),grid,...
title('Step Response for a Second-Order System')
```

which generate the plot shown in Figure 15.15. In this case, the function automatically selects the time values.

Similarly, the statement step(num,den,t) produces a step response plot using a user-specified input time vector t. Using the spring-mass-damper example, the statements

```
num = 10;
den = [1 4 3];
t = 0:0.05:10;
step(num,den,t),grid,...
title('Step Response for a Second-Order System')
```

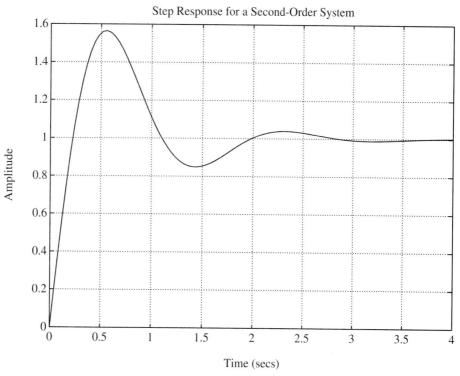

Fig. 15.15 Step response for a second-order system.

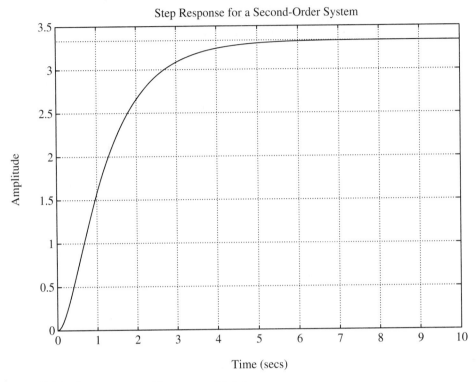

Fig. 15.16   Step response with user-specified times.

result in the plot shown in Figure 15.16 for 201 evenly spaced points between the times of 0 and 10 seconds.

A step response plot can also be generated using arguments that represent the continuous state-space system:

$$x' = Ax + Bu$$
$$y = Cx + Du$$

For example, the state-space equations for the spring-mass-damper example can be obtained by using the tf2ss function

```
[a,b,c,d] = tf2ss(num,den);
```

which results in

$$a = \begin{bmatrix} -4 & -3 \\ 1 & 0 \end{bmatrix} \qquad b = \begin{bmatrix} 1 \\ 0 \end{bmatrix} \qquad c = \begin{bmatrix} 0 & 10 \end{bmatrix} \qquad d = \begin{bmatrix} 0 \end{bmatrix}$$

The following statement will now produce a step response plot that is similar to the plot in Figure 15.16, but using an automatically selected time vector:

```
step(a,b,c,d),grid,...
title('Step Response for a Second-Order System')
```

When used with multi-input systems, `step(a,b,c,d)` produces a plot for each input and output combination of the system. Using `step(a,b,c,d,iu)` will produce a step response plot from a selected input `iu` to all of the outputs of the system. Including a vector of user-specified time values `t` in the argument list will cause the step function to calculate the system response at only those times. Note, however, that if `t` is included in the argument list of a state-space system, `iu` must also be included, even for a single-input system.

Adding the left-side arguments in the `step` function statement causes the function to calculate the output matrices, but does not generate a step response plot. This makes the output matrices `y`, `x`, and `t` available to be plotted using a separate statement. The output variables `x` and `t` are optional and can be omitted from the argument list if not needed. For example, the statements below produce the step response point plot with automatically selected time values, using the values of a, b, c, and d from the previous example:

```
[y,x,t] = step(a,b,c,d);
plot(t,y,'*'),...
title('Unit Step Response'),...
xlabel('Time, seconds'),ylabel('Output, volts'),grid
```

Optimal System Design

linear quadratic estimator

linear quadratic regulator

The `lqe` function is used to solve the continuous-time linear-quadratic estimator design problem, and `lqr` is used to solve the continuous-time linear-quadratic regulator design problem. Both functions solve the associated Riccati equations. These functions are used in the design of optimal estimator and optimal control systems.

The various forms of these functions are shown below:

```
[l,p,e] = lqe(a,g,c,q,r);
[l,p,e] = lqe(a,g,c,q,r,n);
[k,s,e] = lqr(a,b,q,r);
[k,s,e] = lqr(a,b,q,r,n);
```

We now give examples of these forms.

Kalman filter

To calculate the gain matrix L of the continuous, stationary Kalman filter:

$$x' = Ax + Bu + L(y - Cx - Du)$$

for the continuous-time state-space system:

$$x' = Ax + Bu + Gw$$
$$y = Cx + Du + v$$

having zero-mean process and measurement noise vectors w and v, and covariance matrices:

$$E[ww'] = Q, \qquad E[vv'] = R, \qquad E[wv'] = 0$$

we can use the following statement:

```
[l,p,e] = lqe(a,g,c,q,r);
```

The input arguments a, g, c, q, and r correspond to the matrices that appear in the system and covariance equations above. The output matrix l contains the Kalman filter gains. The Riccati equation solution p, and the closed-loop eigenvalues e of the estimator, are optional outputs that can be omitted from the output argument list if they are not needed. This statement

```
[l,p,e] = lqe(a,g,c,q,r,n);
```

similarly calculates l, p, and e when the process and measurement noise is correlated as indicated by

$$E[wv'] = N$$

In both cases, the resulting Kalman filter produces a linear-quadratic-gaussian (LQG) optimal estimate of the state-vector x.

For example, suppose that astronomers wish to estimate very accurately the distance and velocity, relative to the Earth, of a comet that is approaching the sun. Their plan is to use a radar to measure the distance to the comet and to then use a Kalman filter to estimate both the distance and the velocity. The comet is being steadily accelerated by the pull of gravity from the sun, but material that is being ejected from the comet is also causing disturbance forces that are modeled as process noise. In addition, the radar cannot measure the distance perfectly, so sensor noise must be included in the measurement equation. The equations that the astronomers have developed are

$$x' = Ax + Bu + Gw$$
$$y = Cx + Du + v$$

where

$$A = \begin{bmatrix} 0 & 1 \\ 0 & 0 \end{bmatrix}, \qquad G = \begin{bmatrix} 0 \\ 0.7e{-}15 \end{bmatrix}, \qquad C = \begin{bmatrix} 1 & 0 \end{bmatrix}$$
$$Q = 5.0e\,13, \qquad R = 1.5e\,06$$

Thus, to compute the Kalman filter gain L, the Riccati equation solution, and the estimator eigenvalues, the following statements are used:

```
a = [0 1; 0 0];
g = [0 0.7e-15]';
c = [1 0];
```

```
q = 5.0e13;
r = 1.5e6;
format long
[l,p,e] = lqe(a,g,c,q,r);
```

The matrices computed by these statements are the following:

$$l = 1.0e-05 \begin{bmatrix} 0.28430448050527 \\ 0.00000040414519 \end{bmatrix}$$

$$p = \begin{bmatrix} 4.26456720892909 & 0.00000606217783 \\ 0.00000606217783 & 0.00000000001724 \end{bmatrix}$$

$$e = 1.0e-05 \begin{bmatrix} -0.14215224029764 + 0.14215224029764i \\ -0.14215224029764 - 0.14215224029764i \end{bmatrix}$$

The output vector l is the gain L, which is used in the Kalman filter equation above to produce an optimal estimate of the comet's position and velocity.

The lqr function can be used to calculate the optimal gain matrix K of the linear-quadratic regulator feedback equation:

$$u = -Kx$$

that is used to control the system

$$x' = Ax + Bu$$

The gain K minimizes the quadratic performance index

$$J = \int (x'Qx + u'Ru) \, dt$$

where Q and R are specified "weighting" matrices. If the performance index uses a cross-weighting matrix N, such that

$$J = \int (x'Qx + 2u'Nx + u'Ru) \, dt$$

then N is added to the right-side argument list. The function reference is then the following:

```
[k,s,e] = lqr(a,b,q,r,n);
```

Notice that Q, R, and N are not the noise covariance matrices that were defined earlier for the lqe problem. Outputs s and e are the solution to the matrix Ricatti equation and the closed-loop eigenvalues, respectively, where e are the eigenvalues of the matrix $A-B*K$. As before with the lqe statement, s and e are optional outputs that can be omitted from the lqr argument list if they are not needed.

For example, suppose we want to design an optimal regulator to control a system that consists of an inverted pendulum on a movable base. The purpose of the regulator will be to move the base such that the pendulum remains vertical, much the same way as you would balance a stick on your finger. The four states of the system are the position and velocity of the movable base and the angular position and velocity of the pendulum. The only control that is to be used is a horizontal force on the movable base. Assume the state-space equations for this system are

$$x' = Ax + Bu$$
$$y = Cx + Du$$

and the control law is

$$u = -kx$$

where k is the optimal feedback gain matrix that minimizes the cost function

$$J = (x'Qx + u'Ru) \, dt$$

Use lqr to calculate the gain matrix k, assuming that

$$a = \begin{bmatrix} 0 & 1 & 0 & 0 \\ 0 & 0 & -2.2 & 0 \\ 0 & 0 & 0 & 1 \\ 0 & 0 & 22 & 0 \end{bmatrix} \qquad b = \begin{bmatrix} 0 \\ 0.22 \\ 0 \\ -0.22 \end{bmatrix}$$

$$q = \begin{bmatrix} 1 & 0 & 0 & 0 \\ 0 & 1 & 0 & 0 \\ 0 & 0 & 1 & 0 \\ 0 & 0 & 0 & 1 \end{bmatrix} \qquad r = [0.001]$$

The MATLAB statements are:

```
a = [0 1 0 0; 0 0 -2.2 0; 0 0 0 1; 0 0 22 0];
b = [0 0.22 0 -0.22]';
q = eye(4);
r = 0.001;
[k,s,e] = lqr(a,b,q,r);
```

The values of the matrices computed by the lqr function are the following:

$$k = \begin{bmatrix} -31.6228 & -50.8476 & -596.3584 & -135.3104 \end{bmatrix}$$

$$s = \begin{bmatrix} 1.6079 & 0.7927 & 4.2789 & 0.9365 \\ 0.7927 & 1.0586 & 5.9437 & 1.2897 \\ 4.2789 & 5.9437 & 41.6264 & 8.6544 \\ 0.9365 & 1.2897 & 8.6544 & 1.9047 \end{bmatrix}$$

$$e = \begin{bmatrix} -11.0306 \\ -3.2688 + 1.2792i \\ -3.2688 - 1.2792i \\ -1.0135 \end{bmatrix}$$

PROBLEM SOLVING APPLIED: LASER BEAM STEERING MIRROR CONTROL

The Nova laser system that is described in the chapter opener uses 10 individual lasers that are each 137 meters long. Mirrors and lenses are used to guide the beam through the laser and to the target chamber. Many such lasers use "steering mirrors" that can be quickly moved by a control system to redirect the laser beam. Engineers design the control systems starting with the performance requirements that the mirror needs to meet. For example, the requirements might be that the mirror must be able to redirect the beam 5 degrees in less than a second and hold the beam at the new position with an accuracy of better than a thousandth of a degree. Candidate hardware for the steering mirror and its controller may then be selected and modeled so that designs can be developed for the system. The control system designs are analyzed using such things as root locus, Bode, and step response plots to evaluate controller configurations and gain values. When a design is selected, the hardware is assembled and tested to see if the design specifications have been met.

 1. PROBLEM STATEMENT

A lead-lag compensation design is being evaluated for a steering mirror control system. The model of the control system design that is to be evaluated is shown in Figure 15.17. The block diagram shows the gain, the lead-lag compensator, the mirror plant, and the unity feedback path. Select a gain for the control system that will provide a stable, well-damped response.

2. INPUT/OUTPUT DESCRIPTION

The analysis of the transfer function for this system will allow us to determine the desired gain. The root locus plot represents the output of the program, as shown in the input/output diagram in Figure 15.18.

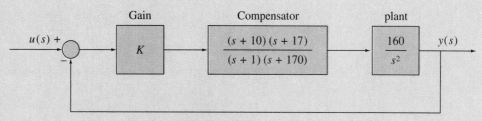

Fig. 15.17 Steering mirror control system diagram.

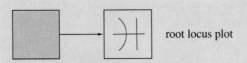

Fig. 15.18   I/O diagram.

 3. HAND EXAMPLE

The best way to select the gain value is to use the root locus plot and choose the closed-loop pole locations of the plant. The location of the plant's poles in the root locus plot provides information on how the system should respond. Having selected the pole locations, we simply determine the associated gain value.

 4. MATLAB SOLUTION

We use the root locus plot to see the locations of the poles as the gain changes. We can then select a location of the plant poles that we believe will provide a stable, well-damped system response and determine the gain that goes with those poles. The version of the root locus function that we use has this form:

```
[r,k] = rlocus(num,den,k);
```

because we would like to be able to see the root locations, select the input gain values, and use transfer functions rather than state-space equations for the inputs. To determine the values for the vectors num and den, we need the open-loop transfer function, which for this system is the product of the compensator and the plant transfer functions. We could multiply the polynomials by hand, but it is quicker to use MATLAB statements to perform the polynomial multiplication. We then use the rlocus statement to plot the root locus to complete this step in the analysis.

```
%
%      This program plots the root locus information
%      for analyzing a steering mirror control system.
%
num = 160*conv([1 10],[1 17]);
den = conv(conv([1 1],[1 170]),[1 0 0]);
%
k = 1:100;
[r,k] = rlocus(num,den,k);
%
plot(r),...
title('Steering Mirror Root Locus Plot'),...
xlabel('Real'),ylabel('Imag')
```

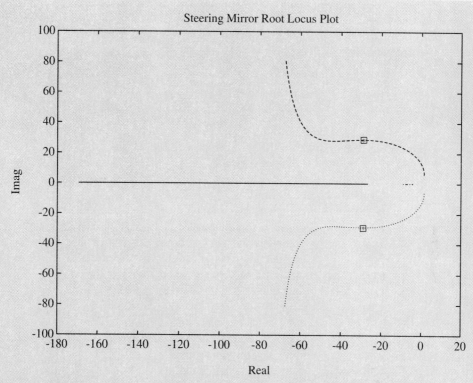

Fig. 15.19 Root locus plot for steering mirror control system.

 5. TESTING

The resulting root locus plot is shown in Figure 15.19. As seen in the plot, all of the closed-loop poles are to the left of the imaginary axis, which ensures that the system will be stable. We have used squares to mark the plant's closed-loop pole locations ($-28.9342 + 28.9028i$ and $-28.9342 - 28.9028i$) that we believe will provide a stable system response with good damping. The gain that is associated with these pole locations is 57.

SUMMARY

Several model conversion and analysis functions for performing linear and control systems design and analysis were presented. Using the conversion functions such as c2d and tf2ss, models can be converted from continuous-time to discrete-time equations and from transfer functions to state-space equations. These conversion functions are useful when designing and analyzing systems using the analysis functions, by making it easy to convert a model that is expressed in one form to another form that the analysis function uses. The design and analysis functions bode, nyquist, rlocus, and step provide useful graphical information about the response of a system. The lqe and lqr functions are used for designing optimal estimators and controllers.

MATLAB SUMMARY

This MATLAB summary lists all the special symbols, commands, and functions that were described in this chapter. A brief description is also included for each one.

Commands and Functions

bode	computes magnitude and phase response
c2d	converts continuous state-space to discrete state-space
lqe	computes a linear-quadratic estimator
lqr	computes a linear-quadratic regulator
nyquist	computes the Nyquist frequency response
residue	computes a partial fraction expansion
rlocus	computes the root-locus
ss2tf	converts state-space to transfer function
ss2zp	converts state-space to zero-pole-gain
step	computes the unit step response
tf2ss	converts transfer function to state-space
tf2zp	converts transfer function to zero-pole-gain
zp2ss	converts zero-pole-gain to state-space
zp2tf	converts zero-pole-gain to transfer function

PROBLEMS

Problems 1 to 6 relate to the engineering application presented in this chapter. Problems 7 to 25 relate to new engineering applications.

Steering Mirror Control System   Assume that the design specifications for the steering mirror control system are

> Bandwidth (-3 dB): 10 Hz
>
> Peak Overshoot: 20%
>
> Settling Time: 0.5 second

1. Using the gain value of 57 determined in the Problem Solving Applied section, generate the open-loop Bode plot of the mirror control system. Find the crossover frequency, which is the frequency where the gain curve crosses the x-axis. Determine the phase margin at the crossover frequency by subtracting 180 degrees from the phase angle.

2. Generate the Nyquist plot and the Nichols plot for the open-loop mirror control system.

3. Compute the closed-loop transfer function for the mirror control system using the conv function. Generate the Bode plot for the closed-loop transfer function, plotting the frequency in Hz, and determine the closed-loop bandwidth. (The closed-loop bandwidth is the frequency where the gain curve drops below −3dB.) Does the control system meet the 10 Hz bandwidth requirement?

4. Plot the closed-loop step response for the mirror control system, and determine the peak overshoot and the settling time. The peak overshoot is the percentage by which the peak value of the step response exceeds the final value. The settling time is the time that it takes for the step response to come within 5% of its final value. Have the peak overshoot and the settling time specifications been met?

5. Convert the closed-loop transfer function to state-space equations and generate the Bode plot and step response of the mirror control system. Are the results the same as in problems 3 and 4?

6. Convert the open-loop transfer function of the mirror plant only to state-space equations. Using these weighting matrices

$$Q = \text{eye}(2); \qquad R = 0.05;$$

solve the linear-quadratic-regulator problem to determine the optimal feedback gain matrix for the mirror controller.

System Conversions   These problems use the conversion functions to convert a system from one form to another form. The problems refer to this set of system equations.

a. $$G(s) = 10\,\frac{3s + 1}{s^5 + 7s^4 + 12\,s^2}$$

b. $$G(s) = \frac{s + 1}{s(s + 2)(s^2 + 4s + 8)}$$

c. $$G(s) = 15\,\frac{s^2 + s + 10}{s^2 + 6s + 10}$$

d. $$G(s) = \frac{4}{s^3 + 6s^2 + 11s + 6}$$

e. $$x' = Ax + Bu$$
$$y = Cx + Du$$

$$A = \begin{bmatrix} 0 & 1 & 0 \\ 0 & 0 & 1 \\ -6 & -11 & -6 \end{bmatrix}, \qquad B = \begin{bmatrix} 0 & 0 & 4 \end{bmatrix}'$$

$$C = \begin{bmatrix} 1 & 0 & 0 \end{bmatrix}, \qquad D = \begin{bmatrix} 0 \end{bmatrix}$$

7. Use the `tf2zp` function to convert transfer function (a) to the zero-pole-gain form.

8. Use the `conv` function and the `tf2zp` function to convert transfer function (b) to the zero-pole-gain form.

9. Use the `residue` function to convert transfer function (a) to partial-fraction form.

10. Use the `residue` function to convert transfer function (c) to partial-fraction form.

11. Use the `c2d` function and a sample period of 0.1 second to convert the continuous-time state-space equations in (e) to discrete-time equations.

12. Use the `tf2ss` function to convert (d) to state-space equations, and the `residue` function to convert (d) to partial-fraction form. Do you see any similarity between the two representations?

13. Use the `ss2tf` function to convert (e) to transfer function form. Do you see any similarity with the results of problem 12?

14. Use the `ss2zp` function to convert (e) to zero-pole-gain form, and then the `zp2ss` function to convert the result back to state-space form. What happened? What are the similarities with problems 12 and 13?

15. Use the `ss2zp` function to convert (e) to zero-pole-gain form, and then the `zp2tf` function to convert the result to transfer function form. Did you get the same results as problem 13?

16. Use the `ss2tf` function to convert the state-space equations obtained in problem 14 to transfer function form. Did you get transfer function (d)?

System Design and Analysis   These problems use the design and analysis functions presented in this chapter to provide further analysis of the system equations presented in the previous set of problems.

17. Generate the Bode and Nyquist plots for (a) using only the right-side arguments.

18. Generate the Bode and Nyquist plots for (b) using left and right-side arguments and a user-specified frequency range.

19. Generate the Bode and Nichols plots for (c) for a user-specified range of frequencies.

20. Generate the Bode plots for the transfer function (d) and the state-space equations (e). Are they the same?

21. Generate the root locus plots for (d) and (e). Are they the same?

22. Generate the root locus plot for (c) using the left-side arguments, a user-specified set of gains and x's to mark the pole locations.

23. Generate the root locus plot for (a), and then use tf2ss to convert (a) to state-space equations. Generate the root locus plot for the state-space equations. Are the plots the same?

24. Convert (b) to state-space equations and generate the root locus plot using the left-side arguments and a user-specified set of gains.

25. Given this second order plant:

$$H(s) = \frac{1}{s^2}$$

convert the transfer function to state-space equations and solve the linear-quadratic-regulator problem using the following values:

$$Q = \begin{bmatrix} 1 & 0 \\ 0 & 1 \end{bmatrix} \qquad R = [0.01]$$

APPENDIX A

A Quick Reference to MATLAB Functions

This summary contains a brief description of all MATLAB functions in the book "The Student Edition of MATLAB." Most of these functions are discussed in this text; for specific page references, refer to the index.

Function Name	Function Description
abs	computes the absolute value
acos	computes the arccosine
acosh	computes the inverse hyperbolic cosine
all	determines if all elements are nonzero
angle	computes phase angles in radians
ans	variable created for expressions
any	determines if any elements are nonzero
asin	computes the arcsine
asinh	computes the inverse hyperbolic sine
atan	computes the arctangent (2 quadrant)
atan2	computes the arctangent (4 quadrant)
atanh	computes the inverse hyperbolic tangent
axis	specifies the manual axis scaling on plots
backsub	uses backsubstitution to solve system of equations
balance	balances a matrix to improve its conditioning
bar	generates a bar graph
bessel	evaluates a Bessel function
bilinear	maps an analog function into a digital function
bode	generates Bode frequency response plots
break	terminates the execution of loops
buttap	designs an analog Butterworth filter
butter	designs a digital Butterworth filter
c2d	converts a continuous system to a discrete system
casesen	selects or deselects case sensitivity
cdf2rdf	converts complex diagonal matrix to real diagonal
ceil	rounds to the nearest integer toward ∞
chdir	changes directories

cheb1ap	designs an analog Chebyshev Type I filter
cheb2ap	designs an analog Chebyshev Type II filter
cheby1	designs a digital Chebyshev Type I filter
cheby2	designs a digital Chebyshev Type II filter
chol	computes the Cholesky factorization of a matrix
clc	clears the command window
clear	clears the workspace
clg	clears the graph window
clock	returns the current time and date
cplxpair	reorders values into complex-conjugate pairs
compan	computes a companion matrix to a vector
computer	returns a string containing the type of computer
cond	computes the condition number of a matrix
conj	computes the complex conjugate
contour	generates a contour plot
conv	multiplies two polynomials (convolution)
corrcoef	computes the correlation coefficients
cos	computes the cosine
cosh	computes the hyperbolic cosine
cov	computes the covariance matrix
cumprod	computes a cumulative product
cumsum	computes a cumulative sum
date	returns a string containing the date
deconv	divides two polynomials (deconvolution)
delete	deletes files
demo	prints a menu of the available demonstrations
det	computes the determinant of a square matrix
diag	generates a diagonal matrix
diary	saves the session in a disk file
diff	computes differences between adjacent elements
dir	prints the current directory
disp	displays text or a matrix
echo	enables echo of M-files during execution
eig	computes the eigenvalues and eigenvectors
ellip	designs a digital elliptic filter
ellipap	designs an analog elliptic filter
ellipj	computes the Jacobian elliptic function

ellipk	computes the elliptic integral of the first kind
else	optional clause in an if structure
elseif	optional clause in an if structure
end	terminates control structures
erf	computes the error function
errorbar	adds errorbars to a plot
eps	returns the floating-point relative accuracy
error	displays error messages
etime	calculates the elapsed time between two times
eval	executes text messages
exist	determines if a variable exists
exit	terminates MATLAB
exp	computes the exponential with base e
expm	computes the matrix exponential with base e
eye	generates an identity matrix
feval	initiates a function evaluation
fft	computes the discrete Fourier transform
fft2	computes the 2D discrete Fourier transform
fftshift	swaps the left half of a matrix with the right half
filter	applies a digital filter to a signal
find	finds the indices of elements
finite	returns 1's where elements are finite
fix	rounds to the nearest integer toward zero
fliplr	flips a matrix left-to-right
flipud	flips a matrix up-to-down
floor	rounds to the nearest integer toward $-\infty$
flops	counts the number of floating-point operations
fmin	determines the minimum of a 1D function
fmins	determines the minimum of a multivariable function
for	generates a loop
format	sets the output display format
fprintf	prints formatted output
freqs	computes the analog frequency response
freqz	computes the digital frequency response
function	defines a user-defined function
funm	computes general matrix functions
fzero	determines the zero of a function of one variable

gallery	generates a gallery of interesting matrices
gamma	evaluates the gamma function
getenv	returns the environment variable
ginput	obtains coordinates from the current graph
global	defines global variables
grid	adds a grid to the current graph
grpdelay	computes the group delay of a digital filter
gtext	adds text on a graph using a mouse
hadamard	computes the Hadamard matrix
hamming	generates values in a Hamming window
hankel	generates a Hankel matrix
hanning	generates values in a Hanning window
help	prints a list of HELP topics
hess	determines the Hessenberg form of a matrix
hex2num	converts hexadecimal string to double precision
hilb	generates a Hilbert matrix
hist	generates a histogram
hold	holds the current graph on the screen
home	moves the command cursor to the home position
i	$\sqrt{-1}$
if	generates a conditional control flow statement
imag	computes the imaginary part of a complex number
Inf	represents infinity
input	allows user input
int2str	converts an integer to a string
inv	computes the inverse of a square matrix
inverf	computes the inverse error function
invhilb	generates an inverse Hilbert matrix
isempty	returns 1 if a matrix is empty
isnan	returns 1's where elements are NaNs
isstr	returns 1's where elements are strings
j	$\sqrt{-1}$
keyboard	invokes the keyboard as an M-file
kron	computes a Kronecker product

length	returns the length of a vector
linspace	generates linearly spaced vectors
load	loads variables saved in a file
log	computes the natural logarithm
log10	computes the logarithm base 10
loglog	generates log-log plots
logm	computes the matrix natural logarithm
logspace	generates logarithmically spaced vectors
lp2bp	converts a lowpass analog filter to bandpass
lp2bs	converts a lowpass analog filter to bandstop
lp2hp	converts a lowpass analog filter to highpass
lp2lp	converts a lowpass analog filter to lowpass
lqe	computes a linear-quadratic estimator design
lqr	computes a linear-quadratic regulator design
lu	computes an LU matrix decomposition
magic	generates a magic square
MATLABPATH	sets the MATLAB search path
max	determines the maximum value
mean	determines the mean or average value
median	determines the median value
menu	generates a menu of choices for user input
mesh	generates a 3D mesh surface
meshdom	evaluates functions of two variables
min	determines the minimum value
NaN	representation for Not-a-Number
nargin	the number of input arguments used in a function
nargout	the number of output arguments used in a function
nnls	computes the nonnegative least squares solution
norm	computes the norm of a matrix
null	finds the null space of a matrix
num2str	converts numbers to strings
nyquist	calculates the Nyquist frequency response
ode23	solves an ordinary differential equation
ode45	solves an ordinary differential equation
ones	generates a matrix of 1's
orth	finds the range space of a matrix

pack	performs memory garbage collection
pascal	generates Pascal's triangle
pause	pauses until a key is pressed
pi	π
pinv	computes a Moore–Penrose pseudoinverse
plot	generates a linear plot
polar	generates a polar plot
poly	computes a polynomial from roots
polyfit	computes a least squares polynomial
polystab	stabilizes a polynomial with respect to the unit circle
polyval	evaluates a polynomial
polyvalm	evaluates a polynomial in a matrix sense
print	sends a high-resolution plot to the printer
prod	computes a product of elements
prtsc	initiates a print screen
qr	computes a QR factorization of a matrix
quad	computes a numerical integration
quad8	computes a numerical integration
quit	terminates MATLAB
qz	gives access to intermediate results
rand	generates random numbers
rank	calculates the rank of a matrix
rat	computes a rational approximation
rcond	returns the condition number of a matrix
real	computes the real part of a complex number
rem	computes the remainder of a division
remez	designs an optimal FIR filter
reshape	reshapes a matrix
residue	computes a partial fraction expansion
return	returns from a function
rlocus	computes the root-locus
roots	determines the roots of a polynomial
rot90	rotates a matrix in 90-degree steps
round	rounds to the nearest integer
rref	produces the reduced row echelon form
rsf2csf	converts the real Schur form to a complex form

save	saves variables in a file
schur	computes the Schur form of a matrix
semilog	generates a semilog plot
setstr	specifies string handling
shg	shows graph window
sign	returns -1, 0, or 1 depending on the sign
sin	computes the sine
sinh	computes the hyperbolic sine
size	returns the size of a matrix
sort	sorts a matrix in ascending order
spline	interpolates using a cubic spline
sprintf	converts strings to numbers for printing
sqrt	computes the square root
sqrtm	computes the matrix square root
ss2tf	converts state-space to transfer function
ss2zp	converts state-space to zero-pole-gain function
stairs	generates a stairstep graph
startup	a file automatically invoked in starting MATLAB
std	computes the standard deviation
step	calculates the unit step response of a system
strcmp	compares two strings
subplot	splits the graph window into subwindows
sum	computes a sum of elements
svd	computes the matrix singular value decomposition
table1	performs linear interpolation with a 1D table
table2	performs linear interpolation with a 2D table
tan	computes the tangent
tanh	computes the hyperbolic tangent
text	adds text to the current graph
title	adds a title to the current graph
tf2ss	converts a transfer function to state-space
tf2zp	converts a transfer function to zero-pole-gain
toeplitz	generates a Toeplitz matrix from vectors
trace	computes the trace of a matrix
tril	generates a lower triangular matrix
triu	generates an upper triangular matrix
type	lists the contents of a file on the screen

unwrap	removes 2π discontinuities
vander	computes a Vandermonde matrix
what	lists a directory of M-files and MAT-files
while	generates a loop
who	lists the variables currently in memory
whos	lists the current variables and sizes
xlabel	adds a label to the x-axis of the current graph
ylabel	adds a label to the y-axis of the current graph
yulewalk	designs an optimal IIR digital filter
zeros	generates a matrix of zeros
zp2ss	converts zero-pole-gain to state-space
zp2tf	converts zero-pole-gain to a transfer function

APPENDIX B
Reading and Writing MAT-Files

This appendix is taken from the student edition of MATLAB [12], and presents several techniques for reading and writing MAT-files using MS-DOS subroutines or Macintosh subroutines.

MS-DOS Subroutines

The save command in MATLAB saves the variables currently in memory into a binary disk file called a MAT-file. (It is called a MAT-file because the filename extension is .MAT.) The load command is the inverse—it reads the MAT-file on disk into MATLAB's memory. There are occasions in which you may find it desirable to directly read and write MAT-files from your own Fortran, C, or Pascal programs. To help you with this task, there is a collection of subroutines for this purpose in the \MATLAB\LOADSAVE directory. In addition, the structure of a MAT-file is documented fully under the load function in the reference section of the student edition of MATLAB.

C Programs There are three files in the \MATLAB\LOADSAVE directory associated with reading and writing MAT-files from C programs. These routines are machine-independent and portable to all different platforms on which MATLAB is currently available.

LOADMAT.C	C routines to read MAT-files
SAVEMAT.C	C routines to write MAT-files
TESTLS.C	A main program illustrating the use of LOADMAT.C and SAVEMAT.C

Pascal Programs There is one file in \MATLAB\LOADSAVE that gives an example of reading and writing MAT-files from Pascal. It is tested with Turbo Pascal.

TESTLS.PAS	Pascal program and procedures showing how to read and write MAT-files

Fortran Programs Fortran poses interesting challenges with respect to reading and writing MAT-files. MAT-files are arbitrary byte streams, for which Fortran does not provide a convenient way of writing. Some Fortrans on some platforms do provide extensions that allow them to be written, but often only in a very inefficient manner.

There are two distinct sets of files in \MATLAB\LOADSAVE for reading and writing MAT-files from Fortran programs. The first set (also the preferred set) implements a machine-independent set of subroutines for working with MAT-files. It does this, however, by coding the subroutines in C. This means that you must have access to the object libraries of a C compiler in order to link your Fortran program.

If you do have a C compiler, these routines are machine-independent and available on all different platforms on which MATLAB is currently running.

The files in \MATLAB\LOADSAVE associated with the preferred set are

FLOADSAV.C	A C module containing the functions MOPEN, MCLOSE, MLOAD, and MSAVE, which open and close MAT-files and do buffered file I/O.
FLOADSAV.OBJ	Compiled version of FLOADSAV.C (small memory model)
FLOADXX.F	An example of a Fortran program using the MLOAD function
FSAVXX.F	An example of a Fortran program using the MSAVE function
FLOADSAV.BAT	A batch file showing how to compile and link the above files

The FLOADSAV.C routines are implemented to support Microsoft Fortran V5.0 and Microsoft C 6.0. You must have the small memory model object code libraries from Microsoft C 6.0 in order to use these files. It should be possible to modify FLOADSAV.C to support other Fortran compilers, if desired.

Near to the top of FLOADSAV.C is the line: #define PC. Change this line to port FLOADSAV.C to other platforms.

If you do not have access to a C compiler, the other files in the \MATLAB\LOADSAVE directory for reading and writing MAT-files from Fortran programs are

LOADMAT.FOR	Fortran subroutines to read MAT-files
SAVEMAT.FOR	Fortran subroutines to write MAT-files
TESTLS1.FOR	A main program illustrating the use of SAVEMAT.FOR
TESTLS2.FOR	A main program illustrating the use of SAVEMAT.FOR and LOADMAT.FOR

These files require Microsoft Fortran, are inefficient because they use single-byte file direct access input/output, and are not portable to other computers. But if you do not have access to a C compiler, and you are not worried about porting your Fortran programs to other computers, they will get the job done.

Macintosh Subroutines

The save command in MATLAB saves the variables currently in memory into a binary disk file called a MAT-files. (It is called a MAT-file because the filename extension is .mat.) The load command is the inverse; it reads the MAT-file on disk into MATLAB's memory. Sometimes you may want to read and write MAT-files directly from your own programs. To help you with this task, there is a collection of subroutines in the LOAD SAVE folder in the MATLAB_Toolbox. In addition, the structure of a MAT-file is documented fully under `load` in the reference section of the student edition of MATLAB.

C Language Programs   There are four files in the LOAD SAVE: C folder associated with reading and writing MAT-files from C programs. These routines are compiler-independent on the Macintosh.

`loadmat.c`	C routines to read MAT-files
`savemat.c`	C routines to write MAT-files
`setmat.c`	C routines to set MAT-file icon
`testls.c`	A main program illustrating the use of `loadmat.c`, `savemat.c`, and `setmat.c`

Fortran Programs   Fortran poses interesting challenges with respect to reading and writing MAT-files. MAT-files are arbitrary byte streams, for which Fortran does not provide a convenient way of writing. Furthermore, different techniques for accessing the Macintosh Toolbox (routines that require you to change a file's creator and type) make it impossible to have compiler-independent routines. As a result, the following folders are in the LOAD SAVE folder:

ABSOFT MF	Contains `matfile.for` and `testls.for` for Absoft's compiler
ABSOFT MF II	Contains `matfile.f` and `testls.f` for Absoft's MacFortran II compiler
LS	Contains `matfile.f` and `testls.f` for Language Systems FORTRAN compiler

APPENDIX C
MATLAB on MS-DOS PCs

This appendix assumes that MATLAB has already been installed on your MS-DOS PC using the installation instructions that came with the student edition of MATLAB software. The material in this section is taken from the student edition of MATLAB reference manual [12], and discusses how to use DOS-specific features of MAT-LAB.

Invoking MATLAB

The command MATLAB invokes MATLAB. MATLAB automatically senses the type of graphics card included in your system.

Command Line Editor

The arrow keys on the keypad can be used to edit mistyped commands or to recall previous command lines. For example, suppose you enter

```
log(sqt(atan2(3+4)))
```

You have misspelled sqrt. MATLAB responds with the error message

```
Undefined variable or function.
Symbol in question sqt.
```

Instead of retyping the entire line, simply hit the Up-Arrow key. The incorrect line will be displayed again and you can move the cursor over using the Left-Arrow key until you can insert the missing r:

```
log(sqrt(atan2(3+4)))
ans =
0.2026
```

The arrow keys on the keypad work on copies of the previous input lines, which have been saved in a moderately sized input buffer. Table C.1 contains a brief description of their functions.

Editors and External Programs

The exclamation point character (!) is used within MATLAB to indicate that the rest of an input line should be issued as a command to the DOS operation system. This is quite useful for invoking DOS utilities or for running other programs without

TABLE C.1   Last-Line Editing and Recall Keys

KEY	FUNCTION
Up Arrow	Recall previous line
Down Arrow	Recall next line
Left Arrow	Move left one character
Right Arrow	Move right one character
Ctrl-Left-Arrow	Move left one word
Ctrl-Right-Arrow	Move right one word
Home	Move to beginning of line
End	Move to end of line
Esc	Cancel current line
Ins	Toggle between insert and overtype mode
Del	Delete character at cursor
Backspace	Delete character left of cursor

quitting MATLAB. For example:

```
!date
```

runs the MS-DOS utility that shows the date, and

```
!edlin darwin.m
```

invokes the `edlin` editor on a file named `darwin.m`. After these programs complete, control is returned to MATLAB.

In MATLAB, the amount of memory available for an external program is limited to the unused portion of the memory available in the MATLAB workspace. The `who` command shows how much free memory is left, and external programs must fit into this space. If you get a message saying there is not enough memory, you can try using the MATLAB `pack` command or you can clear unnecessary variables.

As you become more proficient with MATLAB, you will find yourself working with M-files more and more. M-files are created and modified using an editor or word processor. By design, MATLAB does not have a built-in editor. The choice of editors and word processors is a matter of personal preference. Instead of forcing you to learn a new editor, MATLAB lets you use whichever editor you are already accustomed to using.

There are several ways in which you can intermingle the use of your editor with the use of MATLAB. The most obvious way is to terminate the MATLAB session and to start up the editor. When finished editing, the editor can be terminated and MATLAB reinvoked. Unfortunately, the process of creating and debugging M-files usually involves repeated MATLAB to EDIT to MATLAB cycles, and this slow, manual sequence quickly becomes cumbersome.

The best method of using an editor with MATLAB is accomplished without terminating MATLAB. The exclamation point character (!), introduced in the first part of this section, allows programs to be run directly from MATLAB, without terminating the program. It requires, however, that you have enough free memory in your workspace to hold your editor. Small programming editors work well using this method.

A less desirable method of working with an editor is to use a simple MS-DOS batch file to automate the cycle. The file MATLAB.BAT, used to invoke MATLAB, contains a loop that cycles between the MATLAB program and an editor program. To install your favorite editor, edit MATLAB.BAT and put the command that invokes your editor in the designated area.

If you are running MATLAB and you want to edit a file, simply type edit. This will save your variables in a file, quit MATLAB, and invoke the editor that you specified. When you exit from your editor, control is automatically returned to MATLAB and the variables reloaded. For more information, type help edit in MATLAB or see MATLAB.BAT.

Which method should you use? It depends. Most people who have lots of memory, a small editor, and usually work with small variables will find the ! method is best. If you have limited memory, a large editor, or full workspaces, you may prefer to use the batch file cycling method.

A final note: If your word processor has two ways of writing files, a document mode and a nondocument mode, choose the nondocument mode for editing MATLAB M-files. The other file type may contain strange control characters, perhaps disagreeable to MATLAB.

Environment Parameters

MATLAB has a number of configuration parameters that it obtains from MS-DOS environment variables. The environment is a special global "message-board" area used by the MS-DOS operating system to hold customization information that various programs might want. The MS-DOS operating system prompt you see on your screen is one example of an environment parameter. Two other environment parameters are important to MATLAB, PATH and MATLABPATH.

PATH

The MS-DOS PATH command allows programs to be run from hard disk subdirectories or disk drives other than the current directory. Without a PATH setting, the current directory has to be changed to the directory containing a program before the program can be invoked. PATH give MS-DOS a search path to use when looking for commands. Normally, if you input the name of something to MS-DOS, for example, by typing IHTFP, the MS-DOS interpreter:

1. Searches the current directory for IHTFP
2. Searches the directories specified by the environment symbol PATH for IHTFP

If MS-DOS finds IHTFP, it will try to execute it. The search path is set by listing the directories separated by semicolons. For example:

```
C> PATH=A:\;B:\;C:\BIN;C:\MATLAB\BIN
```

causes MS-DOS to search on drive A, drive B, and then in directories C: \BIN and C: \MATLAB\BIN for commands. The PATH setting is normally put in the AUTOEXEC.BAT file.

MATLABPATH

Like MS-DOS, MATLAB has a search path. The MS-DOS search path is specified with PATH; MATLAB's search path is specified with MATLABPATH. If you input the name of something to MATLAB, for example, by typing test, the MATLAB interpreter:

1. Looks to see if test is a variable.
2. Checks if test is a built-in function.
3. Searches the directories specified by the environment symbol MATLABPATH for test.M. MATLAB will use the first occurrence it finds along the path. This is important if there is more than one version of a file stored with the same name, but in different places on the search path.

MATLABPATH is defined in the file MATLAB.BAT. It contains a semicolon-separated list of the full pathnames to the M-file directories. The list should include standard M-file collections. A typical list might include

\MATLAB\MATLAB
\MATLAB\DEMO
\MATLAB\SIGNAL

You can find out what MATLABPATH is set to on your system by executing help from inside MATLAB.

If you examine MATLAB.BAT, you will see something like

SET MATLABPATH=\MATLAB\MATLAB; \MATLAB\DEMO

The MS-DOS SET command is used to specify environment symbols, and here it is used to set MATLABPATH.

If you want, you can add a new directory to MATLABPATH. This allows you to organize your own libraries of M-files. Suppose a geophysicist has written a set of M-files for analyzing geophysical data and puts them in a directory called \MATLAB\GEO and then changes the SET command in MATLAB.BAT to

SET MATLABPATH=\MATLAB\MATLAB; \MATLAB\DEMO; \MATLAB\GEO

This geophysicist can maintain his or her own private library of M-files. MATLAB will respond to them in the normal way and they won't be mixed in with all the system M-files in \MATLAB\MATLAB.

It is also possible to remove the MATLABPATH definition from MATLAB.BAT and to place it into your AUTOEXEC.BAT file.

If a pathname does not exist or is misspelled, DOS or MATLAB will return the error message, Path not found.

Increasing Environment Space

If you see the message Out of environment space briefly on the screen when MATLAB is invoked, you will have to take special action before MATLAB will respond to MATLABPATH correctly. This message means that you have exceeded the 128 bytes in the DOS environment table and that the MATLABPATH you set has been truncated.

If you issue the SET command from DOS, you will see the various strings currently in the environment table. There may be a long PATH statement or something else using up a lot of space. To remedy the problem, you could clear some strings out of the table by shortening the path or by removing other items from the environment, but you will probably prefer to enlarge the table instead.

To increase the size of the environment table under DOS 3.0 or 3.1, put

```
SHELL=C:\COMMAND.COM C:\ /P /E:30
```

in your CONFIG.SYS file. This reserves $30 \times 16 = 480$ bytes in the table. Make sure you put spaces before /P and /E. The C:\ is the path to COMMAND.COM. Under DOS 3.2 and above, the meaning of the number after the /E: was changed to specify the environment size in bytes, instead of bytes/16 (pages). So to reserve 480 bytes under these versions of DOS, use instead

```
SHELL=C:\COMMAND.COM C:\ /P /E:480
```

In addition to the limitation on the overall size of the environment table, DOS limits each individual environment setting to 128 bytes. Unfortunately, there is no way around this limitation. If you hit this limit, the only choice is to shorten the directory names into which MATLAB is installed, or to use MATLAB's built-in MATLABPATH command. Type help matlabpath for more information.

Printing with Screen Dumps

The graphics screen on your PC can be dumped to dot-matrix or laser printers using the normal PC technique of pressing the Shift-PrtSc keys, provided that you have an appropriate driver loaded into memory. The MATLAB command prtsc does the same thing and allows screen dumps to be triggered under program control.

DOS comes with a driver called GRAPHICS.COM that is appropriate for dumping CGA graphics to Epson dot-matrix printers. It is loaded into memory by executing the command GRAPHICS from the DOS prompt. It must be loaded prior to invoking MATLAB and is usually placed in the AUTOEXEC.BAT file so that it is loaded automatically each time your system is booted.

Hercules graphics cards come with a diskette that contains a driver called HPRINT.COM that is appropriate for dumping Hercules graphics to Epson dot-matrix printers. Using this driver, the key sequence Shift-PrtSc-1 triggers a dump of the graph screen.

Two drivers are provided along with MATLAB for systems with EGA or VGA cards and Epson or Hewlett-Packard LaserJet printers. They can be found in the \MATLAB\BIN directory:

EGAEPSON. COM   Dump EGA and VGA graphics to Epson compatible printers

EGALASER. COM   Dump EGA and VGA graphics to HP LaserJet compatible printers

Some VGA clones won't work properly with these dump utilities when in VGA mode. If you encounter such problems, try starting MATLAB in EGA mode. They are loaded by executing them in the usual way (like GRAPHICS. COM).

Due to a bug in MS-DOS, the second plot you send to an Epson printer using prtsc may have a 1-inch form-feed gap in it. To prevent this, in between sending plots to the printer, hit the button on the printer to take it offline, hit the form-feed button on the printer, and then hit the online button again. Alternatively, the command prtsc ('ff') does a prtsc command with a form-feed patch added automatically.

APPENDIX D
MATLAB on the Macintosh

This appendix assumes that MATLAB has already been installed on your Macintosh using the installation instructions that came with the software for the book "The Student Edition of MATLAB." The material in this section is taken from reference manual [12] for the same book and discusses how to use Macintosh-specific features of MATLAB.

MATLAB Window Types

MATLAB has four different types of windows, which will be discussed separately. If there are several windows open on the Macintosh screen, you can activate a particular window by clicking anywhere within the window. Since the menu bar across the top of the screen is context-sensitive, it changes to reflect the menu choices applicable to the newly selected window after you click in the window.

Except for the Help window, you can grow or shrink all windows using the grow boxes in the lower right-hand corners or zoom boxes in the upper right-hand corners of the window.

Command Window The Command window allows you to enter commands and MATLAB to display its numerical output. Most mathematical operations with MATLAB are performed by issuing commands at the command prompt, or by running scripts, which are stored lists of commands. A few operations, like file management and editing, can be done only from the menus. Other operations can be done from the menu bar or by entering a command in the Command window.

MATLAB keeps a history of your session. You can scroll back through the session history using the scroll bars. To make scrolling through your session history easier, MATLAB automatically positions the insertion point at the end of the command line if you have moved away from the command prompt and pressed a key. If the character that you type is displayable (such as a letter or digit), the character is added to the command line.

Memory for the history is quite large—about 32K bytes. If you exceed this limit, by producing a large amount of output, the beginning of the history is truncated.

To delete the session history, thereby freeing the memory it uses, click on the close box in the upper-left corner of the command window. Select New from the File menu. A dialog box appears. Click on the Command radio button, and then on the OK button to open a new Command window.

When the Command window is active, the menu bar options include the apple (on-line Help), File (choices for managing files and printing), Edit (cut and paste to the Clipboard), Workspace (load or save data from MAT-files), M-File (execute script M-files), Format (set the numeric display format), and Window (bring a hidden window to the front).

Graph Window   The Graph window is where plots appear when graph-producing commands are issued. For example, from the Command window, enter

```
x = 0.1:0.5:4*pi;
y = x.*(sin(x)+1);
plot(y*y)
```

If your Graph window is obscured by other windows, it pops automatically to the top and displays a family of growing sine curves.

When the Graph window is active, the menu bar contains new selections for Gallery (the type of graph paper) and Graph (options for changing the appearance of the graph).

Edit Window   MATLAB has its own built-in editor for creating and modifying M-files, which can be scripts or functions. (Scripts are files containing stored sequences of MATLAB commands. Functions are like scripts, but allow you to pass arguments and have local internal variables.) As you become more experienced with MATLAB, you may find that instead of entering commands in the Command window, you prefer the "what if" capabilities afforded by entering commands into an Edit window, as a script, where they can be changed and reexecuted easily. Suppose you have been entering statements into the Command window to plot Bessel functions of different orders:

```
n = 0;
x = 0:0.5:20;
y = bessel(n,x);
plot(x,y)
```

Rather than typing the same command sequence repetitively for different values of *n*, you can

1.   Select New from the File menu.
2.   Enter the commands into the Edit window.
3.   Select Save And Go from the File menu.
4.   Enter a filename, perhaps Bess, and click on the Save button.

This executes the commands and plots the Bessel function of order zero. To plot the Bessel function of order one:

1.   Click on the Edit window to make it active.
2.   Change n = 0 to n = 1 in the Edit window.
3.   Select Save And Go from the File menu to rerun the script.

Help Window   MATLAB has a Help system that gives you immediate online assistance if you have forgotten how to use a command, or if you need a quick overview of a function that is unfamiliar to you. Online Help is brief; detailed descriptions of the functions and commands can be found in this text or in the reference section of "The Student Edition of MATLAB."

To use Help

1. Select About MATLAB from the apple menu. A Help window appears.
2. Double-click on the category of interest. Except for Built-in commands, the categories correspond to the folders that lie in the MATLAB search path, which by default, includes all the folders beneath the directory in which MATLAB resides. Another list of items appears in the Help window. If you have chosen the Built-in commands category, the items correspond to built-in MATLAB commands. Otherwise, the items correspond to M-files on disk.
3. Scroll to the item of interest and double-click on it with the mouse. A description of the item appears in the Help window.

After you are finished reading the Help entry, you can click on the Topics button to move back up to the list of Help items. From here, you can choose another item by double-clicking on it, or you can click on Topics to return you to the list of Help categories. Clicking on the close box in the upper-left corner closes the Help window.

The Help system is organized hierarchically, according to the MATLAB search path. The Help button and double-clicking on items moves you down the tree out into the folders. The Topics button moves you back up toward the root. Within an individual folder, the actual Help text is the first few commented lines of each M-file.

The Demos button in the Help window lists a set of demonstration scripts for you to run. It is the same as selecting Run Script from the M-File menu and moving to the Demonstrations folder.

You can have several Help windows open at once. Help windows remain on the screen until you close them. They can be kept on the screen concurrently with other windows for easy reference.

Using the Clipboard

You can move information within MATLAB and between MATLAB and other applications using the Macintosh Clipboard. The Cut, Copy, Paste, and Clear commands under the Edit menu let you move information between the active window and the Clipboard. Show Clipboard shows you the current contents of the Clipboard.

In the Command and Edit windows, the Cut, Copy, Paste, and Clear commands work in the normal Macintosh style on text that is selected (highlighted). In the Graph window, the only valid Edit operation is Copy, which copies the plot in the Graph window onto the Clipboard. Internally, the plot is saved in PICT format, and you can paste it into applications that work with PICT files, including Page-Maker and MacDraw. Pasting your graph into MacDraw allows you to rearrange it, add text, change fonts, etc.

If you scroll back through the session history, you can select any previous text with the mouse and copy it to the Clipboard. This allows you to copy a selected command line, and then paste it in order to reexecute it. The Paste command automatically positions the blinking insertion point at the end of the Command window before doing the paste.

When the Graph window is active, Print from the File menu sends a copy of the current graph to the printer. Copy from the Edit menu saves a copy of the graph on the Clipboard in PICT format. Save As from the File menu saves the file on disk in either PICT or Encapsulated PostScript File (EPSF) format.

You can use the Clipboard to import data from other applications. Suppose you have several columns of data saved to the Clipboard from a spreadsheet application. To import this data into a MATLAB variable named Q:

1. Select New from the File menu.
2. Enter Q=[] ; into the Edit window.
3. Position the insertion point between the two brackets.
4. Select Paste from the Edit menu.
5. Select Save And Go from the File menu.
6. Enter a filename, perhaps Q, and click on the Save button.

Your data are now stored in the MATLAB workspace with the name Q.

Invoking MATLAB

To invoke MATLAB, double-click on the MATLAB program icon. After a moment the MATLAB Command window appears. You also see a graph window partially hidden behind it. In the Command window is a prompt that looks like >>. The MATLAB interpreter is awaiting instruction from you.

You can invoke MATLAB by double-clicking on the program file, an M-file, a MAT-file, or a settings file. If you double-click on an M-file, MATLAB opens an Edit window on the designated M-file after starting. If you double-click on a MAT-file, MATLAB loads the data into the workspace after starting.

If you double-click on a settings file, MATLAB loads the user preferences and MATLAB's search path stored in that file. It is possible, therefore, to vary user preferences and MATLAB's search path by having several settings files, one for each configuration.

We recommend that you work in a folder on your Macintosh other than the main MATLAB folder. The MATLAB folder and related subfolders contain many files that are part of the MATLAB system, and we encourage you to keep your own files separate. We suggest that you create a folder called Work in the MATLAB folder or use your own folders elsewhere in your file system.

When working in several folders on the Macintosh, the problem of how to invoke programs that are located elsewhere, without maintaining duplicate copies of programs, arises. With MATLAB this is especially true because in normal use the MATLAB program file remains where it resides in the MATLAB folder. MATLAB expects to find its M-file folders beneath the folder in which it resides.

To invoke MATLAB from another folder, copy a small M-file, MAT-file, or settings file to the places from which you need to invoke MATLAB. Double-clicking on this file automatically starts MATLAB.

Printing

MATLAB allows you to make a hard copy in the normal Macintosh way. Under the File menu are the usual two items for printing:

Page Setup   Select page orientation, reduction, etc.
Print   Print current window

Selecting Print sends to the printer the contents of the currently active window. If you are in a Command or Edit window and you have made a selection, your selection is sent to the printer. Otherwise the entire window is sent to the printer. If a Graph window is active, the graph is printed.

The dialog boxes that appear when you print contain options for number of copies to print, page range, paper size, paper orientation, etc. The specific choices depend upon whether you have a LaserWriter, ImageWriter, or other printer connected to your system.

The dialog box that appears if you are printing a graph to a LaserWriter or other laser printer contains a choice of Best/Fixed Size or Draft/Resizable radio buttons at the bottom of the dialog box. The first option sends PostScript directly to the printer for the highest-quality plot. Unfortunately, the dimensions of the Graph window are not captured by this option. The second option uses QuickDraw and reflects the plot in the Graph window, but produces lower-quality output than the first option.

To print a graph under program control from an M-file, use the MATLAB command `prtsc`, which is short for Print Screen. `prtsc` makes the Graph window active and triggers a Print using the current page setup and job setup options.

APPENDIX E

References

These references were cited at various locations throughout this text.

[1] "10 Outstanding Achievements, 1964–1989," National Academy of Engineering, Washington, DC, 1989.

[2] ETTER, D. M. *Structured Fortran 77 for Engineers and Scientists*, 4th ed., Benjamin/Cummings, Redwood City, CA, 1993.

[3] "The Federal High Performance Computing Program," Executive Office of the President, Office of Science and Technology Policy, Washington, DC, September 8, 1989.

[4] ETTER, D. M. *Fortran 77 with Numerical Methods for Engineers and Scientists*, Benjamin/Cummings, Redwood City, CA, 1992.

[5] BAGGEROER, A. B. "Sonar Signal Processing," in *Applications of Digital Signal Processing*, Prentice Hall, Englewood Cliffs, NJ, 1978.

[6] JONES, E. R., and R. L. CHILDERS. *Contemporary College Physics*, Addison-Wesley, Reading, MA, 1990.

[7] RICHARDSON, M. *College Algebra*, 3rd ed., Prentice Hall, Englewood Cliffs, NJ, 1966.

[8] KREYSZIG, E. *Advanced Engineering Mathematics*, John Wiley & Sons, New York, 1979.

[9] KAHANER, D., CLEVE MOLER, and STEPHEN NASH. *Numerical Methods and Software*, Prentice Hall, Englewood Cliffs, NJ, 1989.

[10] EDWARDS, JR., C. H. and D. E. PENNEY. *Calculus and Analytic Geometry*, 3rd ed., Prentice Hall, Englewood Cliffs, NJ, 1990.

[11] POTTER, M. C., and D. C. WIGGERT. *Mechanics of Fluids*, Prentice Hall, Englewood Cliffs, NJ, 1991.

[12] *The Student Edition of MATLAB*, Prentice Hall, Englewood Cliffs, NJ, 1992.

[13] FRALEIGH, J. B., and R. A. BEAUREGARD. *Linear Algebra*, Addison-Wesley, Reading, MA, 1987.

[14] OPPENHEIM, A. V., and R. W. SCHAFER. *Discrete-Time Signal Processing*, Prentice Hall, Englewood Cliffs, NJ, 1989.

[15] PARKS, T. W., and C. S. BURRUS. *Digital Filter Design*, John Wiley & Sons, New York, 1987.

[16] HALE, F. J. *Introduction to Control System Analysis and Design*, Prentice Hall, Englewood Cliffs, NJ, 1973.

[17] OGATA, K. *Modern Control Engineering*, Prentice Hall, Englewood Cliffs, NJ, 1970.

Solutions to Selected Problems

Chapter 2

1.
```
%
%   This program generates a table of conversions
%   from degrees to radians.
%
deg = 0:10:360;
rad = deg*pi/180;
table(:,1) = deg';
table(:,2) = rad';
disp('Angle Conversion')
disp('Degrees and Radians')
disp(table)
```

9.
```
%
%   This program generates a table of conversions
%   from yen to deutsche marks.
%
yen = 100:9900/24:10000;
dollars = yen*0.0079;
dm = dollars*1.57;
table(:,1) = yen';
table(:,2) = dm';
disp('Currency Conversion')
disp('Yen and Deutsche Marks')
disp(table)
```

Chapter 3

11.
```
%
%   This program predicts the number of bacteria
%   in a colony after 6 hours.
%
yold = 1;
t = 1:6;
ynew = yold*exp(1.386*t);
```

```
        table(:,1) = t';
        table(:,2) = ynew';
        disp('Bacteria Growth')
        disp('Hours and Number of Bacteria')
        disp(table)

20. %
    %  This program converts frequency in rps to Hz.
    %
    rps = input('Enter frequency in rps ');
    hertz = rps/(2*pi);
    fprintf('%8.2f rps = %8.2f Hz \n',rps,hertz)
```

Chapter 4

```
10. %
    %  This program determines the time at which
    %  a rocket begins falling back to the ground
    %  and the time of impact.
    %
    t = 0:2:100;
    height = 60 + 2.13*t.^2 - 0.0013*t.^4...
             + 0.000034*t.^4.751;
    N = length(height);
    %
    %  Compute differences between adjacent heights
    %  to determine when rocket begins to fall.
    %
    difference = height(2:N) - height(1:N-1);
    k_fall = find(difference<0);
    k_impact = find(height<=0);
    if k_fall>0
       fprintf('Rocket begins falling at %5.0f s \n',...
                t(k_fall(1)))
    else
       fprintf('Rocket does not begin falling in 100 s \n')
    end
    if k_impact>0
       fprintf('Rocket impacts at %5.0f s \n',...
                t(k_impact(1)))
    else
       fprintf('Rocket does not impact in 100 s \n')
    end
```

25.
```
%
%  This program prints the number of sensors and
%  the number of seconds covered by the data.
%
load sensor.dat;
[rows,cols] = size(sensor);
fprintf('Number of sensors = %6.0f \n',cols)
fprintf('Seconds of data = %6.0f \n',rows-1)
```

Chapter 5

11.
```
%
%  This program computes the reliability
%  of a series design.
%
rand('uniform')
rand('seed',123)
total = 0;
for n=1:5
    component1 = rand(1000,1);
    component2 = rand(1000,1);
    works = find(component1<=0.8 & component2<=0.92);
    total = total + length(works);
end
fprintf('Reliability is %5.2f \n',...total/5000)
```

17.
```
%
%  This program determines the day and week
%  with maximum and minimum power output.
%
load plant.dat;
%
min_value = min(plant(:));
min_index = find(plant==min_value);
min_day = floor(min_index/8 + 1);
min_week = min_index - (min_day - 1)*8;
%
max_value = max(plant(:));
max_index = find(plant==max_value);
max_day = floor(max_index/8 + 1);
max_week = max_index - (max_day - 1)*8;
%
fprintf('Minimum Power Output = %e \n',min_value);
disp('week and day produced')
```

```
    disp([min_week min_day])
    fprintf('Maximum Power Output = %e \n',max_value);
    disp('week and day produced')
    disp([max_week max_day])
```

Chapter 6

2.
```
%
% This program computes the molecular weights for
% a group of protein molecules using a data file
% that contains the occurrence and number of amino
% acids in each protein molecule.
%
load protein.dat
mw = [ 89 175 132 132 121 146 146  75 156 131 ...
       131 147 149 165 116 105 119 203 181 117 ];
%
weights = protein*mw';
%
[rows cols] = size(protein);
for k=1:rows
    fprintf('protein %3.0f: weight = %5.0f \n',...
            k, weights(k))
end
fprintf('\n average molecular weight = %7.2f \n',...
        mean(weights))
```

16.
```
%
% This program determines if a matrix is diagonal.
% If the matrix is diagonal, then program determines
% if the matrix is also an identity matrix.
%
load array.dat;
top = triu(array,1);
bottom = tril(array,-1);
if top==zeros(top)&bottom==zeros(bottom)
   if diag(array)==diag(eye(array))
      disp('Identity')
   else
      disp('Diagonal')
   end
else
    disp('Not Diagonal')
end
```

Chapter 7

18.
```
%
%  This program generates a 3-D plot
%  of the sinc function.
%
[xgrid,ygrid] = meshdom(-10:2:10,-10:2:10);
r = sqrt(xgrid.^2 + ygrid.^2);
z = sin(r)./r;
index = find(z==NaN);
if length(index) ~=0
   z(index) = 1;
end
mesh(z),title('3-D Sinc Function')
```

25.
```
%
%  This program generates a 3-D plot
%  of a paraboloid.
%
[xgrid,ygrid] = meshdom(-5:10,-5:10);
z = 2*xgrid.^2 + 2*ygrid.^2 +2*xgrid.*ygrid...
    - 14*xgrid - 16*ygrid + 42;
contour(z),title('Paraboloid')
```

Index

+	addition, 54
−	subtraction, 54
*	scalar multiplication, 54; matrix multiplication, 159
/	scalar right division, 54; matrix right division, 221
\	scalar left division, 54; matrix left division, 221
∧	scalar power, 54; matrix power, 161
.*	array multiplication, 58
./	array right division, 58
.\	array left division, 58
.∧	array exponentiation, 58
=	assignment, 94
<	less than, 94
<=	less than or equal, 94
==	equal, 94
~=	not equal, 94
>	greater than, 94
>=	greater than or equal, 94
|	or, 94
&	and, 94
~	not, 94
>>	MATLAB prompt, 25
'	conjugated transpose, 158
.'	unconjugated transpose, 158
!	execute operating system command, 407
%	comments, 26; format specifications, 36
( )	expression precedence, 54; argument enclosure, 68; subscript enclosure, 30
,	separator, 28
.	decimal point, 29
...	ellipsis, 29
\n	new line, 36
:	vector generation, 32
;	end rows, 28; suppress printing, 28
[]	matrices, 29; empty matrix, 33
∧c	local abort, 26

A

Abort command, 26
abs, 68, 83, 396
acos, 70, 396
acosh, 72, 396
Adaptive noise canceling, 311, 316
Aliasing, 324
all, 98, 396
Amino acids, 164, 177
Analog:
  filter, 354
  frequency response, 332, 398
  signal, 322
Analog-to-digital converter (A/D), 61
angle, 83, 396
Angle quadrant, 84
ans, 26, 50, 396
any, 98, 396
Argument, 50
Arithmetic:
  expression, 53
  logic unit (ALU), 22
  operators, 54
  precedence rules, 54
Array (element-by-element):
  division, 58
  exponentiation, 59
  operations, 57
  product, 58
ASCII, 27
  importing and exporting, 31
  text file, 27, 31
asin, 70, 396

ISBN 0-13-280470-0